Writing Reaction Mechanisms in Organic Chemistry

To Silvia

Una mano lava la otra.

Writing Reaction
Mechanisms in Organic Chemistry

AUDREY MILLER

Department of Chemistry
University of Connecticut
Storrs, Connecticut

ACADEMIC PRESS, INC.

Harcourt Brace Jovanovich, Publishers

San Diego   New York   Boston   London   Sydney   Tokyo   Toronto

This book is printed on acid-free paper. ∞

Copyright © 1992 by Academic Press, Inc.

Academic Press, Inc.
1250 Sixth Avenue, San Diego, California 92101-4311

United Kingdom Edition published by
Academic Press Limited
24–28 Oval Road, London NW1 7DX

Library of Congress Cataloging-in-Publication Data

Miller, Audrey.
Writing reaction mechanisms in organic chemistry / Audrey Miller.
p. cm.
Includes index.
ISBN 0-12-496711-6 (pbk.)
1. Chemistry, Organic. 2. Chemical reactions. I. Title.
QD251.2.M53 1992
547.1'39--dc20 91-48080
CIP

PRINTED IN THE UNITED STATES OF AMERICA
92 93 94 95 96 97 BB 9 8 7 6 5 4 3 2 1

Contents

CHAPTER 6

CHAPTER 7

Preface

The ability to write feasible reaction mechanisms in organic chemistry depends on the extent of the individual's preparation. This book assumes the knowledge obtained in a one-year undergraduate course. A course based on this book is suitable for advanced undergraduates or beginning graduate students in chemistry. It can also be used as a supplementary text for a first-year course in organic chemistry.

Because detailed answers are given to all problems, the book can also be used as a tutorial and a review of many important organic reaction mechanisms and concepts. The answers are conveniently located at the end of each chapter. Examples of unlikely mechanistic steps have been drawn from my experience in teaching a course for beginning graduate students. As a result, the book clears up many aspects that are confusing to students. The most benefit will be obtained from the book if an intense effort is made to solve the problem before looking at the answer. It is often helpful to work on a problem in several different blocks of time.

The first chapter, a review of fundamental principles, reflects some of the deficiencies in knowledge often noted in students with the background cited above. The second chapter discusses some helpful techniques that can be utilized in considering possible mechanisms for reactions which may be found in the literature or during the course of laboratory research. The remaining chapters describe several of the common types of organic reactions and their mechanisms and propose mechanisms for a variety of reactions reported in the literature. The book does not cover all types of reactions. Nonetheless, anyone who works all the problems will gain insights that should facilitate the writing of reasonable mechanisms for many organic reactions.

Literature sources for most of the problems are provided. The papers cited do not always supply an answer to the problem but put the problem into a larger context. The answers to problems and examples often consider more than one possible mechanism. Pros and cons for each mechanism are provided. In order to emphasize the fact that frequently more than one reasonable pathway to a product may be written, in some cases experimental evidence supporting a particular mechanism is introduced only at the end of consideration of the problem. It is hoped that this approach will encourage users of this book to also consider more than one mechanistic pathway.

I acknowledge with deep gratitude the help of all the students who have taken the course upon which this book is based. Special thanks to Drs. David Kronenthal, Tae-Woo Kwon, John Freilich, and Professor Hilton Weiss for reading the manuscript and making extremely helpful suggestions. Many thanks to Dr. James Holden for his editing of the entire manuscript and to my editor, Nancy Olsen, for her constant encouragement.

Chapter 1

Introduction

This chapter presents a review and practice of the fundamental principles that are useful tools in the writing of the mechanisms for organic reactions. In order to write reasonable reaction mechanisms you must have an understanding of electron distribution and density and the relationship between the electrons in the starting materials and in the products. Mastering the techniques in this chapter will facilitate writing sensible representations of mechanisms.

1-1 Writing Structures for Organic Compounds

1-1-1 HOW TO WRITE LEWIS STRUCTURES AND CALCULATE FORMAL CHARGES

In order to write or understand reaction mechanisms, it is essential to be able to construct Lewis structures for any organic compound. This is especially true because lone pairs of electrons, which are frequently used when writing the mechanism of a reaction, are often not pictured in the chemical literature.

1

There are two methods commonly used to show Lewis structures. One shows all the electrons as dots. The other shows all bonds (2 shared electrons) as lines and all unshared electrons as dots. This book will use the latter method.

Helpful Hint 1-1. To facilitate the drawing of Lewis structures, estimate the number of bonds in stable structures using the following equation when the number of electrons is even:

(Electron Demand − Electron Supply)/2 = Number of bonds

When neutral molecules are being considered, the contribution of each atom to the electron supply is the number of electrons in the outer shell of that neutral atom. This is the group number of the element in the periodic table. The electron demand is 2 for each hydrogen and 8 for all other atoms usually considered in organic chemistry, **with the exception of elements in group IIIA: B, Al, and Ga.** For these elements the demand is 6. Other exceptions are noted, as they arise, in examples and problems. The electron supply must be decreased by one for each positive charge of a cation and must be increased by one for each negative charge of an anion.

Use the estimated number of bonds to draw that number of two-electron bonds in your structure. This may involve drawing a number of double and triple bonds. Once the bonds are drawn, add the lone pairs of electrons to atoms (like nitrogen, oxygen, sulfur, phosphorus and the halogens) to reach a total of 8 electrons. When this process is complete, there should be 2 electrons for hydrogen; 6 for B, Al or Ga; and 8 for all other atoms. This total number of electrons for each element in the final representation of a molecule is obtained by counting each electron around the element as 1 electron, even if the electron is shared with another atom. (This should not be confused with counting electrons for charges or formal charges.) This total number of electrons is also equal to the electron demand of the element. Thus, when the total number of electrons around each element equals the electron demand, the number of bonds will be as calculated in Helpful Hint 1-1.

Using the molecular formula for a molecule, the total number of rings and/or π bonds can be calculated. This is based on the fact that an acyclic saturated hydrocarbon contains $(2n + 2)$ hydrogens when n is the number of carbons; and each time a ring or π bond is formed, there will be two less hydrogens necessary to complete the structure.

Helpful Hint 1-2. From the number of carbons (n) in the molecular formula, calculate the number of hydrogens for the corresponding saturated hydrocarbon $(2n + 2)$. Take the number of hydrogens in the molecular formula, subtract 1 for every nitrogen or phosphorus and add 1 for every halogen. Subtract this number from $2n + 2$ and divide by two. This number is the number of rings and/or π bonds in the compound. This hint does not hold for molecules in which there are atoms, like sulfur and phosphorus, whose valence shell has expanded beyond 8.

Example 1-1. Calculate the number of rings and/or π bonds corresponding to each of the following molecular formulas.

(a) $C_2H_2Cl_2Br_2$ For this molecular formula n = 2 and 2n + 2 = 6. Taking the number of hydrogens (2) and adding 4, 1 for each halogen, gives a total of 6. Thus the number of rings and/or π bonds is 0/2 or 0.

(b) C_2H_3N For this molecular formula 2n + 2 is also 6. Taking the number of hydrogens (3) and subtracting a hydrogen because of the nitrogen present, gives 2. (6 − 2)/2 = 2. The molecular formula C_2H_3N represents a compound which contains either 2 rings, 2 π bonds or 1 π bond and 1 ring. (See Example 1-2.)

Some of the atoms in some neutral molecules have charges. Because the total charge of the molecule balances to zero, these charges are called formal charges. To calculate formal charges, use the completed Lewis structure. For each atom count the following:

(number of unshared electrons) + [(number of shared electrons)/2]

If this number is equal to the valence shell electrons of the neutral atom (as ascertained from the group number in the periodic table), there is no formal charge. If the number is more than the number of valence shell electrons, the charge is negative (one for each excess electron); if the number is less than the valence shell electrons, the formal charge is positive (one for each excess proton).

Example 1-2. Calculate the formal charges on all atoms in the following structures.

(a)

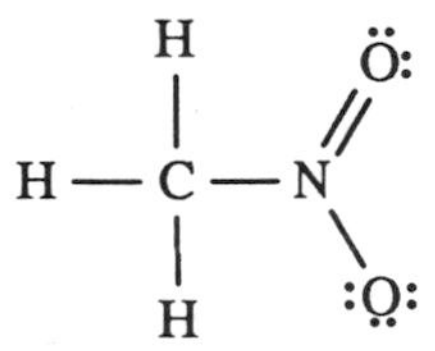

$$
\begin{array}{ccc}
 & H & \ddot{O}: \\
 & | & \nearrow\!\!\!\nearrow \\
H\!-\!\!&C\!-\!N & \\
 & | & \searrow \\
 & H & :\ddot{Q}:
\end{array}
$$

The number of electrons for each hydrogen is 2/2 = 1. This equals the number of valence shell electrons for hydrogen, and hydrogen has 0 formal charge. The number of electrons for carbon is 8 (two for each bond)/2 = 4. This equals the number of valence shell electrons for carbon which has 0 formal charge. The number of electrons for nitrogen is 8/2 = 4. This is one less than the number of valence shell electrons so nitrogen has a charge of +1. The doubly bonded oxygen has 4 + 4/2 = 6 electrons for a formal charge of 0. The singly bonded oxygen has 6 + 2/2 = 7 electrons for a formal charge of −1.

(b)

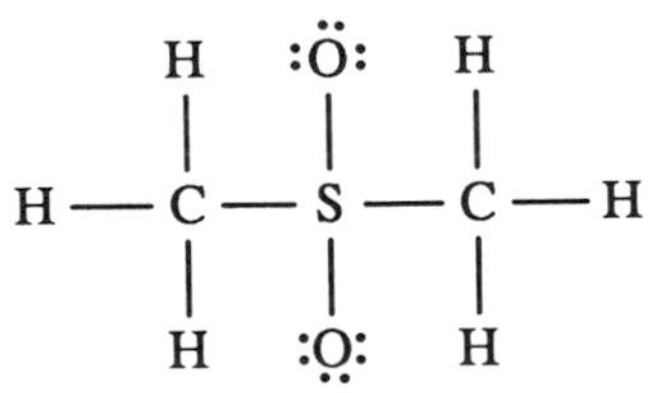

$$
\begin{array}{ccccc}
 & H & :\ddot{O}: & H & \\
 & | & | & | & \\
H\!-\!&C\!-\!&S\!-\!&C\!-\!&H \\
 & | & | & | & \\
 & H & :\ddot{Q}: & H &
\end{array}
$$

The calculations for carbon and hydrogen are the same as those for part (a). Each oxygen has 6 + 2/2 = 7 electrons for a formal charge of −1. The sulfur has 8/2 = 4 electrons for a formal charge of +2.

Helpful Hint 1-3. When drawing Lewis structures, be consistent with the following common structural features of molecules.

1.   Since hydrogen forms only one covalent bond, it must always be on the periphery of a structure.

2. The typical bonding patterns for carbon, oxygen and nitrogen are described below.

   In any of the structures below, the R groups may be hydrogen, alkyl or aryl groups or any combination of these groups. These substituents do not change the bonding pattern depicted.

(a) With few exceptions, which include CO, isonitriles (RNC), and carbenes, carbon in neutral molecules has four bonds. Such molecules include compounds which contain just σ bonds and also compounds which contain a combination of σ and π bonds.

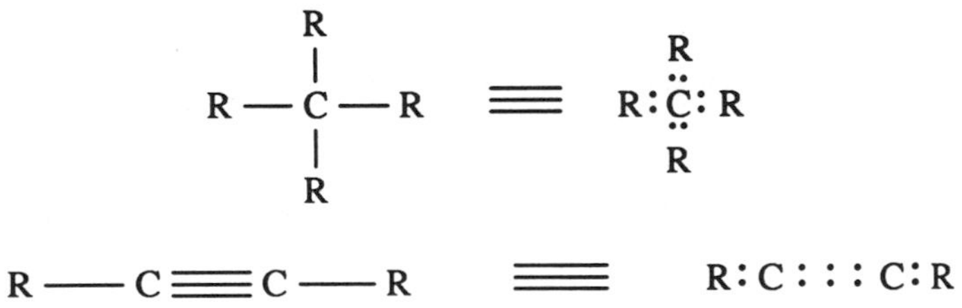

$$R-\overset{\displaystyle R}{\underset{\displaystyle R}{\overset{|}{\underset{|}{C}}}}-R \quad\equiv\quad R\!:\!\overset{..}{\underset{..}{C}}\!:\!R$$

$$R-C\equiv C-R \quad\equiv\quad R\!:\!C\!:\!:\!:\!C\!:\!R$$

(b) Carbon with a single positive or negative charge has three bonds.

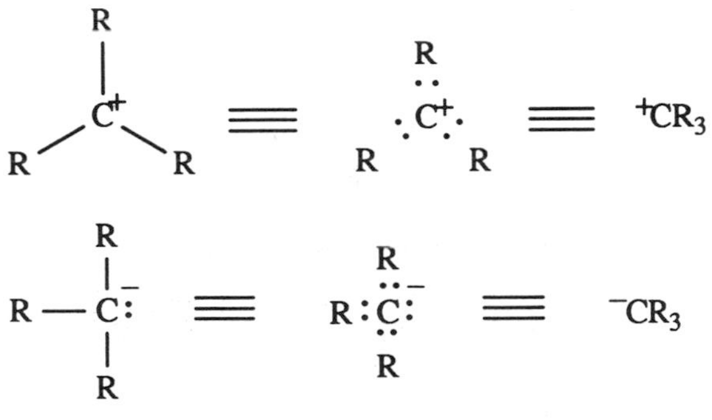

$$\overset{\displaystyle R}{\underset{R\diagdown \diagup R}{\overset{|}{C^+}}} \quad\equiv\quad \overset{\displaystyle R}{\underset{R\quad R}{\overset{..}{.\,C^+.}}} \quad\equiv\quad {}^+CR_3$$

$$R-\overset{\displaystyle R}{\underset{\displaystyle R}{\overset{|}{\underset{|}{C}}}}\!:^{-} \quad\equiv\quad R\!:\!\overset{..}{\underset{..}{C}}\!:^{-} \quad\equiv\quad {}^-CR_3$$

(c) Neutral nitrogen, with the exception of nitrenes (see Chapter 4), has three bonds and a lone pair.

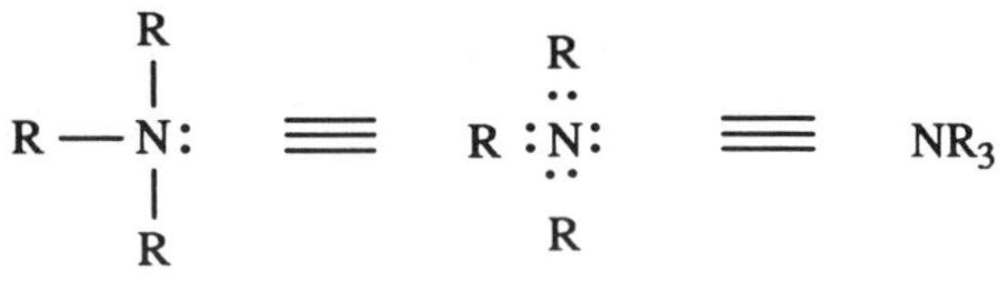

$$R-\overset{\displaystyle R}{\underset{\displaystyle R}{\overset{|}{\underset{|}{N}}}}\!: \quad\equiv\quad R\!:\!\overset{..}{\underset{..}{N}}\!: \quad\equiv\quad NR_3$$

(d)   Positively charged nitrogen has four bonds and a positive
      charge; exceptions are nitrenium ions (see Chapter 4).

$$R-\overset{\overset{\displaystyle R}{|}}{\underset{\underset{\displaystyle R}{|}}{N^{+}}}-R \quad \equiv \quad R:\overset{\overset{\displaystyle R}{..}}{\underset{\underset{\displaystyle R}{..}}{N}}{}^{+}\!\!:R \quad \equiv \quad {}^{+}NR_4$$

(e)   Negatively charged nitrogen has two bonds and two lone
      pairs of electrons.

$$R-\overset{..}{\underset{\underset{\displaystyle R}{|}}{N}}{}^{-}\!: \quad \equiv \quad R:\overset{..}{\underset{..}{N}}{}^{-}\!: \quad \equiv \quad {}^{-}NR_2$$

(f)   Neutral oxygen has two bonds and two lone pairs of elec-
      trons.

$$R\diagdown\!\!\overset{..}{\underset{..}{O}}\!\!\diagup R \quad \equiv \quad R:\overset{..}{\underset{..}{O}}: \quad \equiv \quad R_2O$$

(g)   Oxygen-oxygen bonds are uncommon; they are present
      only in peroxides, hydroperoxides and diacyl peroxides.
      (See Chapter 5.) The formula, RCO_2R, implies the follow-
      ing structure:

$$R\diagdown\!\!\overset{\overset{\displaystyle \overset{..}{O}:}{\|}}{C}\!\!\diagup\!\overset{..}{\underset{..}{O}}\!\!\diagdown R$$

(h)   Positive oxygen usually has three bonds and a lone pair of
      electrons; exceptions are the very unstable oxenium ions
      which contain a single bond to oxygen and two lone pairs
      of electrons.

$$R\diagdown\!\!\overset{+}{\underset{\underset{\displaystyle R}{|}}{\overset{..}{O}}}\!\!\diagup R \quad \equiv \quad R:\overset{..}{\underset{..}{O}}{}^{+}\!\!:R \quad \equiv \quad R_3O^{+}$$

3.   Sometimes a phosphorus or sulfur atom in a molecule is
      depicted with 10 electrons. Since phosphorus and sulfur
      have d orbitals, the outer shell can be expanded to accom-

modate more than 8 electrons. If the shell, and therefore
the demand, is expanded to 10 electrons, one more bond
will be calculated by the equation used to calculate number of bonds. See Example 1-4.

Example 1-3. Write possible Lewis structures for C_2H_3N.

	electron supply	electron demand
3H	3	6
2C	8	16
1N	5	8
	16	30

The estimated number of bonds is (30–16)/2 = 7.

As calculated in Example 1-1, this molecular formula represents molecules which contain 2 rings and/or π bonds. However, since it requires a minimum of 3 atoms to make a ring,
and since hydrogen cannot be part of a ring because each hydrogen forms only one bond, 2 rings are not possible. Thus, all
structures with this formula will have either a ring and a π
bond, or 2 π bonds. Because no information is given on the
order in which the carbons and nitrogen are bonded, all possible bonding arrangements must be considered.

Structures **1-1** through **1-9** depict some possibilities. The
charges shown in the structures are formal charges. When
charges are not shown, the formal charge is zero.

1-1

1-2

1-3

1-4

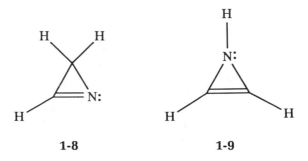

Structure **1-1** contains 7 bonds using 14 of the 16 electrons of the electron supply. The remaining two electrons are supplied as a lone pair of electrons on the carbon so that both carbons and the nitrogen have 8 electrons around them. This structure is unusual because the right hand carbon does not have four bonds to it. Nonetheless, **1-1**, an isonitrile, (see Helpful Hint 1-3) is expected to be isolable. Structure **1-2** is a resonance form of **1-1**. Traditionally, **1-1** is written instead of **1-2**, because in the former, both carbons have an octet. Structures **1-3** and **1-4** represent resonance forms for another isomer. However, when a neutral structure like **1-3** can be written for an isomer, a charged resonance form like **1-4** is usually not written because it doesn't contribute as much to the resonance hybrid (see Section 1-3). (These principles, explained for **1-1** and **1-3**, can also be applied to alternative forms for the subsequent structures **1-5** through **1-9**.) Because of the strain energy of three-membered rings and cumulated double bonds, it is anticipated that **1-6** through **1-9** are quite unstable.

It is always a good idea to check your work by counting the

number of electrons shown in the structure. The number of electrons you have drawn must be equal to the supply of electrons.

In the literature a formula is often written to indicate the bonding skeleton for the molecule. This severely limits, often to just one, the number of possible structures which can be written. For example, if you had been asked for the Lewis structure of CH_3CN in Example 1-3, your answer would have been **1-5**. The way in which the formula is represented, indicates that one carbon is bonded to three hydrogens, and that the other carbon and nitrogen are not bonded to any hydrogen. Furthermore, the formula indicates that the skeleton is C-C-N. Then it remains for you to fill in the bonds and lone pair of electrons.

Example 1-4. Write two possible Lewis structures for dimethyl sulfoxide, $(CH_3)_2SO$, and calculate formal charges for all atoms in each structure.

	electron supply	electron demand
2C	8	16
6H	6	12
1S	6	8
1O	6	8
	26	44

According to Helpful Hint 1-1, the estimated number of bonds in (44–26)/2 = 9. Also, Helpful Hint 1-2 calculates 0 rings and/or π bonds. The way the formula is given indicates that both methyl groups are bonded to the sulfur which is also bonded to oxygen. Drawing the skeleton gives the following:

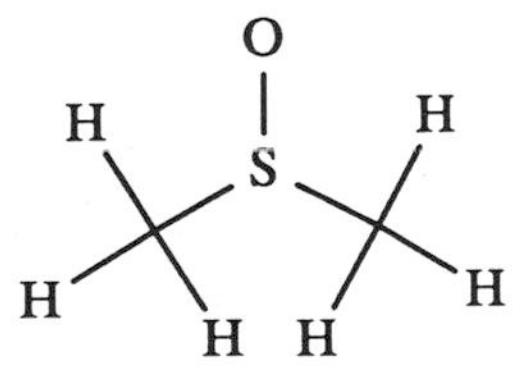

The nine bonds use up 18 electrons from the total supply of 26. Thus there are 8 electrons (four lone pairs) to fill in. In order to have octets at sulfur and oxygen, 3 lone pairs are placed on oxygen and one lone pair on sulfur.

$$:\ddot{O}:$$

1-10

The formal charge on oxygen in **1-10** is −1. There are 6 unshared electrons and $2/2 = 1$ electron from the pair being shared. Thus, the number of electrons is 7, one more than the valence electrons for oxygen.

The formal charge on sulfur in **1-10** is +1. There are 2 unshared electrons and $6/2 = 3$ electrons from the pairs being shared. Thus, the number of electrons is 5, one less than the valence electrons for sulfur.

All of the other atoms in **1-10** have a formal charge of 0.

There is another reasonable structure, **1-11**, for dimethyl sulfoxide which corresponds to an expansion of the valence shell of sulfur to accommodate 10 electrons. Note that our calculation of electron demand counted 8 electrons for sulfur. The 10-electron sulfur has an electron demand of 10 and leads to a total demand of 46 rather than 44 and the calculation of 10 bonds rather than 9 bonds. All atoms in this structure have zero formal charge.

1-11

Helpful Hint 1-2 does not predict the π bond in this molecule, because the valence shell of sulfur has expanded beyond 8. Structures **1-10** and **1-11** correspond to different possible resonance forms for dimethyl sulfoxide (see Section 1-3), and each is a viable possible structure.

Why isn't just one of these two possible structures of dimethyl sulfoxide usually written? Probably because neither structure is totally analogous to a C=O group. For the carbonyl group, a double bond is written between the C and the O, because in that way neither carbon or oxygen has a charge and both have an octet, e.g. **1-12**. Carbon does not have an octet if a single bond is written between the C and the O as in **1-13**.

1-12   **1-13**

By analogy to **1-12**, **1-11** is a structure in which all formal charges are zero and by analogy, **1-10** is a structure in which every atom has an octet.

Helpful Hint 1-4. When the electron supply is an odd number, the resulting unpaired electron will produce a radical; that is, the valence shell of one atom, other than hydrogen, will not be completed. This atom will have 7 electrons instead of 8. Thus, if you get a 1/2 when you calculate the number of bonds, that 1/2 represents a radical in the final structure.

Problem 1-1. Write Lewis structures for each of the following and show any formal charges.

a. *CH_2=CHCHO*

b. *NO_2^+ BF_4^-*

c. *hexamethylphosphorous triamide, [(CH$_3$)$_2$N]$_3$P*

d. *$CH_3N(O)CH_3$*

e. *CH_3SOH (methylsulfenic acid)*

1-1-2 USUAL REPRESENTATIONS OF ORGANIC COMPOUNDS

As illustrated above, the bonds in organic structures are represented by lines. Often, some or all of the lone pairs of electrons are not represented in any way. The reader must fill them in when necessary. To organic chemists, the most important atoms which have lone pairs of electrons are those in groups VA, VIA and VIIA of the periodic table: N, O, P, S and the halogens. The lone pairs on these elements can be of critical concern when writing a reaction mechanism. Thus, you must remember that lone pairs may be present even if they are not shown in the structures as written. For example, the structure of anisole might be written with or without the lone pairs of electrons on oxygen:

Other possible sources of confusion, as far as electron distribution is concerned, are ambiguities you may see in literature representations of cations and anions. Illustrated below are several representations of the resonance forms of the cation produced when anisole is protonated in the *para* position by concentrated sulfuric acid. There are three features to note in the first representation of the product, **1-14**. First, two lone pairs of electrons are shown on the oxygen. Second, the positive charge shown on carbon means that that carbon has one less electron than neutral carbon: electrons on carbon = (6 shared electrons)/2 = 3, while neutral carbon has 4 electrons. Third, both hydrogens are drawn in the *para* position to emphasize that this carbon is now sp^3-hybridized. The second structure for the product, **1-15-1**, represents the overlap of one of the lone pairs of electrons on the oxygen with the rest of the π system. The electrons originally shown as a lone pair are now forming the second bond between oxygen and carbon. Representation **1-15-2**, the kind of structure commonly found in the literature, means exactly the same thing as **1-15-1**, but, *for simplicity, the lone pair on oxygen is not shown.*

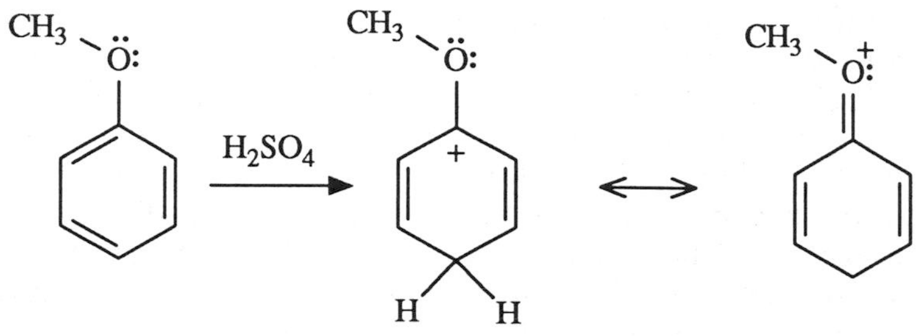

1-14   **1-15-1**

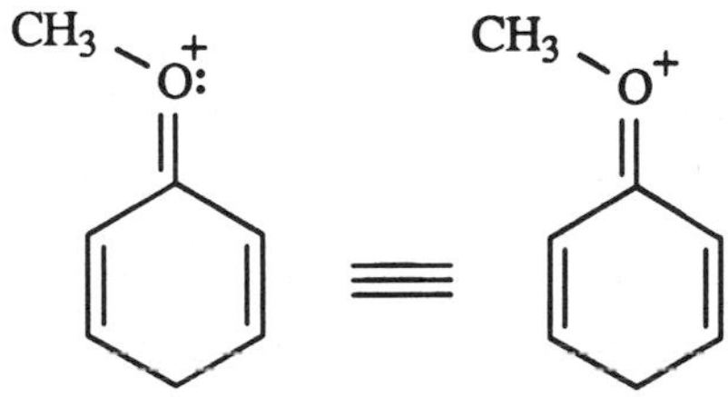

1-15-1   **1-15-2**

There are also several ways in which anions are represented. Sometimes a line represents a pair of electrons (as in bonds or lone pairs of electrons), sometimes a line represents a negative charge, and sometimes a line means both. The following structures represent the anion formed when a proton is removed from the oxygen of isopropyl alcohol.

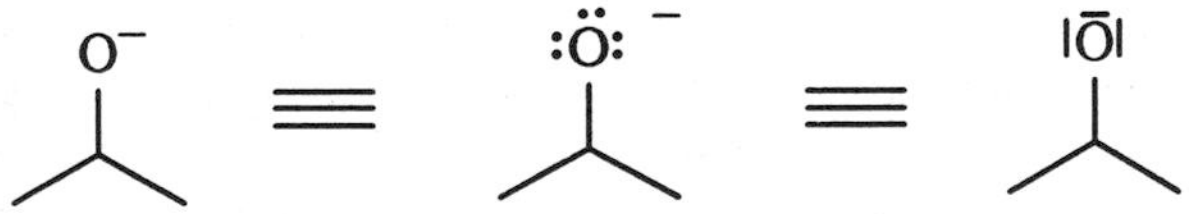

All three representations are equivalent, though the first two are those most commonly used.

1-1-3 GEOMETRY AND HYBRIDIZATION

Particular geometries, spatial orientations of atoms in a molecule, can be related to particular bonding patterns in molecules. These bonding patterns led to the concept of hybridization, derived from a mathematical model. In that model the wave functions for the s and p orbitals in the outermost quantum shell are mixed in ways such that geometries, close to those observed, result. Thus, different combinations of the s orbital and the three p orbitals lead to different geometries. These combinations are called hybridized orbitals.

The hybridization of C, N, O, and P and S are of the most concern because these are the most common atoms besides hydrogen that are encountered in organic compounds. If the expansion of the octet possible for P and S is not included, simple rules can be used to predict the hybridization of any of these atoms in a molecule. If A is a C, N, O, P, or S atom, X is the number of atoms attached to A, and E is the number of lone pairs of electrons on A, then:

1. If $X + E = 4$, A will be sp^3-hybridized and the ideal geometry will have bond angles of 109.5°. The designation, sp^3, means that the character of the s orbital and all three p orbitals are mixed. The sum of the superscripts in a designation of a hybrid orbital is the number of atomic orbitals whose character is mixed and also the number of the hybrid orbitals which is produced. Thus there are four sp^3-hybridized orbitals and according to VSEPR (Valence Shell Electron Pair Repulsion) theory, they will be directed to the corners of a tetrahedron to get the electrons as far apart as possible.

   Notable exceptions are atoms with unshared electrons which may be sp^2 hybridized with the unshared electrons in a p orbital, if this hybridization leads to increased delocalization. Examples are the heteroatoms of structures **1-30** through **1-33** in Example 1-11.

2. If $X + E = 3$, A will be sp^2-hybridized. There will be three hybrid orbitals and an unhybridized p orbital will remain. Again, the hybrid orbitals will be located as far apart as possible. This leads to an ideal geometry with 120° bond angles between the three coplanar hybrid orbitals and 90° between the hybrid orbitals and the remaining p orbital.

3. If $X + E = 2$, A will be sp-hybridized and two unhybridized p

orbitals will remain. The hybrid orbitals will be linear (180°
bond angles), and the p orbitals will be perpendicular to the
linear system and perpendicular to each other.

**Example 1-5. For the following molecule predict the hybridiza-
tions of each carbon and oxygen and discuss the geometries
about each of these atoms.**

The oxygen atom contains two lone pairs of electrons, so X + E
= 3. Thus oxygen is sp²-hybridized. Two of the sp² orbitals are
occupied by the lone pairs of electrons. The third sp² orbital
overlaps with a sp²-hybridized orbital at C-2 to form the C-O σ
bond. The lone pairs and C-2 lie in a plane approximately 120°
from one another. There is a p orbital perpendicular to this
plane.

  C-2 is sp²-hybridized. The three sp²-hybridized orbitals
overlap with orbitals on O-1, C-3 and C-7 to form three σ bonds
which lie in the same plane at approximately 120° from each
other. The p orbital, perpendicular to this plane, is parallel to
the p orbital on O-1 and these p orbitals overlap to produce the
C-O π bond.

  Carbons 3, 4, 5 and 8 are sp³-hybridized. (The presence of
hydrogen atoms is assumed.) Bond angles are approximately
109.5°.

  Carbons 6 and 7 are sp²-hybridized. They are doubly
bonded by a σ bond, produced from hybrid orbitals, and a π
bond produced from their p orbitals.

  Because of the geometrical constraints imposed by the
sp²-hybridized atoms; atoms 1, 2, 3, 5, 6, 7, and 8 all lie in the
same plane.

*Problem 1-2. Discuss the hybridization and geometry for each of
the atoms in the following molecules or intermediates.*

a. $CH_3 - C{\equiv}N$

b. $PhN{=}C{=}S$

c. $(CH_3)_3P$

d. $CH_2{=}CHCH_2^-Li^+$

1-2 Electronegativities and Dipoles

Many organic reactions depend upon the interaction of a molecule which has a positive or fractional positive charge with a molecule which has a negative or fractional negative charge. In neutral organic molecules one can determine the existence of a fractional charge by noting the difference in electronegativity, if any, between the atoms at the ends of a bond. A useful scale of relative electronegativities was established by Linus Pauling. These values are given in Table 1-1, which also gives the relative position of the elements in the periodic table. The larger the electronegativity value, the more electron-attracting the element. Thus, fluorine is the most electronegative element shown in the table.

TABLE 1-1
Relative Values for Electronegativities

H				
2.6				
B	C	N	O	F
2.3	2.7	3.2	3.7	4.0
Al	Si	P	S	Cl
1.7	2.1	2.5	3.0	3.5
				Br
				3.2
				I
				2.8

From Sanderson, R. T. *J. Am. Chem. Soc.* **1983**, *105*, 2259–2261.

An aid to remembering relative electronegativities is to remember that carbon, phosphorus, and iodine have about the same electronegativity and that electronegativity increases as one proceeds either from left to right across a row or from bottom to top up a column in the periodic table.

From the relative electronegativities of the atoms, the relative fractional charges can be ascertained for bonds.

Example 1-6. Relative dipoles in some common bonds.

$$\overset{\delta^+}{C} - \overset{\delta^-}{O} \qquad \overset{\delta^+}{C} = \overset{\delta^-}{O} \qquad \overset{\delta^+}{C} - \overset{\delta^-}{Br} \qquad \overset{\delta^+}{C} - \overset{\delta^-}{N}$$

In all cases the more electronegative element has the fractional negative charge. In the second structure there will be more fractional charge than in the first, because the π electrons in the second structure are held less tightly by the atoms and thus are more mobile. The C-Br bond is expected to have a weaker dipole than the C-O single bond because bromine is not as electronegative as oxygen.

Problem 1-3. Predict the direction of the dipole in the bonds highlighted in the following structures.

a.

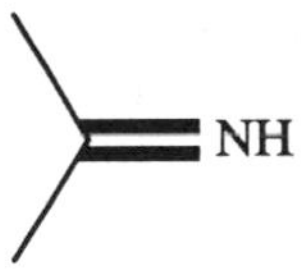

NH

b. Br━━F

c. CH_3 ━━ $N(CH_3)_2$

d. CH_3 ━━ $P(CH_3)_2$

1-3 Resonance Structures and Aromaticity

1-3-1 INTRODUCTION AND DRAWING RESONANCE STRUCTURES

Resonance structures are a formality which help predict the most likely electron distribution in a molecule. A simple method for finding the resonance structures for a given compound or intermediate is to draw one of the resonance structures and then by using arrows to show the movement of electrons, draw a new structure with a different electron distribution. This movement of electrons is formal only; that is, such electron flow does not actually take place in the molecule. The actual molecule is a hybrid of the resonance structures which incorporates some of the characteristics of each resonance structure. Thus, resonance structures themselves are not structures for real molecules or intermediates but are a formality which helps to predict the electron distribution for the real structures. **Resonance structures, and only resonance structures, are separated by a double-headed arrow.**

Example 1-7. Write the resonance structures for naphthalene.

First draw a structure, **1-16,** for naphthalene which shows alternating single and double bonds around the periphery. This is one of the resonance structures which contributes to the character of delocalized naphthalene, a resonance hybrid.

1-16   **1-17**

Each arrow drawn within **1-16** indicates movement of the π electron pair of a double bond to the location shown by the head of the arrow. This gives a new structure **1-17** which can then be manipulated in a similar manner to give a third structure, **1-18.**

1-17   **1-18**

Finally, when all forms have been figured out, they can be presented in the following manner:

How do you know that all possible resonance forms have been written? This is accomplished only by trial and error. If you keep pushing electrons around the naphthalene ring, you will continue to draw structures, but they will be identical to one of the three previously written.

What are some of the pitfalls of this method? If only a single electron pair in **1-17** is moved, **1-19** is obtained. However, this structure doesn't make sense. At the carbon labelled 1, there are five bonds to carbon; this is a carbon with 10 electrons. However, it is not possible to expand the valence shell of carbon. Similar rearrangement of other π bonds in either **1-16, 1-17**, or **1-18** would lead to similar nonsense structures.

1-17   **1-19**

A second possibility would be to move the electrons of a double bond to just one of the terminal carbons; this leads to a structure like **1-20**. However, when more than one neutral resonance structure can be written, doubly charged resonance structures, like **1-20**, and **1-21**, contribute an insignificant amount to the resonance hybrid and are usually not written.

1-17   **1-20**

1-20   **1-21**

Example 1-8. Write resonance forms for the intermediate in the nitration of anisole at the *para* position.

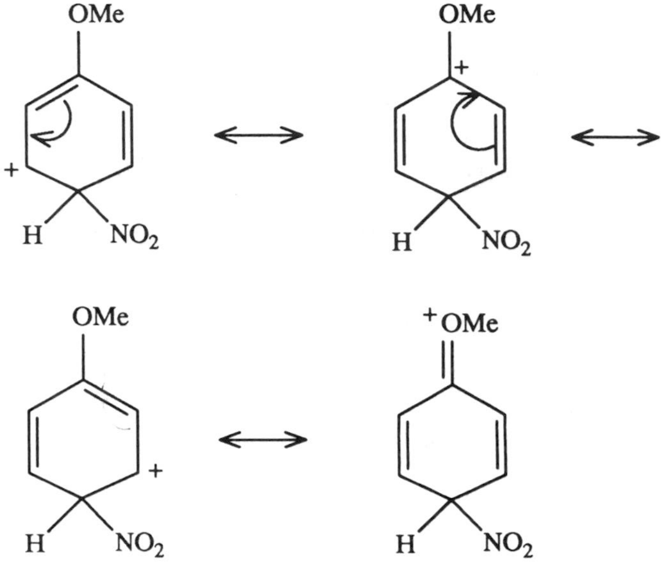

The last structure can be found in the following way:

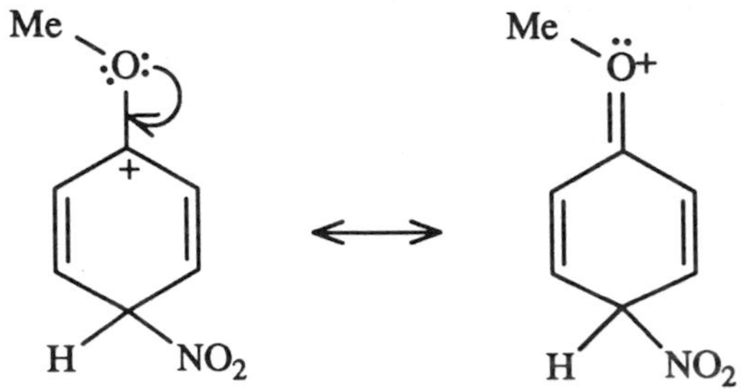

There are actually twice as many resonance forms than those shown since the nitro group is also capable of electron delocalization. Thus, for each resonance form written above, two resonance forms can be substituted in which the nitro group's electron distribution has also been written out:

Because the nitro group is attached to an sp^3-hybridized carbon, it is not conjugated with the electrons in the ring and is
not important to their delocalization. Thus, if resonance forms
were being written to rationalize the stability of the intermediate in the nitration of anisole, the detail in the nitro groups
would not be important, because it does not contribute to the
stabilization of the carbocation intermediate.

A negative charge on atoms in Table 1-1, other than the group III
elements, can also be interpreted as an electron pair. Often a pair of
electrons and a negative sign are used interchangeably. (See Section
1-1-2). For example, the compound depicted in Problem 1-4 e is a
species with a double negative charge; there is an excess of two electrons in the molecule. However, when resonance structures are
drawn for this species, each negative charge shown actually represents two electrons.

Problem 1-4. Draw resonance structures for each of the following.

a. *anthracene*

b.

c. *PhCH$_2$⁺*

d.

e.

f.

(This is the anion radical of 1-iodo-2-benzoylnaphthalene.)

Problem 1-5. Either p-dinitrobenzene or m-dinitrobenzene is commonly used as a radical trap in electron transfer reactions. The compound which forms the most stable radical anion is the better trap. Consider the radical anions formed when either of these starting materials adds an electron and predict which compound is commonly used.

1-3-2 RULES FOR RESONANCE STRUCTURES

1. All of the electrons involved in delocalization are π electrons or electrons, like lone pairs, that can readily be put into p orbitals.

2. Each of the electrons involved in delocalization must have some overlap with the other electrons. This means that if the

orbitals are oriented at a 90° angle, there will be no overlap. The best overlap will occur when the orbitals are oriented at a 0° angle.

3. Each resonance structure must have the same number of π electrons. Count 2 for each π bond; only 2 electrons are counted for a triple bond since only one of the π bonds of a triple bond can overlap with the conjugated π system. Also count 2 for an anion and 0 for a positive charge.

4. The same number of electrons must be paired in each structure. Structures **1-22** and **1-23** are not resonance structures, because they do not have the same number of paired electrons. In **1-22** there are two pairs of paired π electrons: a pair of electrons for the π bond and a pair of electrons for the anion. In **1-23** there is one pair of paired π electrons and two unpaired electrons (shown by the up arrows).

1-22

1-23

5. All resonance structures must have identical geometries. Otherwise they do not represent the same molecule. For example, the following structure is not a resonance form of benzene, because it is not planar. (Note: If it is assumed that the central bond in this structure is a π bond, then it has the same number of π electrons as benzene.)

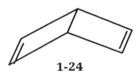

1-24

6. Resonance structures which depend on charge separation are of higher energy and do not contribute as significantly to the resonance hybrid as those structures which do not depend on charge separation.

much more important than

CH_2 ⟋⟍ CH_2 much more important than $\overset{+}{C}H_2$ ⟋⟍ $\overset{-}{C}H_2$

7. Usually, resonance structures are more important when the negative charge is on the most electronegative atom and the positive charge is on the most electropositive atom.

contributes more than

In some cases, aromatic anions or cations are exceptions to this rule. (See below.)

In the example below, 1-26 is less favorable than 1-25, because in 1-26 the more electronegative atom, oxygen, is positive. In other words, although both the positive carbon in 1-25 and the positive oxygen in 1-26 do not have an octet, it is especially destabilizing when the much more electronegative oxygen bears the positive charge.

contributes more than

1-25   1-26

8. Electron stabilization is best when there are two or more structures of lowest energy.

9. The resonance hybrid is more stable than any of the contributing structures.

Problem 1-6 a. Do you think delocalization as shown by the following resonance structures is important? Explain why or why not.

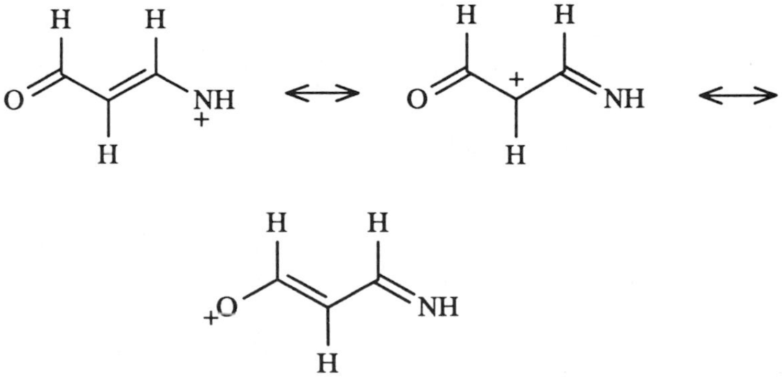

b. If the charges were negative instead of positive would your answer be different? Explain.

Problem 1-7. Write Lewis structures for each of the following and show any formal charges. Also draw all resonance forms for these species.

a. CH₃NO₂

b. PhN₂⁺

c. CH₃COCHN₂ (diazoacetone)

d. N₃CN

1-3-3 AROMATICITY AND ANTIAROMATICITY

Certain cyclic totally conjugated π systems show unusual stability. These systems are said to be aromatic. Hückel originally narrowly described aromatic compounds as those completely conjugated, monocyclic carbon compounds which contain $(4n + 2)$ π electrons. In this designation, n can be 0, 1, 2, 3..., so such systems which contain 2, 6, 10, 14, 18 ... π electrons are aromatic. This description is called Hückel's Rule.

Example 1-9. Some aromatic compounds which strictly obey Hückel's Rule.

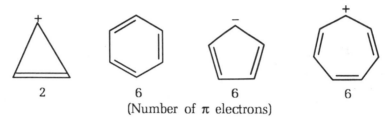

(Number of π electrons)

In these systems, each double bond contributes two electrons, each positive charge on carbon contributes none, and the negative charge (designation of an anion) contributes two electrons. If the first and last two structures did not have a charge on the atom not associated with a double bond, they would not be aromatic, because the π system would not be completely delocalized. That is, if the cyclopropenyl ring is depicted as uncharged, then there are two hydrogens on the carbon with no double bond. This carbon is then sp^3-hybridized and has no p orbital to complete a delocalized system.

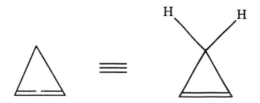

Hückel's Rule has been expanded to cover heterocyclic compounds and fused polycyclic compounds because when these compounds have the requisite number of electrons, they also show unusual stability. In heterocyclic compounds a lone pair of electrons on the heterocyclic atom may be counted as part of the conjugated π system, if necessary or if possible, to attain the correct number for an aromatic system.

Example 1-10. Some aromatic fused ring systems.

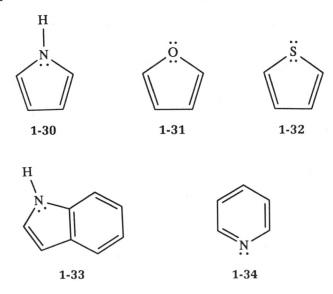

1-27   **1-28**

1-29

Structures **1-27** and **1-28**, which contain 10 totally conjugated π electrons, are examples of Hückel's Rule with n = 2. Structure **1-29** obeys Hückel's Rule with n = 4. In such systems only the electrons located on the periphery of the structure are counted when Hückel's Rule is applied. (See Problem 1-8 e where not all of the electrons in the structure are counted.)

Example 1-11. Some aromatic heterocycles.

1-30   **1-31**   **1-32**

1-33   **1-34**

These examples illustrate how the lone pairs of electrons are considered in determining aromaticity. In each of the examples, the carbons and the heteroatoms are sp²-hybridized, ensuring a planar system with a p orbital perpendicular to this plane at each position in the ring. In **1-30** it is necessary that two electrons are contributed by the nitrogen to give a total of 6π electrons. Thus, the lone pair of electrons would be in the p orbital on the nitrogen. In **1-31** one of the lone pairs of electrons on the oxygen is also in a p orbital parallel with the rest of the π system in order to give a 6π electron aromatic system. Thus, the other lone pair on oxygen must be in an sp²-hybridized orbital which, by definition, is perpendicular to the conjugated π system and therefore cannot contribute to the number of electrons in the overlapping π system. The considerations concerning the sulfur in **1-32** are identical to those for oxygen in **1-31** so this is a 6π electron system. Compound **1-33** is similar to **1-30** with overlap of the additional fused 6-membered ring making this an aromatic 10π electron system. In sharp contrast to the other examples, a totally conjugated 6π electron system is formed in **1-34** with the contribution of only one electron from the nitrogen. The lone pair of electrons will then be in an sp²-hybridized orbital perpendicular to the aromatic 6π electron system. Thus, these two electrons are not part of the delocalized system. In conclusion *one or two electrons from the heteroatom are contributed to its p orbital, as necessary, to reach the number of electrons essential to an aromatic system.*

What about totally conjugated systems which contain 4n (4, 8, 12, 16 ...) π electrons? These systems are actually destabilized by delocalization and are said to be antiaromatic.

Example 1-12. Some antiaromatic systems.

The first three examples contain 4π electrons and the last 8π electrons. All are highly unstable species. On the other hand, cyclooctatetraene, **1-35**, an 8π electron system, is much more stable than any of the above. This is because the π electrons in cyclooctatetraene are not significantly delocalized: the 8-membered ring is bent into a tub-like structure and adjacent π bonds are not parallel.

1-35

Problem 1-8. Classify each of the following compounds as aromatic, antiaromatic, or nonaromatic.

a.

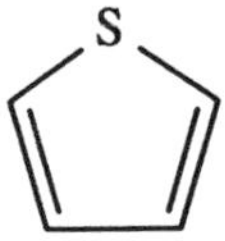

b.

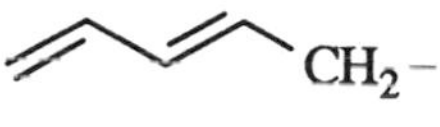

CH$_2$−

c.

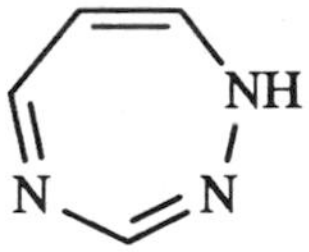

NH

d.

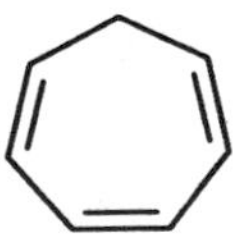

e.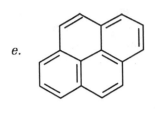

f.

g.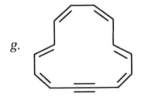

1-4 Equilibrium and Tautomerism

Equilibrium exists when there are equal rates for both the forward and reverse processes of a reaction. Equilibrium is usually designated by half-headed arrows shown for both the forward and reverse reaction. If it is known that one side of the equilibrium is favored, this may be indicated by a longer arrow pointing to the side that is favored.

Example 1-13. An acid-base equilibrium.

$$CH_3CO_2H \ + \ H_2O \ \rightleftharpoons \ CH_3CO_2^- \ + \ H_3O^+$$

Tautomers are isomers which differ in the arrangement of single and double bonds and a small atom, usually hydrogen. Under appropriate reaction conditions, such isomers can equilibrate by a simple mechanism.

Example 1-14. Tautomeric equilibria of ketones.

A common example of tautomers are the keto and enol forms of aldehydes and ketones. This special case of equilibrium is called tautomerism.

NOTE: Students have a tendency to confuse the concept of resonance with that of tautomerism. The following three facts must be clearly distinguished:

(1) Tautomerism is an equilibrium between **isomers**, which differ in the location of a double bond and a hydrogen. Such equilibration is shown with a pair of half-headed arrows as are other equilibria.

(2) Resonance structures, shown with a double headed arrow between them, **ARE NOT** different chemical species in equilibrium; they **DO** represent different π bonding patterns.

(3) Unlike tautomeric structures, in all resonance structures for a given species, the σ bonding pattern (with a few unusual exceptions) and geometry are identical.

Example 1-15. Tautomerism versus resonance.

Compounds **1-36** and **1-37** are tautomers; they are isomers and are in equilibrium with each other:

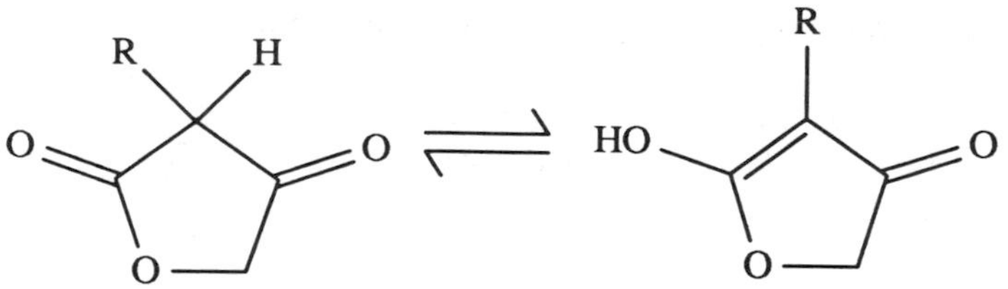

1-36   **1-37**

On the other hand, **1-38**, **1-39**, and **1-40** are resonance forms. The hybrid of these structures can be formed from **1-36** or **1-37** by removing the acidic proton.

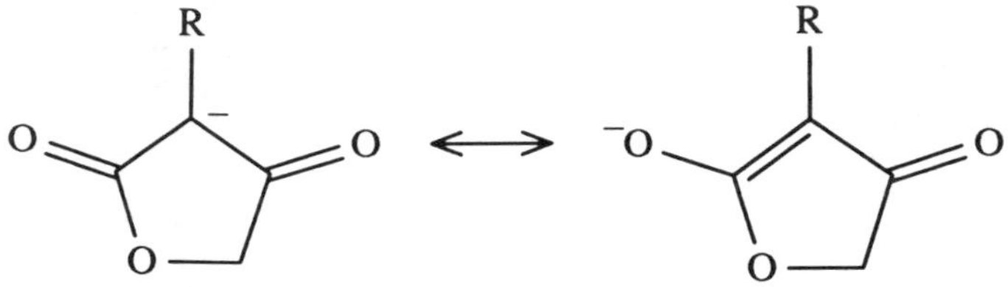

1-38   **1-39**

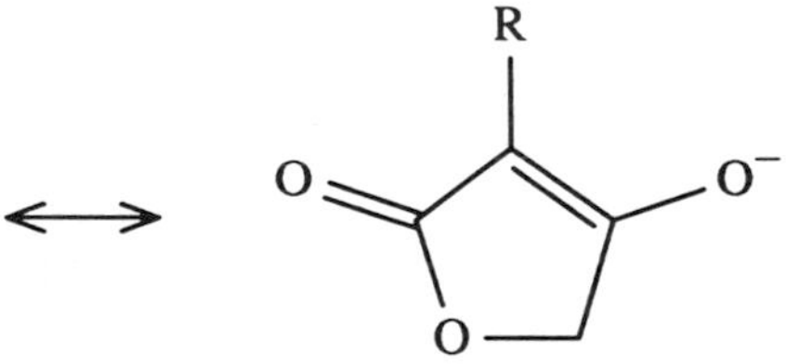

1-40

Note that **1-38**, **1-39**, and **1-40** have the same atoms attached to the same atoms, while the tautomers **1-36** and **1-37** differ in the position of a proton.

Problem 1-9. Write tautomeric structures for each of the following. The number of tautomers you should write, in addition to the original structure, is shown in parentheses.

a.

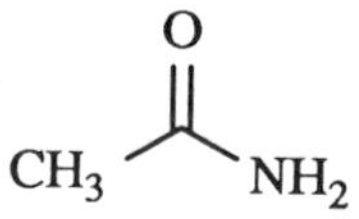

(2)

b. CH_3CHO (1)

c. $CH_3CH=CHCHO$ (1)

d.

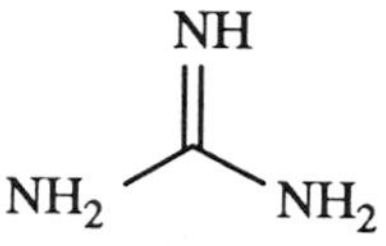

(2)

e.

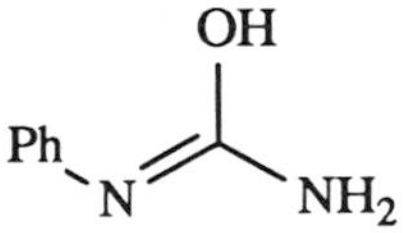

(5)

Problem 1-10. For each of the following sets of structures indicate whether they are tautomers, resonance forms, or the same molecule.

a.

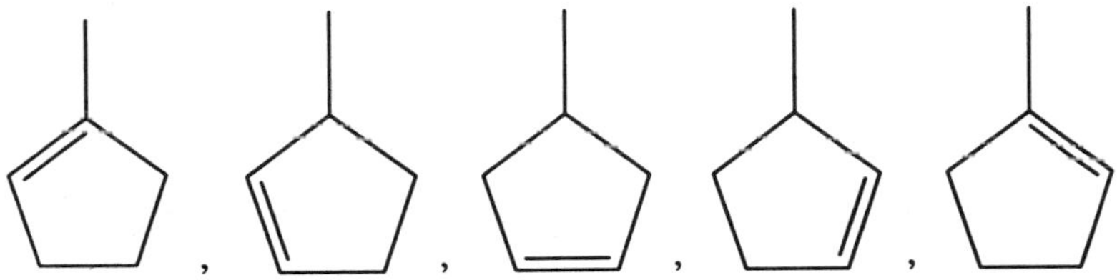

b.

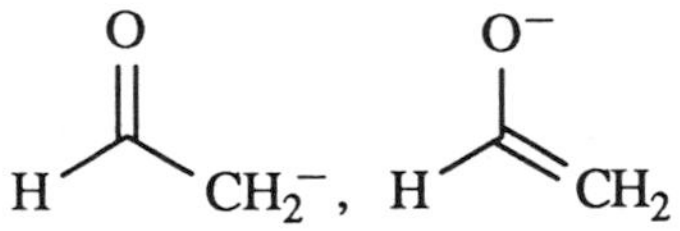

c.

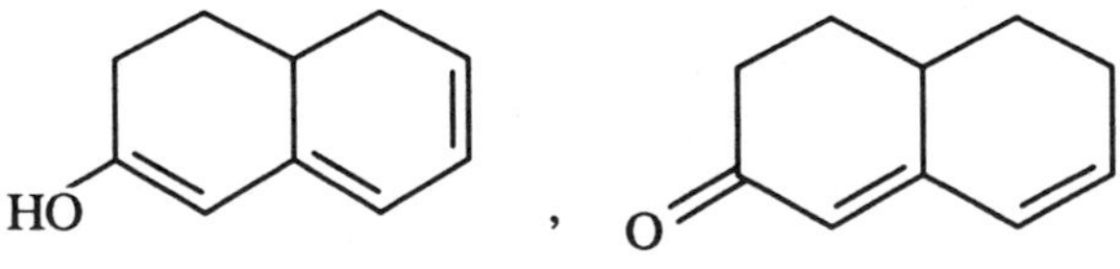

Problem 1-11. *In a published paper, two structures were presented in the following manner and referred to as resonance
forms. Are the structures shown actually resonance forms? If not,
what are they and what change(s) need to be made to make the
picture correct?*

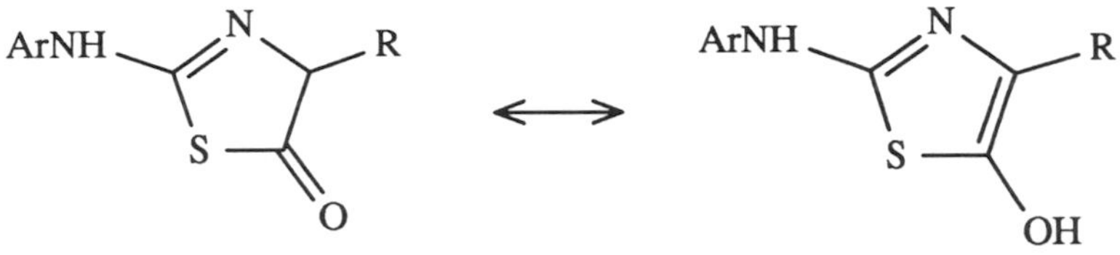

1-5 Acidity and Basicity

A Bronsted acid is a proton donor. A Bronsted base is a proton ac-
ceptor.

$$CH_3CO_2H + CH_3NH_2 \rightleftharpoons CH_3CO_2^- + CH_3\overset{+}{N}H_3$$

   acid   base   conjugate base   conjugate acid

If this equation were reversed, the definitions would be similar:

$$CH_3\overset{+}{N}H_3 + CH_3CO_2^- \rightleftharpoons CH_3NH_2 + CH_3CO_2H$$

   acid   base   conjugate base   conjugate acid

In each equation the acids are the proton donors and the bases are
proton acceptors.

There is an inverse relationship between the acidity of an acid
and the basicity of its conjugate base. That is, the more acidic the
acid, the weaker the basicity of the conjugate base and *vice versa*.
For example, if the acid is very weak like methane, the conjugate
base, the methyl carbanion, is a very strong base. On the other hand,
if the acid is very strong, like sulfuric acid, the conjugate base, the
HSO_4- ion is a very weak base. Because of this reciprocal relationship between acidity and basicity, most references to acidity and
basicity use a single scale of pK_a values, and relative basicities are
obtained from the relative acidities of the conjugate acids. Table 1-2
lists the pK_a values for a variety of acids. Especially at very high pK_a
values, the chart may be inaccurate since various approximations
have to be made in the measurements of such values. This is often
the reason why different pK_a values are given for the same acid in
the literature.

TABLE 1-2

Relative Acidities of Common Organic and Inorganic Substances[a]

Acid	Solvent	pK_a	Conjugate Base	Reference
HI	Ref b	-10	I^-	c
HBr	Aq. H_2SO_4	-9	Br^-	d
HCl	Aq. H_2SO_4	-8	Cl^-	d
$(CH_3)_2\overset{+}{S}H$	Aq. H_2SO_4	-6.99	$(CH_3)_2S$	e
$C_6H_5\text{-}\overset{+}{O}\overset{H}{\underset{CH_3}{\cdot}}$	Aq. H_2SO_4	-6.5	$C_6H_5\text{-}OCH_3$	f
$Ph\text{-}C(\overset{\overset{H\cdot}{O^+}}{\|})\text{-}OEt$	Aq. H_2SO_4	-6.2	$PhCO_2Et$	g
CF_3SO_3H	Ref h	-5.1(-5.9)	$CF_3SO_3{}^-$	c,i
$HClO_4$	Ref h	-5.0	$ClO_4{}^-$	i
FSO_3H	Ref h	-4.8(-6.4)	$FSO_3{}^-$	c,i
$Ph\text{-}C(\overset{\overset{H\cdot}{O^+}}{\|})\text{-}OH$	Aq. H_2SO_4	-4.7	$PhCO_2H$	g
$Ph\text{-}C(\overset{\overset{H\cdot}{O^+}}{\|})\text{-}CH_3$	Aq. H_2SO_4	-4.3	$Ph\text{-}C(\overset{O}{\|})\text{-}CH_3$	g
$PhCH=\overset{+}{O}H$	Aq. H_2SO_4	-3.9	$PhCH=O$	g
$(CH_3)_2\overset{+}{O}H$	Aq. H_2SO_4	-3.8	$(CH_3)_2O$	j
$Ph\text{-}C(\overset{\overset{H\cdot}{S^+}}{\|})\text{-}NH_2$	Aq. H_2SO_4	-3.20	$Ph\text{-}C(\overset{S}{\|})\text{-}NH_2$	k
$(CH_3)_2C=\overset{+}{O}H$	Aq. H_2SO_4	-2.85	$(CH_3)_2C=O$	e
$PhSO_3H$	Ref h	-2.8	$PhSO_3{}^-$	i
H_2SO_4	Ref h	-2.8	$HSO_4{}^-$	i

TABLE 1-2 (continued)

Acid	Solvent	pK_a	Conjugate Base	Reference
(dihydropyranyl oxocarbenium) $\overset{+}{O}\!-\!H$	Aq. H_2SO_4	-2.8	(dihydropyran) O	j
$CH_3\!-\!\overset{\overset{\displaystyle \overset{H\cdot}{S^+}}{\|}}{C}\!-\!NH_2$	Aq. H_2SO_4	-2.51	$CH_3\!-\!\overset{\overset{\displaystyle S}{\|}}{C}\!-\!NH_2$	k
$CH_3\overset{+}{O}H_2$	Aq. H_2SO_4	-2.5	CH_3OH	l
CH_3SO_3H	Ref m	-1.9	$CH_3SO_3^-$	c
$Ph\!-\!\overset{\overset{\displaystyle \overset{H\cdot}{O^+}}{\|}}{C}\!-\!NH_2$	Aq. H_2SO_4	-1.74	$Ph\!-\!\overset{\overset{\displaystyle O}{\|}}{C}\!-\!NH_2$	n
H_3O^+	Ref h	-1.7	H_2O	i
$Ph\!-\!\overset{\overset{\displaystyle \overset{H\cdot}{O^+}}{\|}}{C}\!-\!NHCH_3$	Aq. H_2SO_4	-1.7	$Ph\!-\!\overset{\overset{\displaystyle O}{\|}}{C}\!-\!NHCH_3$	n
$CH_3\!-\!\overset{\overset{\displaystyle \overset{+OH}{S}}{\|\|}}{}\!-\!CH_3$	Aq. H_2SO_4	-1.5	$CH_3\!-\!\overset{\overset{\displaystyle O}{\|\|}}{S}\!-\!CH_3$	k
HNO_3	Ref h	-1.3	NO_3^-	i
$NH_2\!-\!\overset{\overset{\displaystyle \overset{H\cdot}{S^+}}{\|}}{C}\!-\!NH_2$	Aq. H_2SO_4	-1.26	$NH_2\!-\!\overset{\overset{\displaystyle S}{\|}}{C}\!-\!NH_2$	o
$CH_3\!-\!\overset{\overset{\displaystyle \overset{H\cdot}{O^+}}{\|}}{C}\!-\!NH_2$	Aq. H_2SO_4	-0.6	$CH_3\!-\!\overset{\overset{\displaystyle O}{\|}}{C}\!-\!NH_2$	n
CF_3CO_2H	H_2O	-0.6	$CF_3CO_2^-$	c
Cl_3CCO_2H	H_2O	-0.5	$Cl_3CO_2^-$	c
$H\!-\!\overset{\overset{\displaystyle \overset{H\cdot}{O^+}}{\|}}{C}\!-\!NH_2$	Ref p	-0.48	$H\!-\!\overset{\overset{\displaystyle O}{\|}}{C}\!-\!NH_2$	q

TABLE 1-2 (continued)

Acid	Solvent	pK_a	Conjugate Base	Reference
$\overset{H\cdot_{O^+}}{\underset{CH_3}{\overset{\|}{\diagdown}}}N(CH_3)_2$	H_2O, CH_3NO_2	0.1	$\overset{O}{\underset{CH_3}{\overset{\|}{\diagdown}}}N(CH_3)_2$	r
$\overset{H\cdot_{O^+}}{\underset{NH_2}{\overset{\|}{\diagdown}}}NH_2$	Ref p	0.5	$\overset{O}{\underset{NH_2}{\overset{\|}{\diagdown}}}NH_2$	q
$(Ph)_2\overset{+}{N}H_2$	H_2O	0.8	$(Ph)_2NH$	s
$PhSO_2H$	H_2O	1.2	$PhSO_2^-$	c
HO_2CCO_2H	H_2O	1.25	$HO_2CCO_2^-$	c
Cl_2CHCO_2H	H_2O	1.35	$Cl_2CHCO_2^-$	c
$PhCH=\overset{+}{N}HOH$	H_2O	2.0	$PhCH=NOH$	c
H_3PO_4	H_2O	2.1	$H_2PO_4^-$	s
CH_3SO_2H	H_2O	2.3	$CH_3SO_2^-$	c
$\overset{+}{N}H_3CH_2CO_2H$	H_2O	2.35	$\overset{+}{N}H_3CH_2CO_2^-$	c
FCH_2CO_2H	H_2O	2.6	$FCH_2CO_2^-$	c
$ClCH_2CO_2H$	H_2O	2.86	$ClCH_2CO_2^-$	c
HF	H_2O	3.2	F^-	d
HNO_2	H_2O	3.4	NO_2^-	i
CH_3COSH	H_2O	3.4	CH_3COS^-	t
$O_2N-\!\!\!\diagup\!\!\!\diagdown\!\!\!-CO_2H$	H_2O	3.44	$O_2N-\!\!\!\diagup\!\!\!\diagdown\!\!\!-CO_2^-$	c
H_2CO_3	H_2O	3.7	HCO_3^-	u
HCO_2H	H_2O	3.75	HCO_2^-	v
$HOCH_2CO_2H$	H_2O	3.8	$HOCH_2CO_2^-$	v

TABLE 1-2 (continued)

Acid	Solvent	pK_a	Conjugate Base	Reference
Cl–C₆H₄–$\overset{+}{N}H_3$	H_2O	4.0	Cl–C₆H₄–NH_2	w
O_2N–C₆H₃(NO₂)–OH	H_2O	4.1	O_2N–C₆H₃(NO₂)–O^-	v
$PhCO_2H$	H_2O	4.2	$PhCO_2^-$	v
Ph–$\overset{+}{N}H_3$	H_2O	4.6	Ph–NH_2	w
CH_3CO_2H	H_2O	4.76	$CH_3CO_2^-$	i
$PhCH_2\overset{+}{N}H_2OH$	H_2O	4.9	$PhCH_2NHOH$	c
NH_2–C₆H₄–CO_2H	H_2O	4.92	NH_2–C₆H₄–CO_2^-	s
$Ph\overset{+}{N}H(CH_3)_2$	H_2O	5.1	$PhN(CH_3)_2$	x
C₅H₅$\overset{+}{N}H$	H_2O	5.2	C₅H₅N	x
$CH_3\overset{+}{N}H_2OH$	H_2O	6.0	CH_3NHOH	c
$\overset{+}{N}H_3OH$	H_2O	6.0	NH_2OH	c
Ph–SH	H_2O	6.5	Ph–S^-	y
H_2S	H_2O	7.0	HS^-	s
O_2N–C₆H₄–OH	H_2O	7.2	O_2N–C₆H₄–O^-	v
$\overset{+}{N}H_3OH$	H_2O	8.0	NH_2OH	yy
O_2N–C₆H₄–OH	H_2O	8.3	O_2N–C₆H₄–O^-	v

TABLE 1-2 (continued)

Acid	Solvent	pK$_a$	Conjugate Base	Reference
(phthalimide structure: benzene ring fused to a five-membered ring with two C=O and NH)	H$_2$O	8.3	(phthalimide anion: benzene ring fused to a five-membered ring with two C=O and N$^-$)	v
CH$_3$COCH$_2$COCH$_3$	H$_2$O	9.0	CH$_3$COC̄HCOCH$_3$	z
NH$_4^+$	H$_2$O	9.2	NH$_3$	l
(succinimide: five-membered ring with two C=O and NH)	H$_2$O	9.6	(succinimide anion: five-membered ring with two C=O and N̄)	aa
$^+$NH$_3$CH$_2$CO$_2^-$	H$_2$O	9.8	NH$_2$CO$_2^-$	s
(phenol: benzene ring with OH)	H$_2$O	10.0	(phenoxide: benzene ring with O$^-$)	v
CH$_3$NO$_2$	H$_2$O	10.0	$^-$CH$_2$NO$_2$	d
HCO$_3$ $^-$	H$_2$O	10.2	CO$_3$ $^=$	s
PhSH	DMSO	10.3	PhS$^-$	bb
EtSH	H$_2$O	10.6	EtS$^-$	c
(cyclohexyl–$^+$NH$_3$)	H$_2$O	10.7	(cyclohexyl–NH$_2$)	x
Et$_3$$^+$NH	H$_2$O	10.8	Et$_3$N	x
(2-carbethoxycyclohexanone, CO$_2$Et)	H$_2$O	10.9	(2-carbethoxycyclohexanone anion, CO$_2$Et)	v
(CH$_2$=C(OH)CH$_3$)	H$_2$O *	11.0	(CH$_2$=C(O$^-$)CH$_3$)	cc
PhCO$_2$H	DMSO	11.0	PhCO$_2^-$	d
Et$_2$$^+NH_2$	H$_2$O	11.0	Et$_2$NH	x
CH$_2$(CN)$_2$	DMSO	11.1	C̄H(CN)$_2$	dd
(piperidinium, $^+$NH$_2$)	H$_2$O	11.1	(piperidine, NH)	x

TABLE 1-2 (continued)

Acid	Solvent	pK_a	Conjugate Base	Reference
$CH_2(CN)_2$	H_2O	11.4	$\bar{C}H(CN)_2$	ee
HOOH	H_2O	11.6	HOO^-	ff
	H_2O	12.2		gg
$PhCH_2NO_2$	DMSO	12.3	$Ph\bar{C}HNO_2$	d
CH_3CO_2H	DMSO	12.3	$CH_3CO_2^-$	d
CF_3CH_2OH	H_2O	12.4	$CF_3CH_2O^-$	hh
$NCCH_2CO_2Me$	DMSO	12.8	$NC\bar{C}HCO_2Me$	ii
$CH_3COCH_2COCH_3$	DMSO	13.4	$CH_3COCHCOCH_3$	dd
	H_2O	13.4		c
$CH_3COCH_2CO_2Et$	DMSO	14.2	$CH_3CO\bar{C}HCO_2Et$	ii
	H_2O	14.5		jj
CH_3OH	H_2O	15.5	CH_3O^-	hh
H_2O	H_2O	15.7	HO^-	kk
CH_3CH_2OH	H_2O	15.9	$CH_3CH_2O^-$	kk
	DMSO	15.9		ll
	H_2O	15.9		gg
CH_3CHO	H_2O	16.5	$\bar{C}H_2CHO$	mm
$(CH_3)_2CHNO_2$	DMSO	16.9	$(CH_3)_2\bar{C}NO_2$	jj

TABLE 1-2 (continued)

Acid	Solvent	pK_a	Conjugate Base	Reference
$(CH_3)_2CHOH$	H_2O	17.1	$(CH_3)_2CHO^-$	kk
CH_3NO_2	DMSO	17.2	$\bar{C}H_2NO_2$	nn
$(CH_3)_3COH$	H_2O	18	$(CH_3)_3CO^-$	kk
(cyclopentadiene)	DMSO	18.1	(cyclopentadienyl anion)	dd
CH_3CSNH_2	DMSO	18.5	$CH_3CS\bar{N}H$	oo
CH_3COCH_3	H_2O	19.2	$\bar{C}H_2COCH_3$	cc
$O_2N-\langle\rangle-NH_2$	DMSO	20.9	$O_2N-\langle\rangle-\bar{N}H$	ll
$PhCH_2CN$	DMSO	21.9	$Ph\bar{C}HCN$	nn
Ph_2NH	$H_2O/DMSO$	22.4	$Ph_2\bar{N}$	pp
(fluorene)	DMSO	22.6	(fluorenyl anion)	nn
Ph_2NH	DMSO	23.5	Ph_2N^-	ll
$CHCl_3$	H_2O, tt	24	$^-CCl_3$	zz
CH_3COPh	DMSO	24.7	$\bar{C}H_2COPh$	nn
$HC\equiv CH$		25	$HC\equiv C^-$	qq
CH_3CONH_2	DMSO	25.5	$CH_3CO\bar{N}H$	oo
(Ph-C(=O)-CH(CH_3)_2 with H)	DMSO	26.3	(Ph-C(=O)-C(CH_3)_2 anion)	nn
CH_3COCH_3	DMSO	26.5	$\bar{C}H_2COCH_3$	nn
(3-Cl-C_6H_4-NH_2)	DMSO	26.7	(3-Cl-C_6H_4-\bar{N}H)	ll
NH_2CONH_2	DMSO	26.9	$NH_2CO\bar{N}H$	e
$EtCOEt$	DMSO	27.1	$EtCO\bar{C}HCH_3$	nn

TABLE 1-2 (continued)

Acid	Solvent	pK$_a$	Conjugate Base	Reference
Ph—C≡CH	DMSO	28.8	Ph—C≡C$^-$	nn
CH$_3$SO$_2$Ph	DMSO	29.0	$\bar{C}$H$_2$SO$_2$Ph	nn
Cl—⟨⟩—NH$_2$	DMSO	29.4	Cl—⟨⟩—$\bar{N}$H	nn
CH$_3$CO$_2$Et	DMSO	30.5	$\bar{C}$H$_2$CO$_2$Et	ii
(Ph)$_3$CH	DMSO	30.6	Ph$_3$C$^-$	nn
PhNH$_2$	DMSO	30.7	Ph$\bar{N}$H	ll
(dithiane-H)	DMSO	30.6	(dithiane anion)	d
(dithiane-Ph)	DMSO	30.7	(dithiane-Ph anion)	d
CH$_3$SO$_2$CH$_3$	DMSO	31.1	$\bar{C}$H$_2$SO$_2$CH$_3$	nn
CH$_3$CN	DMSO	31.2	$\bar{C}$H$_2$CN	dd
(Ph)$_2$CH$_2$	DMSO	32.3	(Ph)$_2$$\bar{C}$H	dd
CH$_3$CON(CH$_2$CH$_3$)$_2$	DMSO	34.5	$\bar{C}$H$_2$CON(CH$_2$CH$_3$)$_2$	ii
[(CH$_3$)$_2$CH]$_2$NH	THF	35.7	[(CH$_3$)$_2$CH]$_2$$\bar{N}$	rr
[(CH$_3$)$_2$CH]$_2$NH	THF	39	[(CH$_3$)$_2$CH]$_2$$\bar{N}$	ss
NH$_3$	nn	41	$\bar{N}$H$_2$	dd
PhCH$_3$	DMSO	43	$\bar{C}$H$_2$Ph	dd
benzene	CHA	43	⟨⟩$^-$	uu
CH$_2$=CH$_2$	nn	44	CH$_2$=$\bar{C}$H	vv
CH$_3$CH=CH$_2$	nn	47.1-48.0	$\bar{C}$H$_2$CH=CH$_2$	ww
CH$_3$CH$_3$	nn	approx. 50	$\bar{C}$H$_2$CH$_3$	xx
CH$_4$	nn	58 ± 5	$\bar{C}$H$_3$	ww

(a) Abbreviations: DMSO, dimethylsulfoxide; THF, tetrahydrofuran, CHA, cyclo-hexylamine. Most acidities were measured at 25°C. Some are extrapolated values; some are values from kinetic studies. Errors in some cases are several pK units. The further the pK value is from 0–14, the larger the errors because of estimates and assumptions made when water is not the solvent. Values of pK's for the same substance in different solvents differ because of differences in solvation. While Table 1-2 is careful to list the acids' actual structure, not all references do this. Thus you may find lists of the pK_a values for organic amines which refer to the pK_a of the protonated amine rather than of the amine itself. A good rule of thumb is that if the pK_a value given for an amine is <15, it must be the pK_a of the protonated amine rather than of the amine itself.

(b) Calculated from vapor pressure over a concentrated aqueous solution extrapolated to infinite dilution.

(c) Stewart, R. "The Proton: Applications to Organic Chemistry" **1985**, New York: Academic Press.

(d) Bordwell, F. G. *Acc. Chem. Res.* **1988**, *21*, 456–463.

(e) Perdoncin, G.; Scorrano, G. *J. Am. Chem. Soc.* **1977**, *99*, 6983–6986.

(f) Arnett, E. M.; Wu, C. Y. *Chem Ind.* **1959**, 1488.

(g) Edward, J. T.; Wong, S. C. *J. Am. Chem. Soc.* **1977**, *99*, 4229–4232.

(h) Estimated from model kinetic studies, extrapolated to aqueous media.

(i) Guthrie, J. P. *Can. J. Chem.* **1978**, *56*, 2342–2354.

(j) Arnett, E. M.; Wu, C. Y. *J. Am. Chem. Soc.* **1960**, *82*, 4999–5000.

(k) Lemetais, P.; Charpentier, J-M. *J. Chem. Res. (S)* 1981, 282–3.

(l) Deno, N. C.; Turner, J. O. *J. Org. Chem.* **1966**, *31*, 1969–1970.

(m) Highly concentrated solutions extrapolated to dilute aqueous media.

(n) Yates, K.; Stevens, J. B. *Can. J. Chem.* **1965**, *43*, 529–537.

(o) Janssen, M. J. *Rec. Trav. Chim. Pays-Bas* **1962**, *81*, 650–660.

(p) Titrated in acetic acid and corrected to H_2O at 20°C.

(q) Huisgen, R.; Brade, H. *Chem. Ber.* **1957**, *90*, 1432–1436.

(r) Adelman, R. L. *J. Org. Chem.* **1964**, *29*, 1837–1844.

(s) Weast, R. C., Ed. "CRC Handbook of Chemistry and Physics" **1982-1983**, Boca Raton: CRC Press.

(t) Kreevoy, M. M.; Eichinger, B. E.; Stary, F. E.; Katz, E. A.; Sellstedt, J. H., *J. Org. Chem.* **1964**, *29*, 1641–42.

(u) Bell, R. P. "The Proton in Chemistry", 2nd Ed. **1973**, Ithaca: Cornell U. Press.

(v) Bell, R. P.; Higginson, W. C. E. *Proc. Roy. Soc. (London)* **1949**, *197A*, 141–159.

(w) Biggs, A. E.; Robinson, R. A. *J. Chem. Soc.* **1961**, 388–393

(x) Perrin, D. D. "Dissociation Constants of Organic Bases in Aqueous Solution" **1965**, London: Butterworths.

(y) Liotta, C. L.; Perdue, E. M.; Hopkins, Jr. H. P. *J. Am. Chem. Soc.* **1974**, *96*, 7981–7985.

(z) Pearson, R. G.; Dillon, R. L. *ibid.* **1953**, *75*, 2439–2443.

(aa) Pine, S. H. "Organic Chemistry", 5th Ed. **1987**, New York: McGraw-Hill.

(bb) Bordwell, F. G.; Hughes, D. J. *J. Am. Chem. Soc.* **1985**, *107*, 4737–44.

(cc) Chiang, Y.; Kresge, A. J.; Tang, Y. S.; Wirz, J. *ibid.* **1984**, *106*, 460–462.

(dd) Bordwell, F. G.; Bartness, J. E.; Drucker, G. E.; Margolin, Z.; Matthews, W. S. *ibid.* **1975**, *97*, 3226–3227.

(ee) Hojatti, M.; Kresge, A. J.; Wang, W. H. *ibid.* **1987**, *109*, 4023–8.

(ff) Everett, A. J.; Minkoff, G. J. *Trans Far. Soc.* **1953**, *49*, 410–414.

(gg) Ross, A. M.; Whalen, D. L.; Eldin, S.; Pollack, R. M. *J. Am. Chem. Soc.* **1988**, *110*, 1981–1982.

(hh) Ballinger, P.; Long, F. A. *ibid.* **1959**, *81*, 1050–1053.

(ii) Bordwell, F. G.; Fried, H. E. *J. Org. Chem.* **1981**, *46*, 4327–4331.

(jj) Walba, H.; Isensee, R. W. *ibid.* **1956**, *21*, 702–704.

(kk) Murto, J. *Acta Chem. Scand.* **1964**, *18*, 1043–1053.

(ll) Bordwell, F. G.; Algrim, D. J. *J. Am. Chem. Soc.* **1988**, *110*, 2964–2968.

(mm) Guthrie, J. P. *Can. J. Chem.* **1979**, *57*, 1177–1185.

(nn) Matthews, W. S.; Bares, J. E.; Bartmess, J. E.; Bordwell, F. G.; Cornforth, F. J.; Drucker, G. E.; Margolin, Z.; McCallum, R. J.; McCollum, G. J.; Vanier, N. R. *J. Am. Chem. Soc.* **1975**, *97*, 7006–7014.

(oo) Bordwell, F. G.; Algrim, D. J. *J. Org. Chem.* **1976**, *41*, 2507–2508.

(pp) Dolman, D.; Stewart, R. *Can. J. Chem.* **1967**, *45*, 911–925 and 925–928.

(qq) Cram, D. J. "Fundamentals of Carbanion Chemistry" **1965**, New York: Academic Press.

(rr) Fraser, R. T.; Mansour, T. S. *J. Org. Chem.* **1984**, *49*, 3442–3443.

(ss) Chevrot, C.; Perichon, *J. Bull. Soc. Chim. Fr.* **1977**, 421–7.

(tt) Acidities of very weak acids are measured and/or calculated by a variety of indirect methods and may contain large errors.

(uu) Streitwieser, Jr. A.; Scannon, P. J.; Neimeyer, H. H. *J. Am. Chem. Soc.* **1972**, *94*, 7936–7937.

(vv) Maskornick, M. J.; Streitwieser, A. Jr. *Tetrahedron Lett.* **1972**, 1625–1628.

(ww) Juan, B.; Schwar, J.; Breslow, R. *J. Am. Chem. Soc.* **1980**, *102*, 5741–5748.

(xx) Streitwieser, Jr. A.; Heathcock, C. H. "Introduction to Organic Chemistry", 3rd Ed. **1985**, New York: Macmillan.

(yy) Bissot, T. C.; Parry, R. W.; Campbell, D. H. *J. Am. Chem. Soc.* **1957**, *79*, 796–800.

(zz) Margolin, Z.; Long, F. A. *ibid.* **1973**, *95*, 2757–2762.

Problem 1-12. For each of the following pairs indicate which is the strongest base. For b. and d. use resonance structures to rationalize the relative basicities.

a. $H\bar{C} = CH_2$, $\bar{C} \equiv CH$

b. $\bar{C}H_2CON(Et)_2$, $[(CH_3)_2CH]_2\bar{N}$

c. CH_3O^-, $(CH_3)_3CO^-$

d. *p-nitrophenolate, m-nitrophenolate*

Problem 1-13. For each of the following compounds indicate which proton is more likely to be removed when the compound is treated with base and present a rationale for your answer. Assume that equilibria are involved in each case.

a. $CH_3COCH_2COCH_3$

b. $H_2NCH_2CH_2OH$

c.

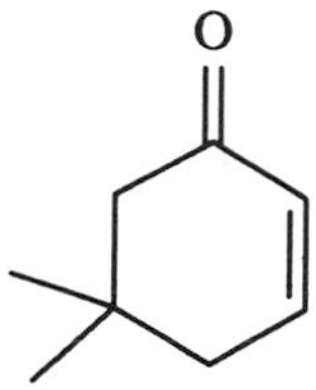

d.

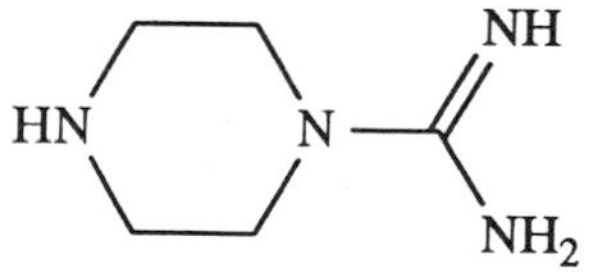

Example 1-16. Calculate the equilibrium constant for the following reaction, and predict if this is a likely reaction to the right.

$$Br^- \;+\; EtOH \;\rightleftharpoons\; HBr \;+\; EtO^-$$

For this reaction the equilibrium constant is:

$$K_a \;=\; \frac{[HBr]\,[EtO^-]}{[Br^-]\,[EtOH]}$$

This equilibrium constant can be calculated by an appropriate combination of the equilibrium constant for the ionization of ethanol and the equilibrium constant for the ionization of HBr. The equilibrium constant for the ionization of ethanol is:

$$10^{-15.9} \;=\; \frac{[EtO^-]\,[H^+]}{[EtOH]}$$

The equilibrium constant for the ionization of HBr is:

$$10^9 = \frac{[Br^-]\,[H^+]}{[HBr]}$$

If the equilibrium constant for ethanol ionization is divided by that for HBr ionization, the equilibrium constant for the reaction of bromide ion with ethanol is obtained:

$$K_a = \frac{10^{-15.9}}{10^9} = \frac{[HBr]\,[EtO^-]\,[H^+]}{[H^+]\,[Br^-]\,[EtOH]} = \frac{[HBr]\,[EtO^-]}{[Br^-]\,[EtOH]}$$

K_a equals $10^{-24.9}$. Thus, it can be concluded that bromide ion does not react with ethanol.

Problem 1-14. Calculate the equilibrium constants for each of the following reactions and predict which direction, forward or reverse, is favored.

a. CH_3CO_2Et + EtO^- $\rightleftharpoons$ $^-CH_2CO_2Et$ + $EtOH$

b.
$$Ph-\overset{\overset{\displaystyle O}{\|}}{C}-CH_3 \;+\; HO^- \;\rightleftharpoons\; Ph-\overset{\overset{\displaystyle O}{\|}}{C}-CH_2^- \;+\; H_2O$$

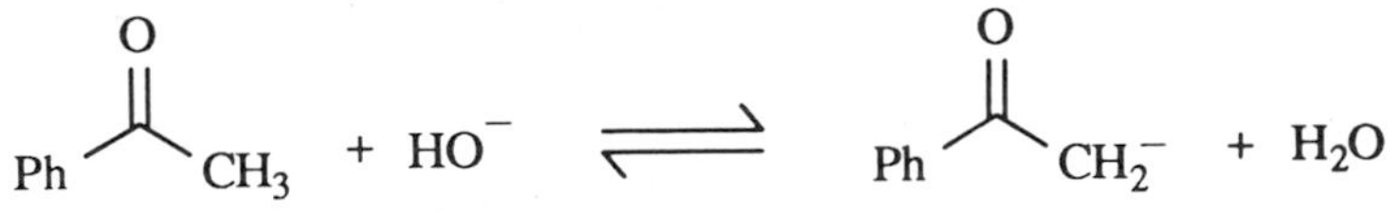

c. Ph_3CH + $^-N(i\text{-}Pr)_2$ $\rightleftharpoons$ Ph_3C^- + $HN(i\text{-}Pr)_2$

1-6 Nucleophiles and Electrophiles

A chemical species that is seeking an electron poor center is called a nucleophile. Such a molecule will have a negative or partially negative charge at the nucleophilic atom. The electron poor center will be an atom which has a positive or partially positive charge. Nucleophilicity is not to be confused with basicity, which refers to reactivity with a proton, though often there are parallels.

Table 1-3 shows one measure of nucleophilicity of nucleophiles toward carbon with larger n values meaning greater nucleophilicity. Each nucleophilicity is a comparison of the reactivity of the nucleophile with methyl bromide at 25°C in water relative to the reactivity of water with methyl bromide under the same conditions. The nucleophilicity, n, is calculated according to the following equation:

$$\log \frac{k}{k_o} = sn$$

where k is the rate constant for reaction with the nucleophile, k_o is the rate constant for reaction with water as the nucleophile, s = 1.00 for methyl bromide (as substrate) and n is the relative nucleophilicity.

TABLE 1-3

Nucleophilicities Toward Carbon[a]

Nucleophile	n	Nucleophile	n
$S_2O_3^{2-}$	6.4	pyridine	3.6
SH^-	5.1	Br^-	3.5
CN^-	5.1	PhO^-	3.5
SO_3^{2-}	5.1	$CH_3CO_2^-$	2.7
I^-	5.0	Cl^-	2.7
$PhNH_2$	4.5	$HOCH_2CO_2^-$	2.5
SCN^-	4.4	SO_4^{2-}	2.5
OH^-	4.2	$ClCH_2CO_2^-$	2.2
$(NH_2)_2CS$	4.1	F^-	2.0
N_3^-	4.0	NO_3^-	1.0
HCO_3^-	3.8	H_2O	0.0
$H_2PO_4^-$	3.9		

(a) From Wells, P. R. *Chem. Rev.*, **1963**, *63*, 171–219.

Relative reactivity in nucleophilic reactions also depends on the substrate and leaving group. While the s values for methyl bromide and chloroacetate are 1.00, they are 1.33 for iodoacetate, 0.87 for benzyl chloride and 1.43 for benzoyl chloride.

The nucleophilicity of a nucleophile is often solvent dependent. Expecially illustrative are the relative nucleophilicities, N_+, measured by Ritchie and coworkers utilizing the equation:

$$\log \frac{k_n}{k_{H_2O}} = N_+$$

where k_n = rate constant for reaction of a cation with a nucleophile in a given solvent, and k_{H_2O} = the rate constant for reaction of the same cation with water in water. Some N_+ values are given in Table 1-4. Note that nucleophilicity is greater in dipolar aprotic solvents like dimethyl sulfoxide and dimethyl formamide than in protic solvents like water or alcohols. Sometimes relative nucleophilicities change upon going from a protic to aprotic solvent. For example, the relative nucleophilicities of the halide ions in water are $I^->Br^->Cl^-$, while in dimethyl formamide the nucleophilicities are reversed; i.e., $Cl^->Br^->I^-$.

TABLE 1-4
Relative Nucleophilicities in Common Solvents[a]

Nucleophile (solvent)	N_+	Nucleophile (solvent)	N_+
H_2O (H_2O)	0.0	PhS^- (CH_3OH)	10.51
CH_3OH (CH_3OH)	1.18	PhS^- [$(CH_3)_2SO$]	12.83
CN^- (H_2O)	3.67	N_3^- (H_2O)	7.6
CN^- (CH_3OH)	5.94	N_3^- (CH_3OH)	8.85
CN^- [$(CH_3)_2SO$]	8.60	N_3^- [$(CH_3)_2SO$]	10.07
CN^- [$(CH_3)_2NCHO$]	9.33		

(a) From Ritchie, C. D. *J. Am. Chem. Soc.*, **1975**, *97*, 1170–1179.

Other common nucleophiles include ROH, RO^-, RCO_2^-, $ArCO_2^-$, RSH, RS^-, ArSH, ArS^-, RSR', ^-S-S^-, amines of all sorts, R_3P, $(RO)_3P$, hydrazine, ^-N=C=O, phthalimidate ion, $LiAlH_4$, $NaBH_4$, $LiEt_3BH$, RMgX, ArMgX, R_2CuLi, and salts of terminal alkynes.

The electron-deficient species which reacts with a nucleophile is called an electrophile. Typical electrophiles include Lewis acids like $ZnCl_2$, $AlCl_3$, and BF_3; PBr_3; $SOCl_2$; the carbon atom attached to X in RX where X = Cl, Br, or I; the α-carbon of α-haloacids, esters and ketones; the carbon atom of R which is attached to oxygen in $ROSO_2R'$, where R' = p-tolyl, CF_3, CH_3, etc.; CH_2N_2 (after protonation of carbon by acid); H_2O_2; epoxides; and nitrous acid (as its anhydride O=N-O-N=O).

Problem 1-15. In each of the following reactions label the electrophilic or nucleophilic center in each reactant. In a, b, and c different parts of acetophenone are behaving as either nucleophile or electrophile.

a.

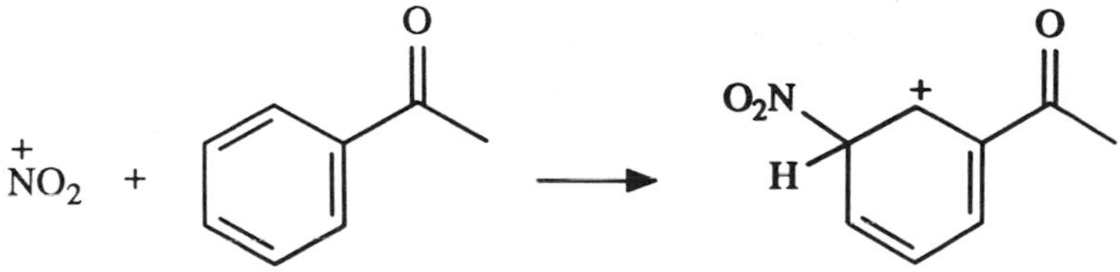

b.

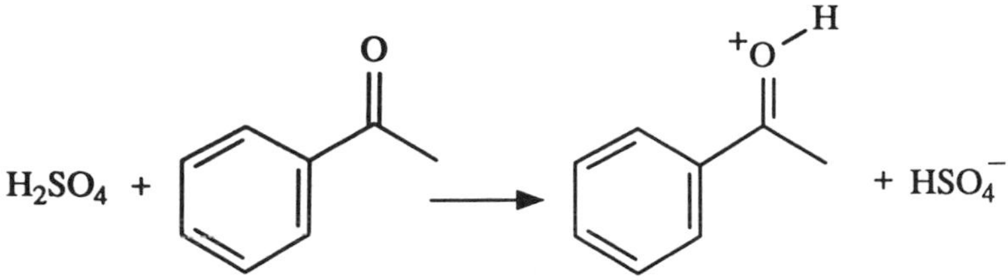

c.

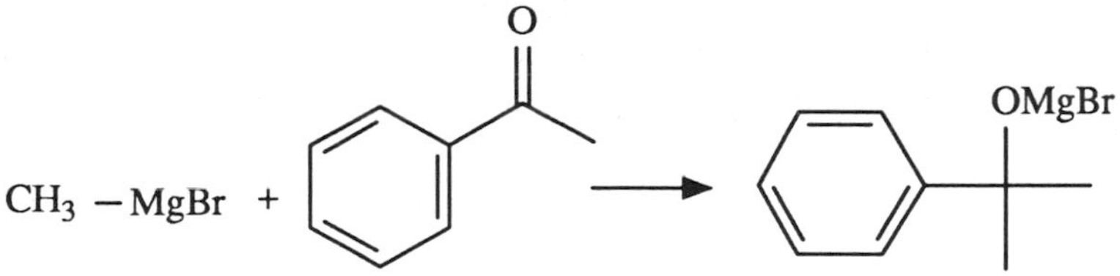

d.

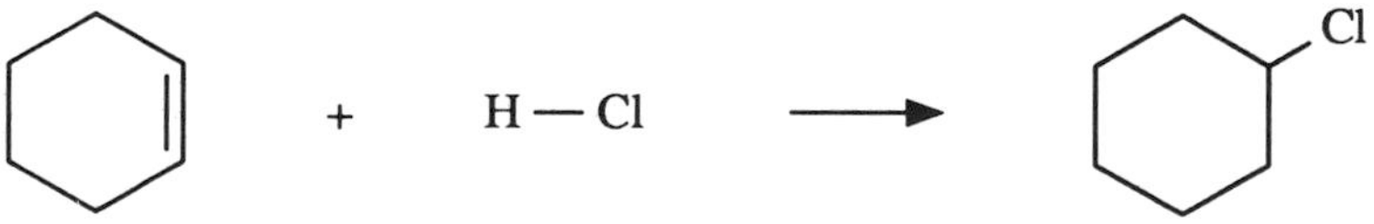

Answers to Problems

1-1.   For all parts of this problem, the overall carbon skeleton is given. Therefore, a good approach is to draw the skeleton of the molecule with single bonds and fill in with extra bonds, if necessary, to complete the octet of atoms other than hydrogen.

1-1.a.   Electron supply = (3x4)(C)+(1x6)(O)+(4x1)(H) =22. Electron demand = (3x8)(C) + (1x8)(O) + (4x2)(H) = 40. Estimate of bonds = (40–22)/2 = 9. This leaves two bonds left over after all atoms are joined by single bonds; thus, there is a double bond (as shown) between the CH_2 and CH group and a double bond between the second CH and the oxygen to give the following skeleton:

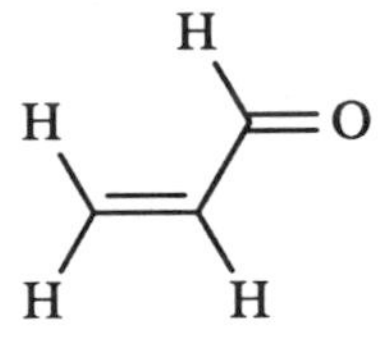

Calculating the number of rings and/or π bonds also shows that 2 π bonds are present. The molecular formula is C_3H_4O. The number of hydrogens for a saturated hydrocarbon is (2 x 3) + 2 = 8. (8–4)/2 = 2 rings and/or π bonds.

Eighteen electrons are used in making the 9 bonds in the molecule. There are 4 electrons left (from the original supply of 22); these fill in the octet on oxygen by adding two lone pairs of electrons to it.

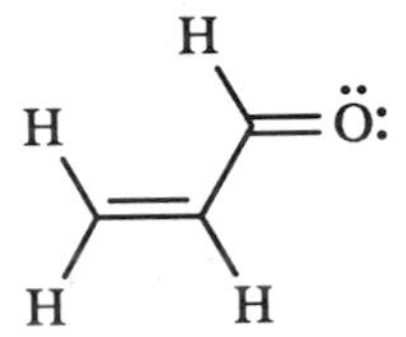

The structure depicted exclusively with dots would be:

$$\begin{array}{c} \text{H}\cdot \quad \cdot\text{H} \\ \ddot{}\text{C}::\text{C}\ddot{} \\ \text{H} \qquad \ddot{}\text{C}::\ddot{\text{O}}: \\ \text{H} \end{array}$$

1-1.b. The way the charges are written in the structure indicates that the NO_2^+ and the BF_4^- are separate entities. For NO_2^+: electron supply = (1x5)(N)+(2x6)(O)–1(positive charge) = 16. Electron demand = 3x8 = 24. Estimate of bonds = (24 –16)/2 = 4. (Electron supply – electrons used for bonds) = 8. These 8 electrons are used to complete the octets of the two oxygens.

$$:\ddot{\text{O}}::\overset{+}{\text{N}}::\ddot{\text{O}}: \quad \text{or} \quad :\ddot{\text{O}}=\!=\!\text{N}^+=\!=\ddot{\text{O}}:$$

For BF_4^- electron supply = (4 x 7)(F) + (1 x 3)(B) +1 (negative charge) = 32. Electron demand = 4 x 8 = 24. Estimate of bonds = (32–24)/2 = 4. (Electron supply – electrons used for bonds) = 24. These electrons are used to complete the octets of the four fluorines.

Since Helpful Hint 1-1 says to count the electron demand of boron as 6, why is a demand of 8 used for boron in this case? When boron is neutral, its electron demand is 6. However, when the empty orbital on boron has accepted an extra pair of electrons, as in the case of tetrafluoroborate, the boron has an octet of electrons, and the electron demand is 8.

1-1.c. Electron supply = (6x4)(C)+(18x1)(H)+(3x5)(N)+(1x5)(P) = 62. Electron demand = (10x8)+(18x2) = 116. Approximate number of bonds = (116–62)/2 = 27. After the bonds are placed in the molecule, there are eight electrons left [62 – 2(27)] which are used to complete the octets of phosphorus and nitrogen.

All charges are zero. If you drew a structure with an oxygen attached to phosphorus, that is hexamethylphosphoric triamide. The "Handbook of Chemistry and Physics" is a good source for formulas for common organic compounds as is the Aldrich catalog.

1-1.d.   Electron supply = (2x4)(C)+(6x1)(H)+(1x5)(N)+(1x6)(O) = 25. Electron demand = (4x8)+(6x2) = 44. Estimate of bonds = (44 − 25) = 9.5. This result means there must be 9 bonds and an odd electron somewhere in the molecule. There are two likely structures which agree with the carbon skeleton given, 1-41 and 1-42.

1-41   1-42

Because of the chemistry of the nitroxide functional group, the odd electron is usually written on oxygen, i.e. 1-41, even though this means that the most electronegative element in the molecule is the one that doesn't have an octet. None of the atoms in 1-41 have a formal charge. In 1-42 there is a +1 charge on nitrogen and a −1 charge on oxygen.

If you wrote **1-43** for the structure, you added an extra electron.

1-43

This gives an overall negative charge which is placed on the oxygen. Since this compound is not neutral, it will not be stable. On the other hand, if you didn't have an odd electron in your structure and there was an atom which didn't have an octet, you didn't put enough electrons in the structure. You should always count the number of electrons in your finished Lewis structure, to make sure you have exactly the number calculated for the electron supply.

Unlike the sulfur in dimethyl sulfoxide (Example 1-2), nitrogen and oxygen do not have d orbitals. Thus, it is not possible for this compound to have a double bond between nitrogen and oxygen, because this will leave either nitrogen or oxygen with more than 8 electrons.

1-1.e. Electron supply = 4 (C)+4 (H)+12 (S and O) = 20. Electron demand = 24 (C,S,O)+8 (H) = 32. Estimate of bonds = (32 − 20)/2 = 6. The molecular formula CH_4OS, suggests no rings or π bonds when sulfur has an octet.

1-44

After putting in the bonds, the eight electrons left over are used to complete the octets of sulfur and oxygen by filling in two lone pairs on each atom. There are no charges on any of the atoms in **1-44**.

The way the structure is given in the problem indicates that the hydrogen is on oxygen. Another possible structure which used to be written for sulfenic acids puts the acidic hydrogen on sulfur rather than oxygen:

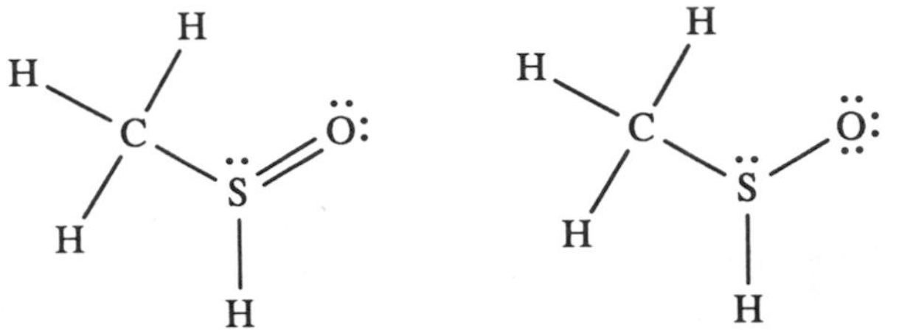

1-45   **1-46**

In **1-45** there are 7 bonds because the sulfur has expanded its valence shell to accommodate 10 electrons. In **1-46** the S has a charge of +1 and the oxygen has a charge of −1. In **1-45** there are no formal charges.

1-2.   It is always a good idea to write a complete Lewis structure before deciding on hybridization so that lone pairs are not missed.

1-2.a.   The CH_3 carbon is sp³-hybridized. The carbon is attached to four distinct groups, 3 hydrogens and a carbon so X=4. The other carbon and the nitrogen are both sp-hybridized because X+E=2. This sp-hybridized carbon is attached to the CH_3 carbon and nitrogen, and the nitrogen is attached to carbon and also contains a lone pair. The C-C-N skeleton is linear because of the central sp-hybridized atom. The H–C–H bond angles of the methyl group are approximately 109°, the tetrahedral bond angle.

1-2.b.   All of the carbons in the phenyl ring are sp²-hybridized. This ring is also conjugated with the external π system, so the entire molecule is planar. The nitrogen and sulfur are sp²-hybridized: nitrogen is bonded to the phenyl group and the external carbon and also contains a lone pair so X+E=3; the sulfur is attached to carbon and contains two lone pairs so X+E=3. The external carbon is sp-hybridized: it is attached to only two atoms, the nitrogen and the sulfur. The N=C=S group is linear.

1-2.c. For phosphorus X+E=4, 3 ligands and one lone pair. Thus, the carbons and the phosphorus are all sp³-hybridized with approximate tetrahedral bond angles.

1-2.d. Each of the carbons is sp²-hybridized. Even though X + E = 4 for the carbon which bears the negative charge, this carbon is sp² and not sp³-hybridized. The three bonds utilize the sp²-hybridized orbitals. This places the two non-bonding electrons in a p orbital parallel to the π bond between the other two carbons. That is, these electrons are delocalized. The decrease in energy, due to delocalization of the negative charge, more than makes up for the increase in energy associated with two electrons being only 90° from those of the σ bond. This angle would be 109.5° if the carbon were sp³-hybridized. All the carbons and hydrogens lie in the same plane; all H-C-H and C-C-C bond angles are approximately 120°.

1-3.a. $\overset{\delta^+}{\diagup}\!\!=\!\!\overset{\delta^-}{NH}$ b. $\overset{\delta^+}{Br}\!\!-\!\!\overset{\delta^-}{F}$ c. $\overset{\delta^+}{H_3C}\!\!-\!\!\overset{\delta^-}{N(CH_3)_2}$

d. $\overset{\delta^-}{H_3C}\!\!-\!\!\overset{\delta^+}{P(CH_3)_2}$

1-4.a.

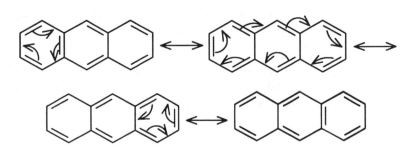

1-47

Is the resonance form **1-48** different from those written above? The answer is no.

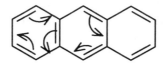

1-48

This represents exactly the same electron distribution as **1-47**. While the double bond between the two right hand rings is written on the right in **1-47** and on the left in **1-48**, the double bond is still between the same two carbons. Thus **1-47** and **1-48** are identical.

Would the following redistribution of electrons lead to a correct resonance form? The answer is no.

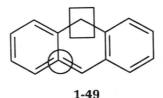

If a double bond is written at every single bond to which an arrow is drawn, **1-49** results:

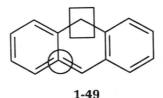

1-49

In **1-49** the carbon which is circled has 10 electrons which is not possible for carbon. Also, the carbon which is boxed doesn't have enough electrons. (This carbon is only attached to one hydrogen, not two.) Thus, this is an impossible structure.

1-4.b.   The nitro group is conjugated with the π-system. Hence there are two forms for the nitro group whenever the second negative charge is located on an atom external to the nitro group.

A structure like **1-50** would be incorrect because it represents a different compound; in fact this compound has one less hydrogen at the terminal carbon. Remember that all resonance forms must have identical geometry.

1-50

1-4.c.

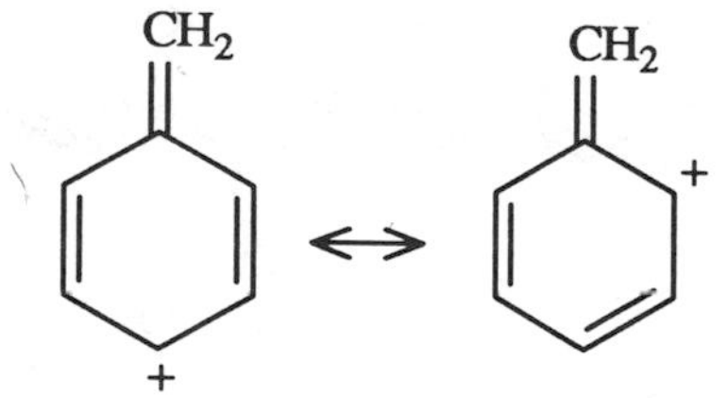

58   *Writing Reaction Mechanisms in Organic Chemistry*

1-4.d.

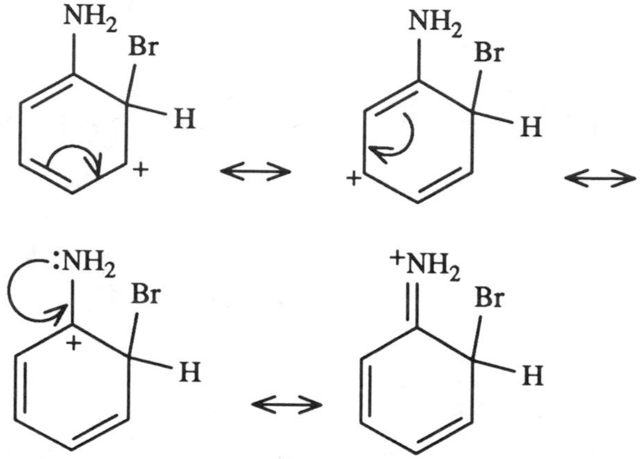

1-4.e.   A circle drawn inside a ring means that there is a totally conjugated π-system. The dinegative charge means that this conjugated system has two extra electrons. Thus two electrons have been added to the neutral, totally conjugated system (a planar cyclooctatetraene). The resonance forms drawn below are a few of the many possible forms that can be drawn.

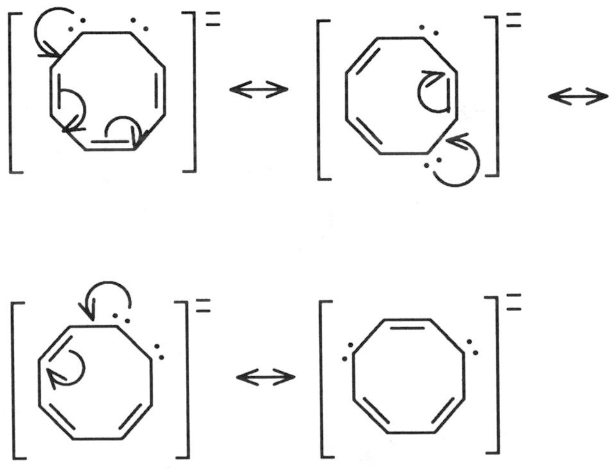

1-4.f.   There are many resonance forms possible. Just a few are
shown below.

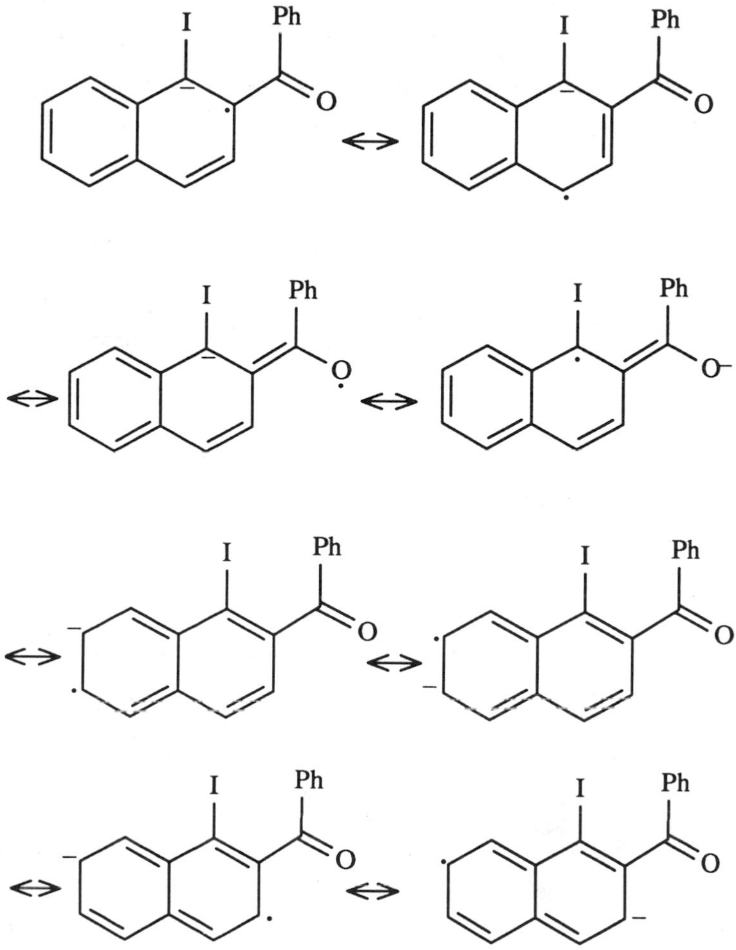

1-5.   Only a few of the possible resonance forms are drawn.
Nonetheless, it can be seen that the anion and radical can
be delocalized into both nitro groups simultaneously for
the *p*-dinitrobenzene and this leads to more possible resonance forms. Because there is more delocalization in the

intermediate from the *para* compound, it should be easier to transfer an electron to *p*-dinitrobenzene.

1-6.a.   The resonance forms with positive charge on nitrogen or oxygen do not contribute significantly to stabilization of the positive charge, because the nitrogen and oxygen do not have an octet when they bear the positive charge. The fact that oxygen or nitrogen does not have an octet is a critical point. When oxygen and nitrogen do have an octet, they can contribute significantly to a resonance hybrid in which they are positively charged. For example, the 1-methoxyethyl carbocation is significantly stabilized by the adjacent oxygen:

1-6.b.   When nitrogen and oxygen are negatively charged, they do have an octet. A negative charge on these atoms, both more electronegative than carbon, is significantly stabilizing; thus this delocalization would be very important.

1-7.a.   Electron supply = (1x4)(C)+(3x1)(H)+(1x5)(N)+(2x6)(O)=24. Electron demand = (4x8)+(3x2) = 38. Estimate of bonds = (38 − 24)/2=7. Using Helpful Hint 1-2, estimates 1 ring or 1 π bond. Two possible skeleton structures are **1-51** and **1-52**.

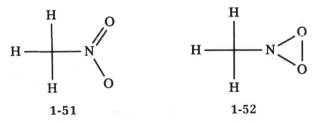

The combination of a highly strained three-membered ring and a weak O-O bond make **1-52** unlikely. When you see the grouping NO_2 in a molecule, you can assume that it is a nitro group. Filling in the 5 lone pairs of electrons gives **1-53** or the resonance structure, **1-54**.

In a neutral molecule, the formal charges must balance to 0. So if oxygen is given a formal charge of −1, some other atom in the molecule must have a formal charge of +1. Calculation shows that this atom is the nitrogen.

Structure **1-55** is incorrect because the nitrogen has ten electrons; nitrogen does not have orbitals available to expand its valence shell.

1-55

1-7.b. Electron supply = (6x4)(C)+(5x1)(H)+(2x5)(N)−1 = 38. Electron demand = (8 x 8) + (5 x 2) = 74. Estimate of bonds = (74 −38)/2 = 18. This cation does not have an odd electron. One electron was removed from the electron supply because the species is positively charged. The resonance structures that contribute most to the resonance hybrid are **1-56** and **1-57**.

1-56   **1-57**

Resonance forms, **1-58** and **1-59**, contribute much less than **1-56** and **1-57** to the resonance hybrid, because the positively charged nitrogen in these forms does not have an octet.

1-58   **1-59**

1-7.c.   Electron supply = (3x4)(C) + (4x1)(H) + (2x5)(N) + (1x6)(O) = 32. Electron demand = (6x8) + (4x2) = 56. Estimate of total bonds = (56 − 32)/2 = 12. Several possible structures, all resonance structures, follow:

A resonance structure which will not contribute as significantly to the hybrid is **1-60**.

1-60

In this structure, the positively charged nitrogen does not have an octet; the structures in which positive nitrogen has an octet are more stable. Notice also that the number of bonds is one less than that estimated.

You might have interpreted the molecule's skeleton to be that shown in resonance structures **1-61** through **1-63**. However, none of these structures is expected to be stable: **1-61** has too much charge; **1-62** has a carbene carbon; and **1-63** has a carbene carbon and a nitrogen which does not have an octet. The first answers drawn for this problem are much more likely.

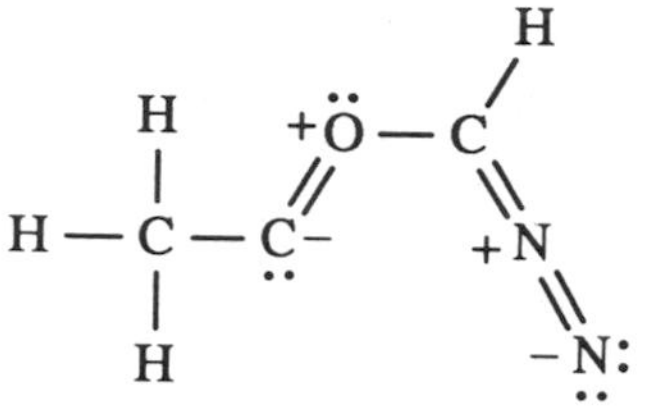

1-61 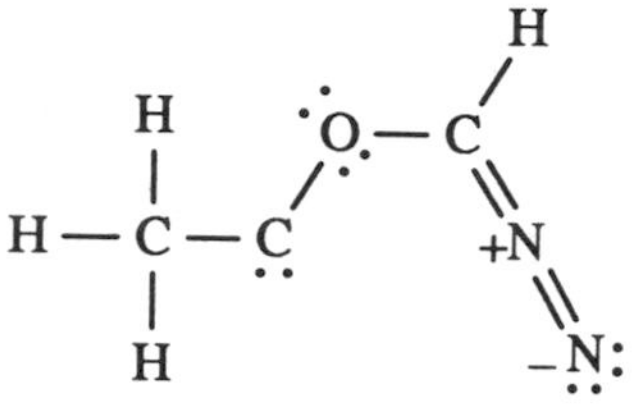   **1-62**

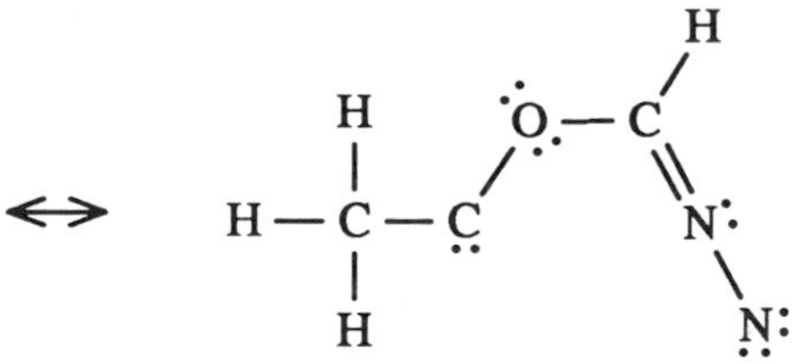

1-63

1-7.d.   Electron supply = (1x4)(C)+(4x5)(N) = 24. Electron demand = 5x8= 40. Estimate of bonds = (40−24)/2 = 8. Because they are resonance structures, all of the following are correct:

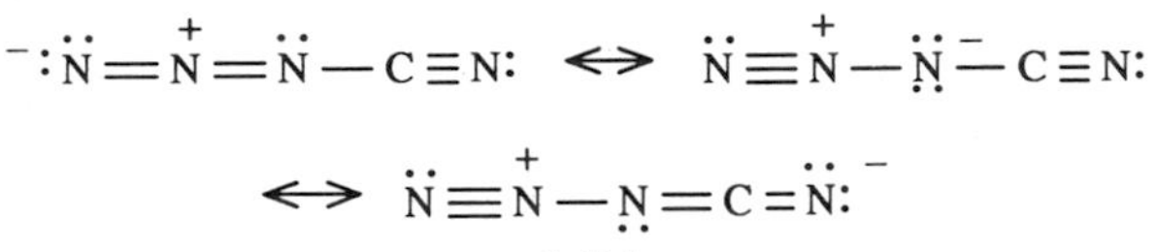

1-64

In a neutral molecule, the formal charges must balance to 0. So if one nitrogen is given a formal charge of +1, some other atom in the molecule must have a formal charge of −1.

Another acceptable structure is **1-65**.

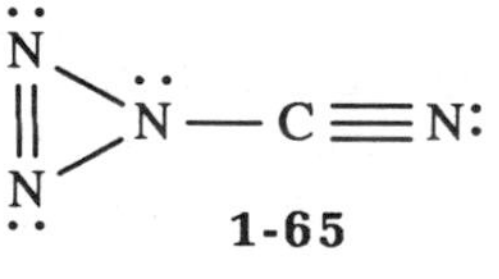

1-65 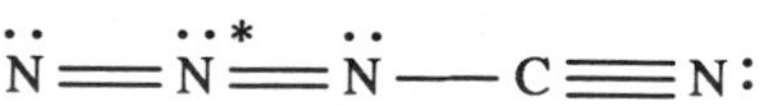   **1-66**

On the other hand, **1-66** is unacceptable. The starred nitrogen has 10 electrons which is not possible for any element in period 2 of the periodic table. Also, the terminal nitrogen next to this nitrogen has only 6 electrons instead of an octet. Other structures, **1-67** and **1-68**, in which one of the nitrogens does not have an octet, are less stable than the resonance structures **1-64**.

$$:N = \ddot{N} - \ddot{N} - C \equiv N: \qquad :\ddot{N} - \ddot{N} = \ddot{N} - C \equiv N:$$

1-67   **1-68**

Structures **1-69** and **1-70**, possible alternatives to a linear nitrogen array for the azide group, are not acceptable. In **1-69** there is a 10 electron nitrogen and in **1-70** the nitrogen at the top doesn't have an octet and is located next to a positive nitrogen, an extremely unstable electron arrangement.

1-69   **1-70**

1-8.a.   This is aromatic. The sulfur is sp^2: one of the lone pairs of electrons is in an sp^2-hybridized orbital, and the other lone pair is in a p orbital parallel with the p orbitals on each of the carbons. Thus, this is a 6π electron system.

1-8.b.   This is nonaromatic. Although the structure contains 6π electrons which are conjugated, it is not cyclic.

1-8.c.   Antiaromatic or nonaromatic. If the NH nitrogen is sp^2-hybridized, the lone pair of electrons on the nitrogen can be in a p orbital which overlaps with the other six p electrons in the ring. This would be an antiaromatic system. However, because of the destabilization which delocalization would cause, it is unlikely that good overlap of the nitrogen lone pair electrons with the π system would occur.

1-8.d.   Nonaromatic. The carbon at the top of the structure is sp^3-hybridized. Thus the delocalization of the 6π electron system is interrupted by this carbon. For aromaticity, not only must the compound have the correct number of electrons, but those electrons must be completely delocalized.

1-8.e.   Aromatic. This compound contains 16π electrons; so, strictly speaking, it does not obey Hückel's Rule. However, if a compound has fused 6-membered rings, all of which are totally conjugated, it is considered to be aromatic. A less simplistic approach is to count only the electrons in the peripheral ring. This can be seen more easily by looking at another resonance structure for the given compound:

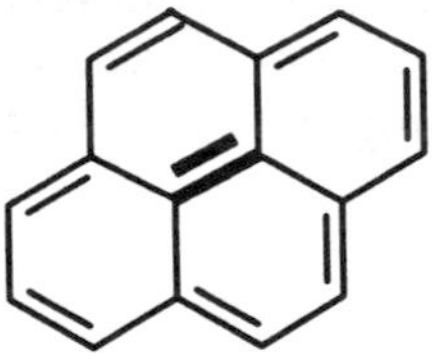

Note that the highlighted bond is not part of the outside ring. Thus, the outside ring has 14π electrons, obeys Hückel's Rule, and is aromatic.

1-8.f.   Aromatic. This is a totally conjugated, cyclic, 10π electron system. The ring is planar. However, the situation is not so simple. It has been calculated that the energy reduction, due to aromaticity, is not large enough to compensate for the electron-electron repulsion energy. Apparently, such anionic compounds exist primarily because the metallic counterions are strongly solvated by the solvents used.

1-8.g.   Aromatic. The totally conjugated system contains 14π electrons. Only two of the π electrons of the triple bond are counted, because the other two π electrons are perpendicular to the first two and cannot conjugate with the rest of the π system.

1-9.a.

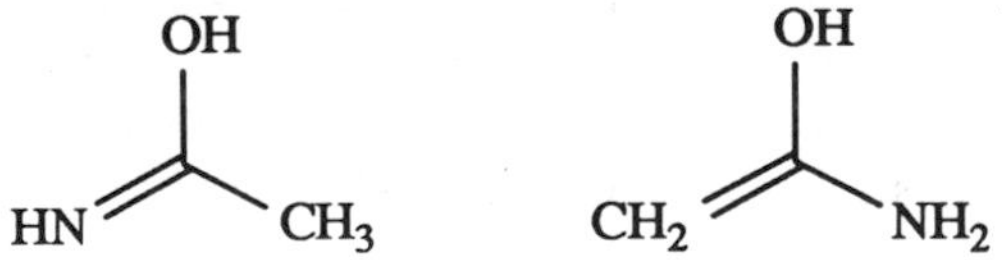

1-9.b. CH$_2$=CHOH

1-9.c. CH$_2$=CH-CH=CHOH

1-9.d.

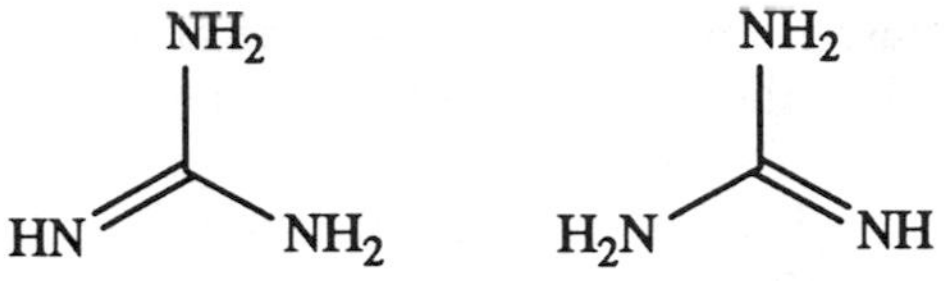

1-9.e.

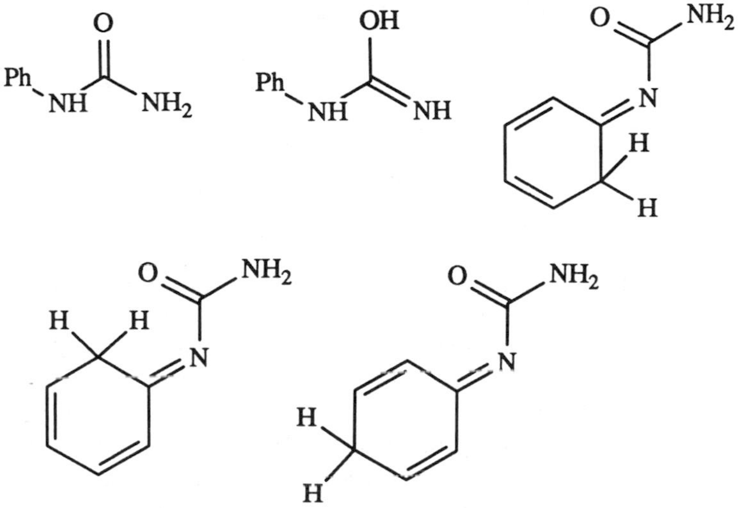

There are two more possible structures in which there is isomerism about the imine nitrogen:

1-10.a. The first structure is the same compound as the last; the second is the same as the fourth. The first, second, and third structures meet the basic requirement for tautomers; interconversion involves only movement of a double bond and one hydrogen atom. However, most chemists would not call them tautomers, because the allylic proton which moves is not very acidic. Its pK$_a$ can be estimated from the

pK$_a$ of the allylic proton of 1-propene, 47.1–48.0 (from
Table 1-2). Thus, the intermediate anion necessary for the
interconversion of these isomers would only be formed
under extremely basic conditions. When the equilibration
is this difficult to effect, the different isomers are usually
not called tautomers. On the other hand, if another double
bond were added to the structure, the compound would be
very acidic (from Table 1-2, the related cyclopentadiene
has a pK$_a$ = 18.1 in DMSO, more acidic than the α protons
in acetone, pK$_a$ = 19.2). Thus, most chemists would call
compounds **1-71** and **1-72** tautomers.

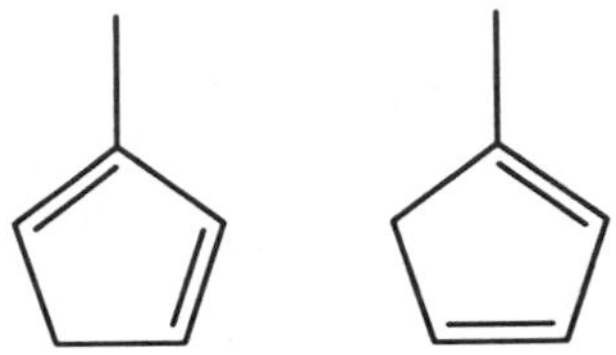

1-71   **1-72**

The structures in the problem are not reasonable resonance
forms, because, for this purpose, a hydrogen atom cannot
relocate and the hybridization of the carbon atoms cannot
change. Therefore, if the double bond were moved in either
direction to the next ring location, it would leave one carbon with only three bonds (carbon 5 in the second structure below) and place five bonds on another (carbon 2 in
the second structure below). These other structures would
have extremely high energies and contribute nothing to a
resonance hybrid.

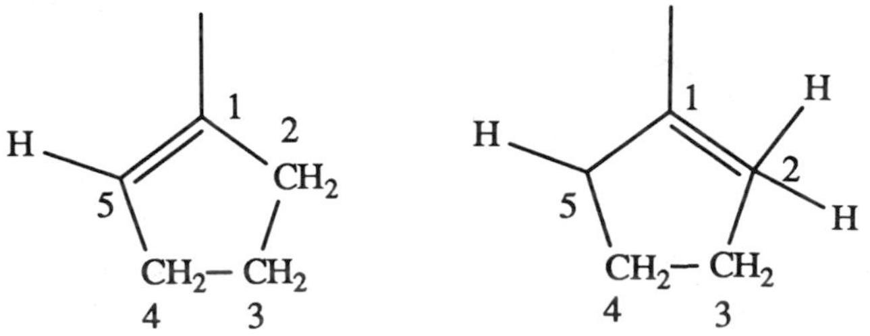

1-10.b.   These two structures are resonance forms. They represent
the simplest enolate ion that can exist.

1-10.c.   These compounds are tautomers. They differ in the placement of a double bond and a proton, **AND** they could readily be in equilibrium with each other.

1-11.   The two structures are tautomers because they differ in the placement of a double bond and a proton, **AND** they could readily be in equilibrium with each other. If the arrow is changed to $\rightleftharpoons$, the picture will be correct.

1-12.a. $CH_2=CH^-$ is more basic than $HC\equiv C^-$. That is, according to Table 1-2, ethene is less acidic than ethyne, so the conjugate base of ethene is more basic than the conjugate base of ethyne.

1-12.b. $[(CH_3)_2CH]_2N^-$ is more basic than $\bar{C}H_2CON(Et)_2$. The diisopropylamide anion cannot be stabilized by resonance, because the N is bonded only to sp^3-hybridized carbons. On the other hand, $\bar{C}H_2CON(Et)_2$, is stabilized by resonance. The negative charge is delocalized on both carbon and oxygen:

This anion is more stable than the diisopropylamide ion, and thus diisopropylamide anion is the more basic.

1-12.c. $(CH_3)_3CO^-$ is more basic than CH_3O^-. The t-butoxide anion is not as well solvated in solution as methoxide, and thus is more basic. Table 1-2 also indicates that methanol is more acidic than t-butyl alcohol, so the relative basicity of the conjugate bases is the reverse.

1-12.d. The m-nitrophenolate anion is less stable than the p-nitrophenolate anion, and therefore is more basic. The resonance forms for m-nitrophenolate are:

The resonance forms for *p*-nitrophenolate are:

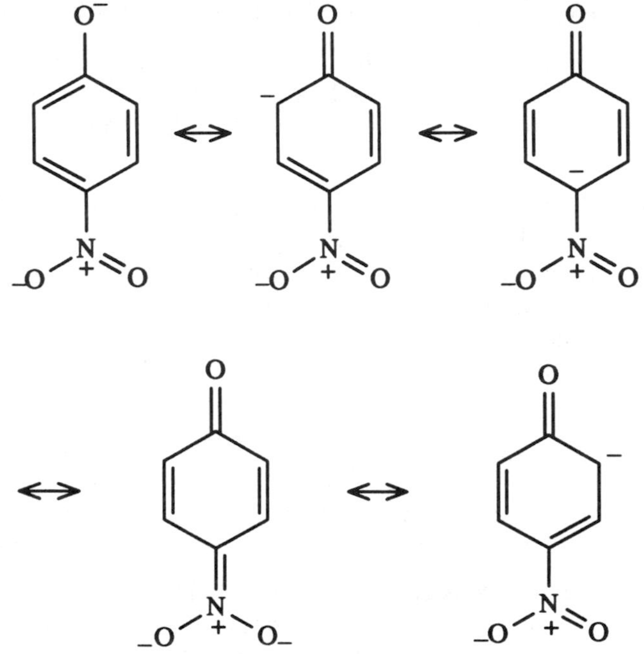

The *p*-nitrophenolate anion is more stable because the negative charge can be delocalized into the nitro group. Because it is more stable, it will be less basic. Note that in the drawings, the other Kekule form of the ring and the other form of the nitro group were omitted where appropriate. Since these forms also stabilize the starting materials, they do not explain a difference in energy between the starting phenol and product phenolate. It is the size of this energy difference which determines the acidity of the starting material.

1-13.a. The carbanion produced by removal of a proton on the CH_2 (methylene) group is the most effectively stabilized by resonance since it is conjugated with two carbonyl groups.

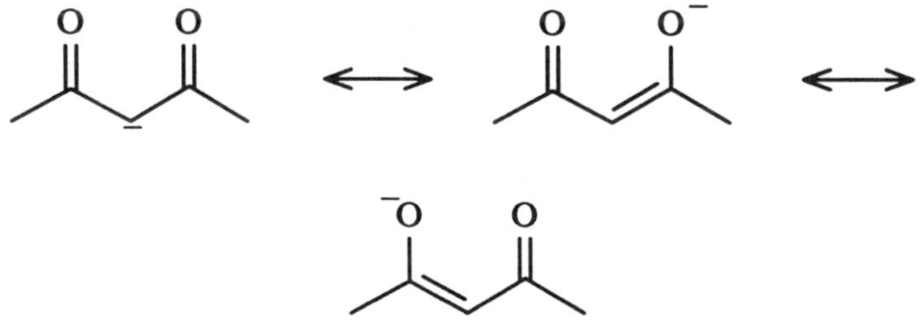

Removal of a proton from a methyl group will give a carbanion which is stabilized by conjugation with only one carbonyl group.

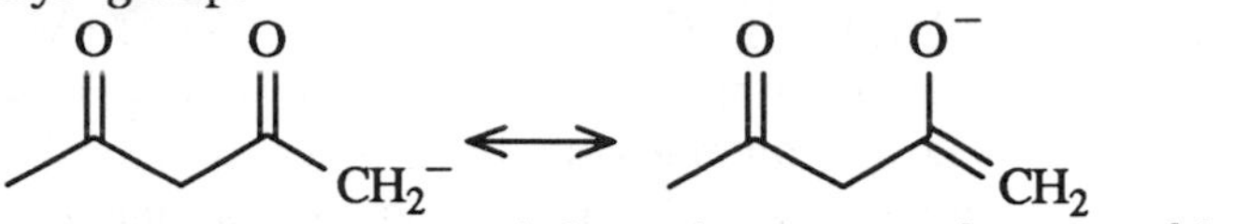

1-13.b. The proton will be removed from the oxygen, because this is the most electronegative element in the molecule and can stabilize the negative charge most effectively.

1-13.c. Removal of a proton from the γ-position gives the most stabilized anion because more delocalization is possible.

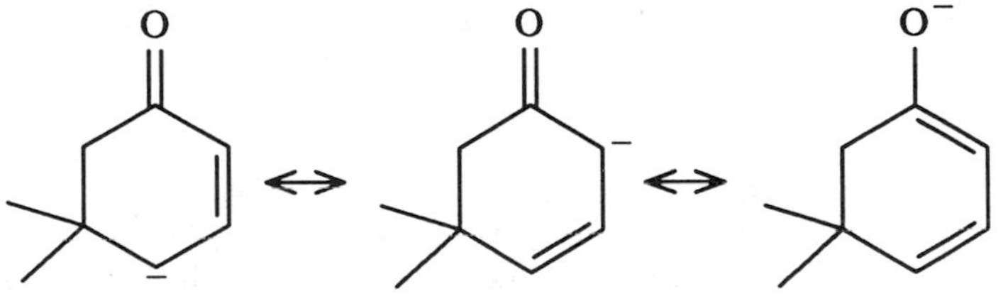

Removal of the α-proton on the sp³-hybridized carbon gives an anion which is not as stabilized, because the charge is located on only two centers, not three.

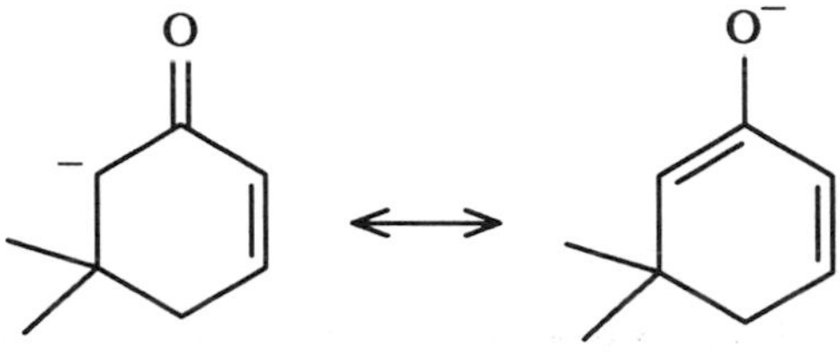

1-13.d. The proton will be removed from the exocyclic amino nitrogen. Consider the anion, **1-73**, produced if the proton on the imino nitrogen were removed:

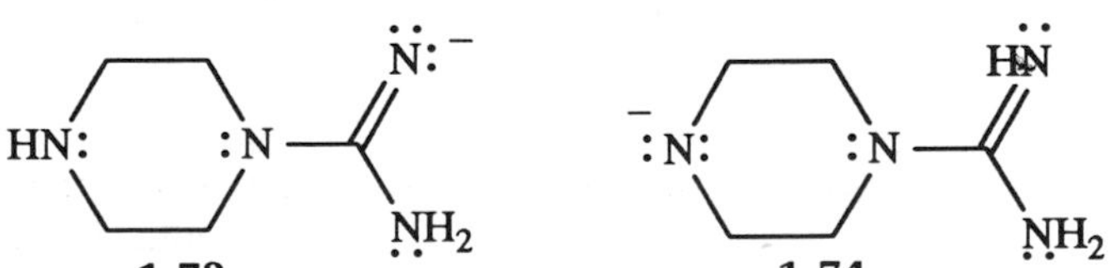

1-73   **1-74**

While this anion is stabilized by the electronegative nitrogen, it is localized on just the one nitrogen. Both lone pairs of electrons on the negative nitrogen will be in sp^2-hybridized orbitals. If a proton is removed from the endocyclic nitrogen of the starting material, giving **1-74**, the lone pair electrons on the anionic nitrogen will be in sp^3-hybridized orbitals. Thus, **1-74** will not be as stable as **1-73**. On the other hand, when a proton is removed from the amino group, a resonance stabilized anion, **1-75**, is produced:

1-75

Since there are two forms of equal energy, and the negative charge is distributed on two nitrogens, the resonance hybrid corresponding to removal of a proton from the amino group is more stable. Thus that proton will be preferentially removed.

1-14.a.

$$K_a = \frac{10^{-30.5}}{10^{-15.9}} = 10^{-14.6}$$

This value tells us that there is very little ester anion present at equilibrium.

1-14.b.

$$K_a = \frac{10^{-24.7}}{10^{-15.7}} = 10^{-9}$$

This value is only an approximation, since the two acidities are not measured in the same solvent. Nonetheless, the value indicates that equilibrium favors starting material to a large extent.

1-14.c.

$$K_a = \frac{10^{-30.6}}{10^{-35.7}} = 10^{5.1}$$

In this case the reaction goes substantially to the right. Using the alternative pK_a of 39 for the amine, the answer would be 10^{84}.

1-15.a. $^+NO_2$ is the electrophile. Writing a Lewis structure for this species indicates that the nitrogen is positive and will be the atom that reacts with the nucleophile: $(:\ddot{O}::\overset{+}{N}::\ddot{O}:)$. In acetophenone, the π electrons of the ring act as the nucleophile. In keeping track of electrons, it is easier to think of the nucleophile as the electron pair of the π bond to the carbon where the nitrogen becomes attached:

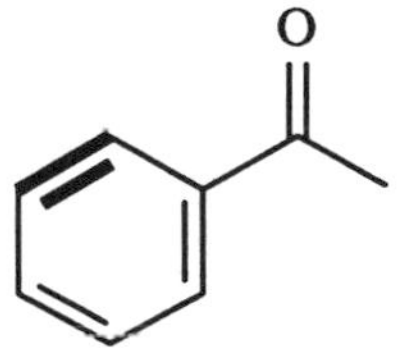

1-15.b. One of the protons of sulfuric acid is the electrophile which is transferred to the oxygen of acetophenone. Thus, this oxygen is acting as the nucleophile in this reaction. The product shown would be a direct result of a lone pair of electrons on the oxygen acting as a nucleophile. One could also show the π electrons of the C=O group acting as the nucleophile. This would give the following structure:

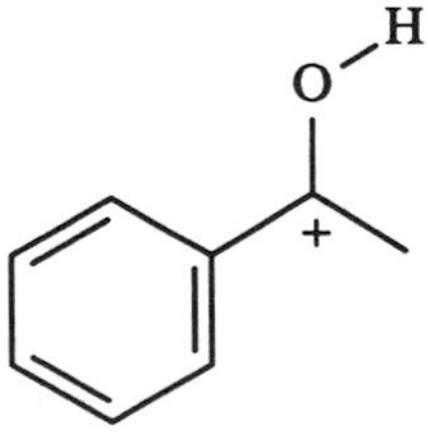

This structure and the structure drawn in the problem are the same, because they are both resonance forms which contribute to the same resonance hybrid.

This reaction could also be described as an acid-base reaction where the acid is the sulfuric acid and the oxygen of the carbonyl group is the base.

1-15.c. The electron pair of the bond between the methyl group and the magnesium is the nucleophile. This bond of the Grignard reagent is a highly polarized covalent bond with carbon being very negative. This carbon becomes attached to the carbon of the carbonyl group in acetophenone with the latter carbon being the electrophile.

1-15.d. The electron pair, constituting the π bond in cyclohexene, is the nucleophile, and a proton from H-Cl is the electrophile.

Chapter 2

General Principles for Writing Organic Mechanisms

In writing a reaction mechanism, we give a step-by-step account of the bond (electron) reorganizations that take place in the course of a reaction. This book will help you to figure out a number of pathways for a new reaction by showing you some of the steps which often take place under a particular set of reaction conditions. This chapter is devoted to some general principles, derived from the results of many experiments by organic chemists, that can be applied to writing organic mechanisms. Subsequent chapters will develop the ideas further under more specific reaction conditions.

It is often difficult to predict what will actually happen in the course of a reaction. If you were planning to run a reaction that's never been done before, you would plan the experiment based on previously run reactions that look similar. You would assume that the steps of bond reorganization that take place in the new reaction are analogous to the reactions previously run. However, you might find that there is a step or steps in your reaction which were not anticipated. In other words, though there are a number of general ideas about the course of reactions which have been developed on the basis of experiments, it is sometimes difficult to choose which ideas apply to a particular reaction. This book will help you to develop the ability to make some of those choices. Nonetheless, you will often be left with more than one possible pathway for a new reaction that you run.

Helpful Hint 2-1. It can be assumed, unless otherwise stated, that when an organic reaction is written, the products shown have undergone any required aqueous workup, which may involve acid or base, to give a neutral organic molecule (unless salts are shown as the product). In other words, often when an equation for a reaction is written in the literature or on an exam, an aqueous workup is assumed, and intermediates, salts, etc. are not shown.

Helpful Hint 2-2. When given an equation for a reaction, first balance all atoms on both sides of the equation. For any such balanced equation, the charges must also balance. Be aware that when equations are written in the organic literature, they are often not balanced.

Example 2-1. Balance atoms first, then charge.

If an atom on the left side of the equation has a positive charge, some atom on the right side must also bear a positive charge.

Example 2-2. Balancing an equation with a net negative charge.

There is a net negative charge on the left side of the equation; thus there must be a net negative charge on the right side of the equation.

Problem 2-1. In the following steps, supply the missing charges. Assume that no molecules with unpaired electrons are produced.

a.

b.

c.

In writing mechanisms bond making and bond breaking processes are shown by curved arrows. The arrows are a convenient tool for thinking about and illustrating what the actual electron redistribution for a reaction may be.

Helpful Hint 2-3. The arrows, which are used to show redistribution of electron density, are drawn from a position of high electron density to a position which is electron deficient. Thus, arrows are drawn leading away from negative charges or lone pairs and toward positive charges or the positive end of a dipole. In other words, they are drawn leading away from nucleophiles and toward electrophiles. Furthermore, it is only in **unusual** reaction mechanisms that two arrows will either lead away from or toward the same atom.

Example 2-3. Arrows show redistribution of electron density.

The following equations show the electron flow for the transformations of Examples 2-1 and 2-2.

For an understanding of why the neutral oxygen in the first equation reacts with a proton of the hydronium ion rather than the positively charged oxygen, see Helpful Hint 2-9.

Helpful Hint 2-4. When you first start drawing reaction mechanisms, rewrite any intermediate structure before you try to manipulate it further. This avoids confusing the arrows associated with electron flow for one step with the arrows associated with electron flow for a subsequent step. As you gain experience, you will not need to do this. It will also be helpful to write the Lewis structure for at least the reacting atom and to write lone pairs on atoms such as nitrogen, oxygen, halogen, phosphorus, and sulfur.

Helpful Hint 2-5. If a reaction is run in a strongly basic medium, any *positively* charged species must be weak acids. If a reaction is run in a strongly acidic medium, any *negatively* charged species must be weak bases. In a weak acid or base (like water), both strong acids and strong bases may be written as part of the mechanism.

Example 2-4. Writing a mechanism in strong base.

The mechanism written below for the hydrolysis of methyl acetate in a strong base is consistent with the experimental data for the reaction.

In accordance with Helpful Hint 2-5, since this is a reaction in a strong base, all the charges in this mechanism are negative. Thus the following steps would be incorrect for a reaction in base, because they involve the formation of ROH_2 (the intermediate) and H_3O^+ which are both strong acids. Another way of looking at this is to realize that, in base, hydroxide ion has a significant concentration. Thus, hydroxide, a much better nucleophile than water, will act as the nucleophile in the first step of the reaction.

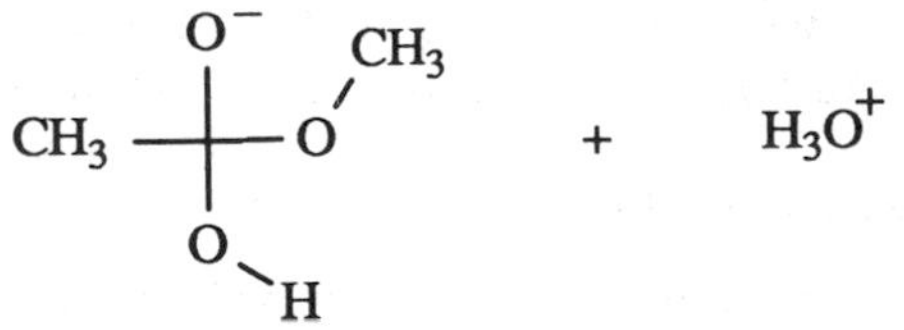

$$CH_3 \underset{\underset{H}{O}}{\overset{O^-}{\underset{|}{\overset{|}{C}}}} O{\overset{CH_3}{}} \quad + \quad H_3O^+$$

Example 2-5. Writing a mechanism in strong acid.

The following step is consistent with the facts known about the esterification of acetic acid with methanol in strong acid.

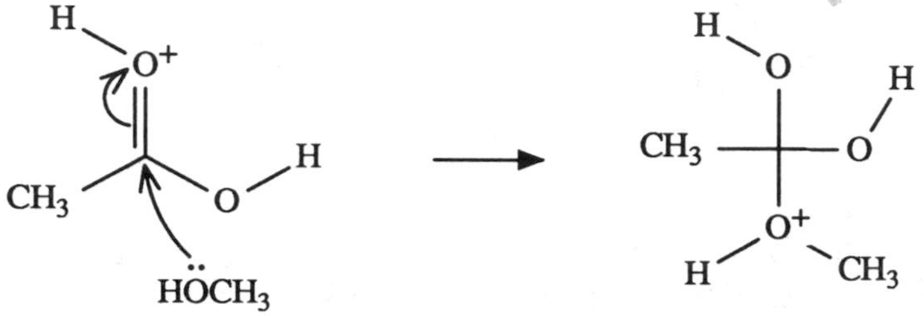

Because the fastest reaction for a strong base, CH_3O^-, in acid is protonation, the following mechanistic step would be highly improbable for esterification in acid:

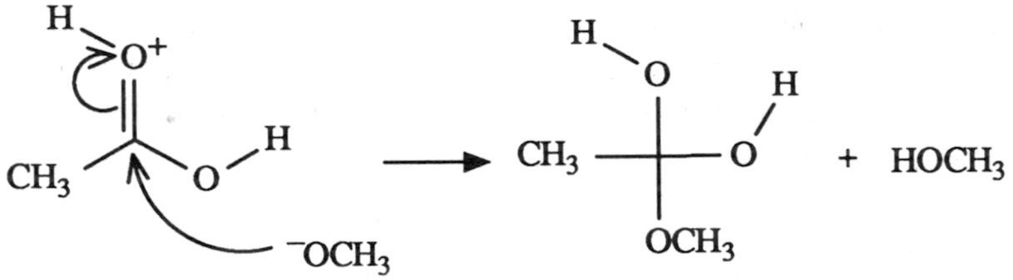

Example 2-6. Writing a mechanism in a weak base or weak acid.

Below is a mechanism often written for the hydrolysis of acetyl chloride in water. (Most molecules of acetyl chloride are probably protonated on oxygen before reaction with a nucleophile, because acid is produced as the reaction proceeds.)

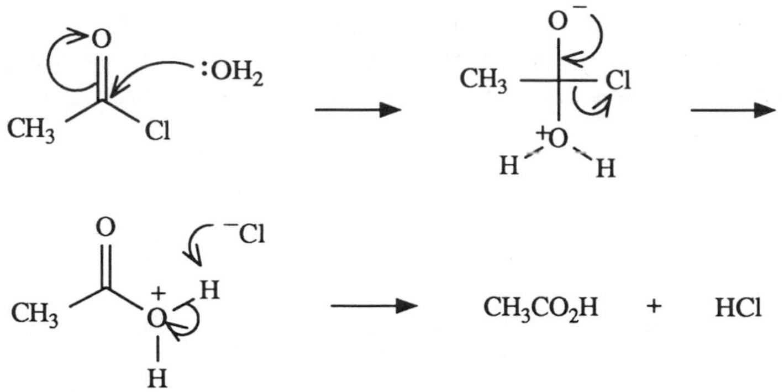

Example 2-7. The mechanism for the tautomerism of enols in water (neutral conditions) involves both strong acids and strong bases as intermediates:

In words, the first step can be described as a proton transfer from the enol to a molecule of water. However, in the figure the arrow must proceed from the nucleophile to the electrophile.

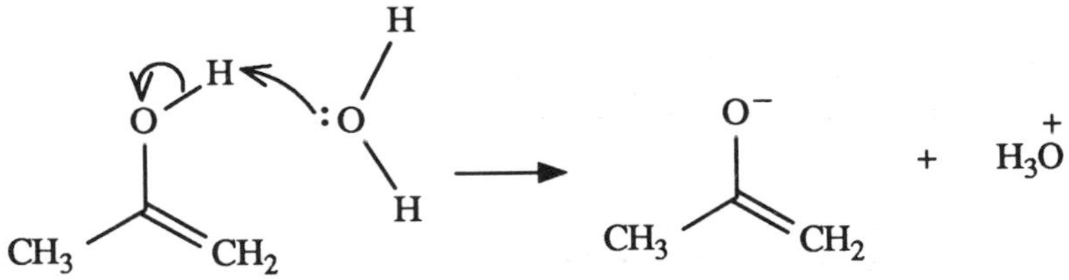

2-1

The anion produced is a resonance hybrid of structures **2-1** and **2-2**.

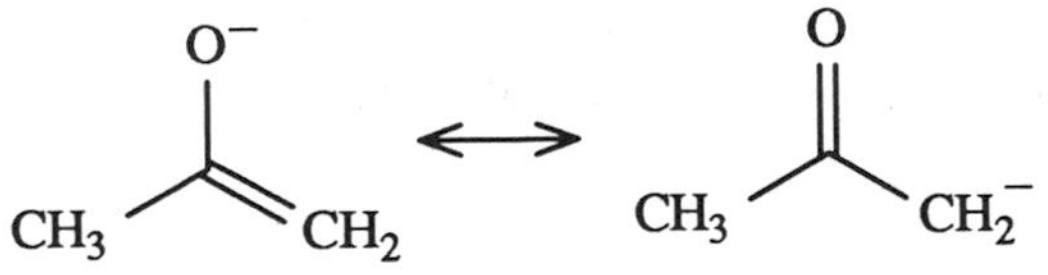

2-1   **2-2**

The hybrid can remove a proton from the hydronium ion to give the ketone form of the tautomers. Although not strictly correct (because resonance structures don't exist), such reactions are commonly depicted as arising from the resonance structure which bears the charge on the atom which is adding the proton.

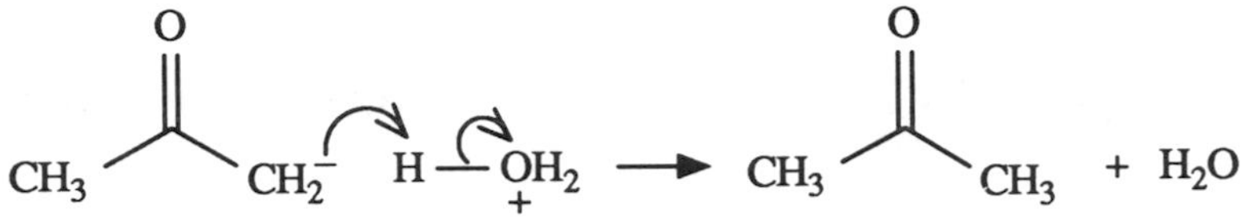

2-2

Helpful Hint 2-6. If the reagent is a strong base, look for acidic protons in the substrate. If a proton is removed from the substrate, look for a reasonable reaction for the anion produced. Alternatively, there may be a leaving group which leaves at the same time that the proton is removed; this would be an example of a concerted elimination. The anion may be a good nucleophile. If so, look for a suitable electrophilic center to which the nucleophile can add. If the anion is both a good nucleophile and a good base, all these possibilities must be considered. (For further detail, see Chapter 3.)

When writing mechanisms in acid and base, keep in mind that protons are removed by bases; even **very weak bases** like HSO_4^-, the conjugate base of sulfuric acid can do this. Protons do not just leave a substrate as H^+, because the bare proton is very unstable! Nonetheless the designation $-H^+$ is often used when a proton is removed from a molecule. (Whether this designation is acceptable is up to individual taste. If you're using this book in a course, you'll have to find out what your instructor will accept.) A corollary is that when protons are added to a substrate, they originate from an acid; that is, protons are not added to substrates as freely floating (unsolvated) protons.

Problem 2-2. In each of the following reactions, the first step in the mechanism is removal of a proton. In each case, put the most likely proton to be removed in a box. Table 1-2 will be helpful.

a.

Cohen, T.; Bhupathy, M.; J. R. *J. Am. Chem. Soc.* **1983**, *105*, 520–525.

b.

Kirby, G. W.; McGuigan, H.; Mackinnon, J. W. M.; Mallinson, R. R. *J. Chem. Soc. Perkin I* **1985**, 405–408; Kirby, G. W.; Mackinnon, J. W. M.; Elliott, S.; Uff, B. C. *ibid.* **1979**, 1298–1302.

c.

In this example, removal of the most acidic proton does not lead to the product. Which is the most acidic proton and which is the one that must be removed in order to give the product?

From Bernier, J. L.; Henichart, J. P.; Warin, V.; Trentesaux, C.; Jardillier, J. C. *J. Med. Chem.* **1985**, *28*, 497–502.

d.

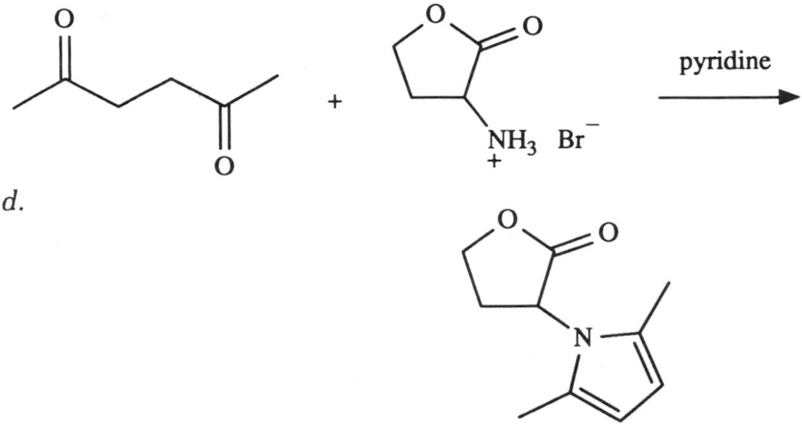

Jaffe, K.; Cornwell, M.; Walker, S.; Lynn, D. G. 190th National Meeting of American Chemical Society, Chicago, **1985**, ORGN 267.

Problem 2-3. For each of the following reactions, the first step in the mechanism is protonation. In each case, put the atom most likely to be protonated in a box.

a.

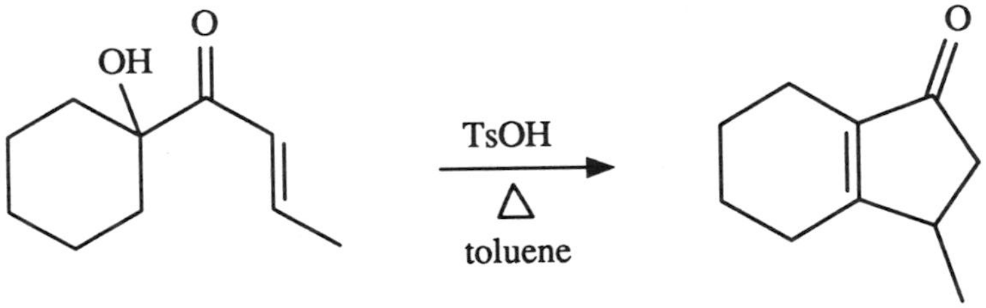

TsOH = *p*-toluenesulfonic acid

Jacobson, R. M.; Lahm, G. P. *J. Org. Chem.* **1979**, *44*, 462–464.

b.

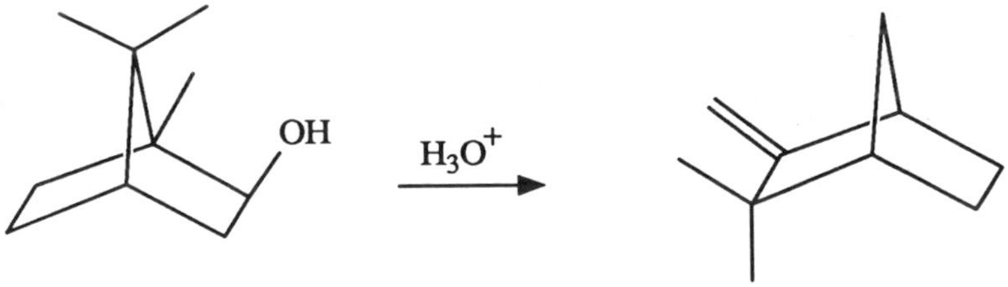

c.

$$H_2O, HCl$$
$$\xrightarrow{\hspace{2cm}}$$
$$DMSO$$
$$76\text{-}81\%$$

House, H. O., "Modern Synthetic Reactions", 2nd ed., 1972, Benjamin: Menlo Park, p. 726.

Helpful Hint 2-7. When a mechanism involves removal of a proton, it is not always removal of the most acidic proton that leads to the product. An example is Problem 2-2c in which removal of the most acidic proton by base does not lead to the product. (A mechanism for this reaction is proposed in the answer to 2-2c.) Similarly, when a mechanism involves protonation, it is not always protonation of the most basic atom that leads to product. Such reactions are called unproductive steps. When equilibria are involved, they are called unproductive equilibria.

Example 2-8. Does a reaction pathway involve removal of a proton or nucleophilic addition?

Consider the reaction of methylmagnesium bromide with 2,4-pentanedione. This substrate contains carbonyl groups which might undergo nucleophilic attack by the Grignard reagent. However, it also contains very acidic protons (see Table 1-2), one of which reacts considerably faster with the Grignard reagent than the carbonyl groups. Thus the reaction of methylmagnesium bromide with 2,4-pentanedione leads to methane, and after aqueous workup, starting ketone.

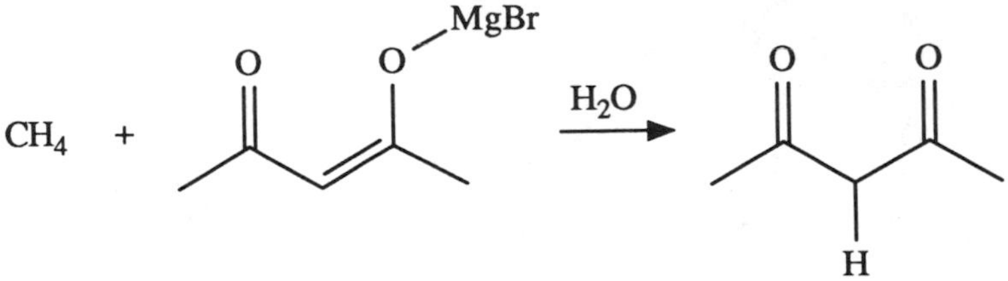

Helpful Hint 2-8. Trimolecular steps are rare, because of the large decrease in entropy associated with three molecules simultaneously assuming the proper orientation for reaction. Try to break a tri-molecular step into two or more bimolecular steps.

Example 2-9. Breaking a trimolecular step into several bimolecular steps.

When mechanisms for the following reaction are considered,

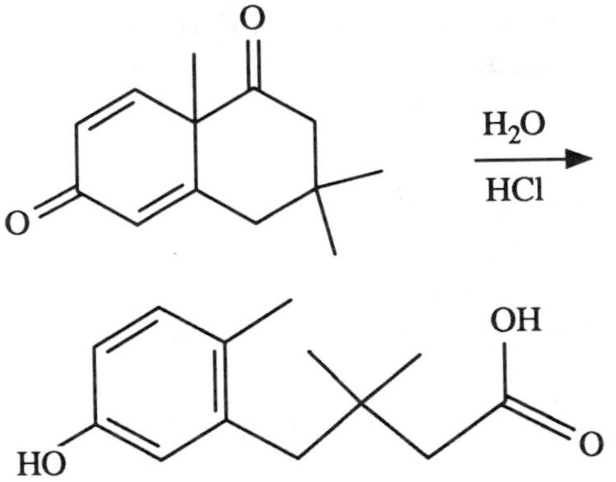

one of the steps could be written as a trimolecular reaction:

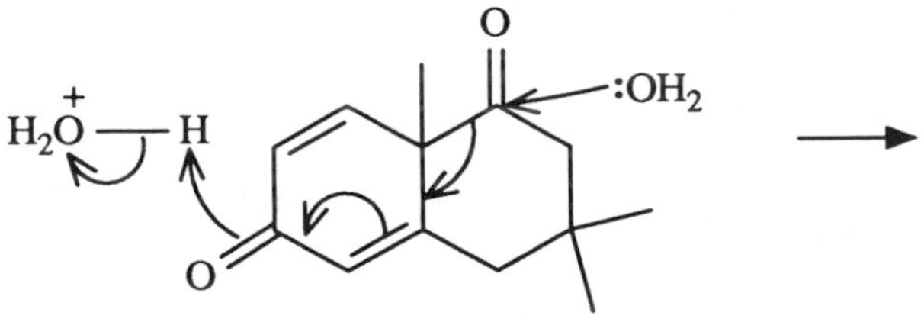

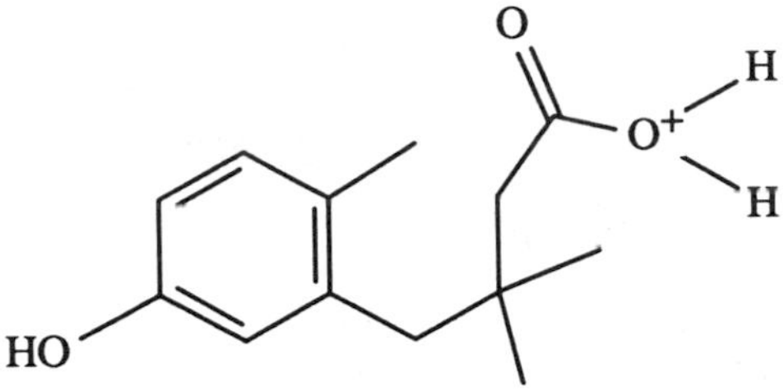

2-3

Intermediate **2-3** can lose a proton to water to give the product.

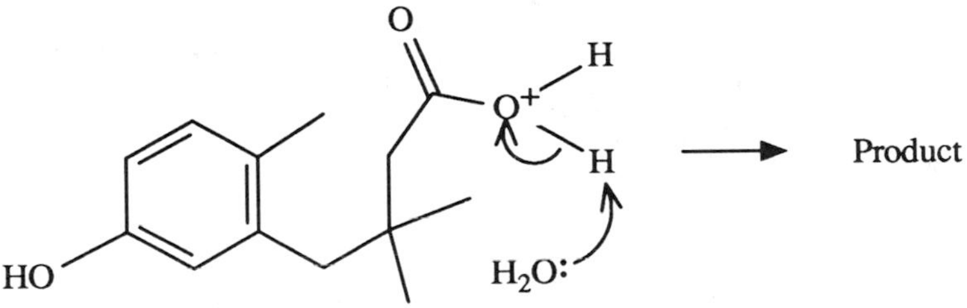

Product

However, another mechanism, which avoids the trimolecular step, can also be written:

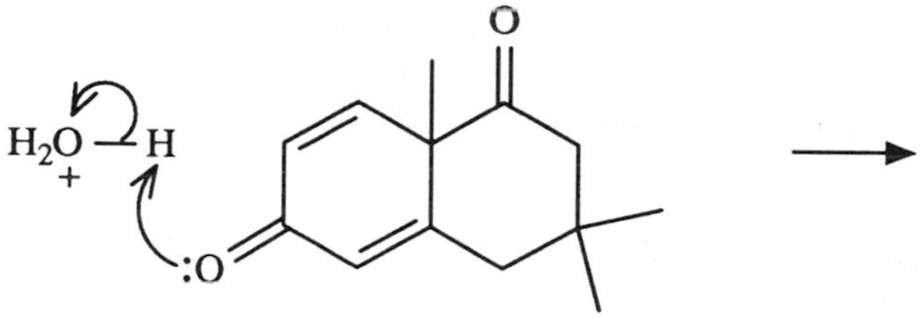

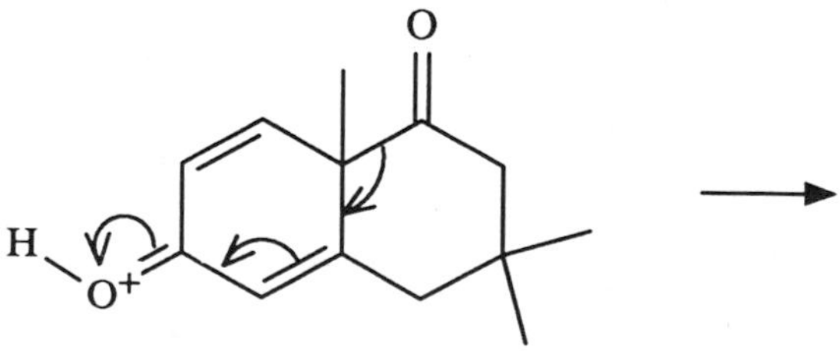

See Problem 4-11 a for an alternative mechanism for this reaction.

Helpful Hint 2-9. Nucleophilic attack cannot occur at a positive oxygen or nitrogen which has a filled valence shell. Only 8 electrons can be accommodated by elements in the second period of the periodic table. However, third period elements, like sulfur and phosphorus, can (and do) expand their valence shells to accommodate 10 (occasionally more) electrons.

Example 2-10. Nitrogen cannot accommodate more than 8 electrons.

The following step is inappropriate because, in the product, nitrogen has expanded its valence shell to 10 electrons.

To avoid this situation, the π electrons in the double bond could move to the adjacent carbon, giving an internal salt called an ylide. Although they are rather unstable, ylides are intermediates in some well-known reactions.

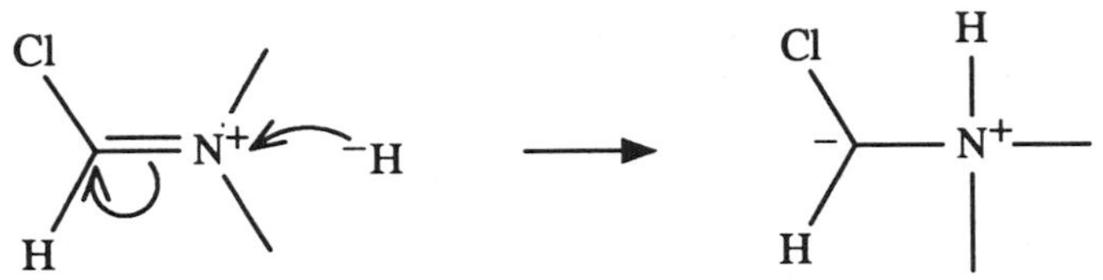

If a neutral nucleophile attacked the nitrogen in a similar manner, there would be three charges on the product. The two positive charges on adjacent atoms would make this a very unstable intermediate.

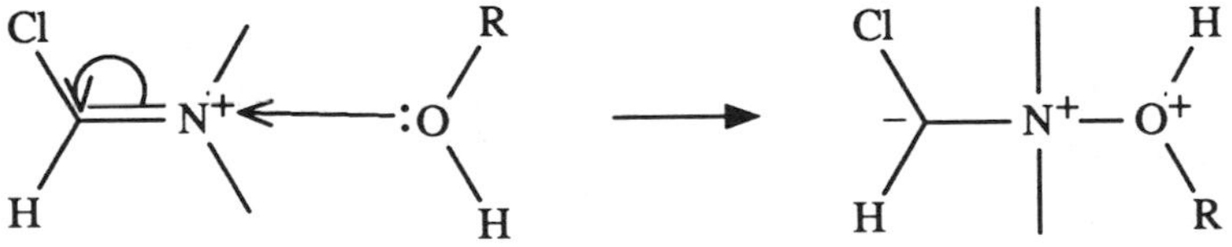

With the above reagents, a more appropriate reaction would be:

Helpful Hint 2-10. Any species containing a positive nitrogen or oxygen, without an octet of electrons, will have high energy, because of the high electronegativity of oxygen and nitrogen. Formation of such species, especially with oxygen, occurs only under very unusual circumstances. An example of an electron-deficient nitrogen species, the nitrenium ion, will be given in Chapter 4. *Neither* of the following are viable steps:

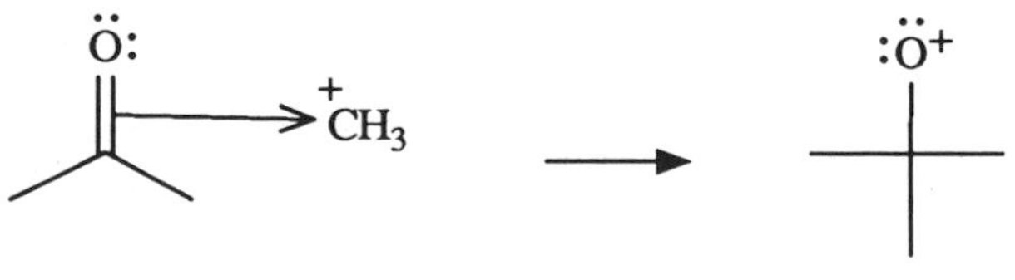

Helpful Hint 2-11. A viable reaction should have some energetic driving force. Examples are: formation of a stable inorganic compound; formation of a stable double bond or aromatic system; formation of a stable carbocation, anion, or radical from a less stable one; or formation of a stable small molecule (see Helpful Hint 2-12). A reaction may be driven by a decrease of enthalpy, increase of entropy, or a combination. Reactions driven by entropy often involve forming more product molecules from fewer starting molecules. Reactions which form more stable bonds are primarily enthalpy driven. When writing a mechanism, constantly ask the questions: Why would this reaction go this way? What is favorable about this particular step?

Example 2-11. A rationale for different leaving groups in the acid and base-promoted hydrolysis of amides.

Ammonia is the leaving group in the acid-promoted hydrolysis of amides. Amide ion, $^-NH_2$, is the leaving group in the base-promoted hydrolysis. The difference can be explained by the driving force of the intramolecular nucleophile relative to the ability of amide ion or ammonia to act as leaving groups. In acid, the intramolecular nucleophile is the oxygen of one of the hydroxyl groups of the tetrahedral intermediate.

In base, the intramolecular nucleophile is the oxyanion of the tetrahedral intermediate:

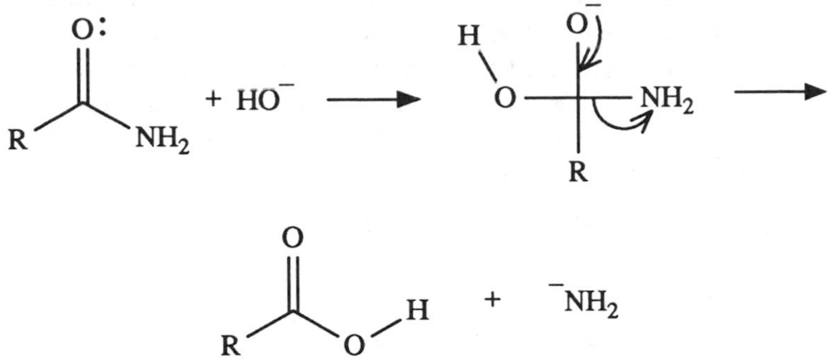

In acid, the hydroxyl group is not a strong enough intramolecular nucleophile to drive the loss of an amide ion, so the nitrogen is protonated first. The leaving group is now the more stable ammonia molecule. In base, the oxyanion, formed by reaction of the original amide with a hydroxyl ion, is a strong enough nucleophile to drive the loss of an amide ion.

Helpful Hint 2-12. Formation of a small stable molecule can be a significant driving force for a reaction. Such molecules include nitrogen, carbon monoxide, carbon dioxide, water, and sulfur dioxide.

Example 2-12. Loss of carbon monoxide in the thermal reaction of tetraphenylcyclopentadienone with maleic anhydride.

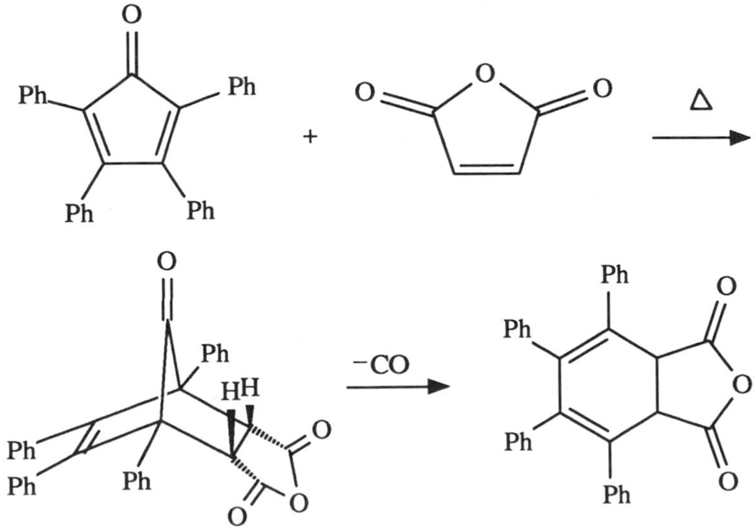

Helpful Hint 2-13. Figuring out the relationship between the atoms in the starting material and the product can be facilitated by numbering the atoms in the starting material and then numbering the same atoms in the product. To do this, first number the atoms of the starting materials in any logical order. Next, identify by number those same atoms discerned in the products, using sequences of atom types and bonding patterns. Then, **using the fewest bond changes, fill in the rest of the numbers.**

Example 2-13. Write a mechanism for the following reaction:

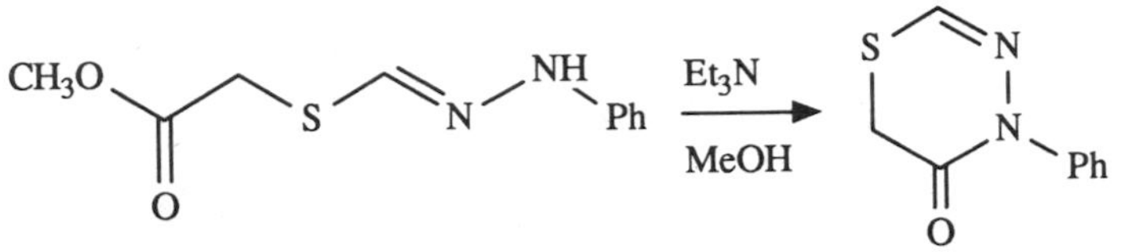

Numbering of the atoms in the starting material and product makes it clear that nitrogen-1 becomes attached to carbon-6.

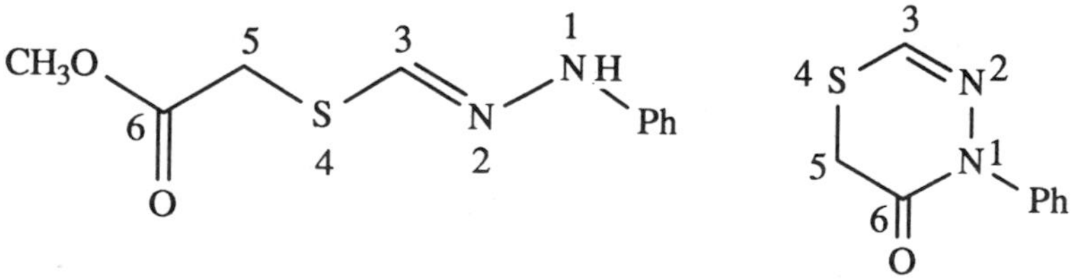

Now it is quite straightforward to write the mechanism as follows:

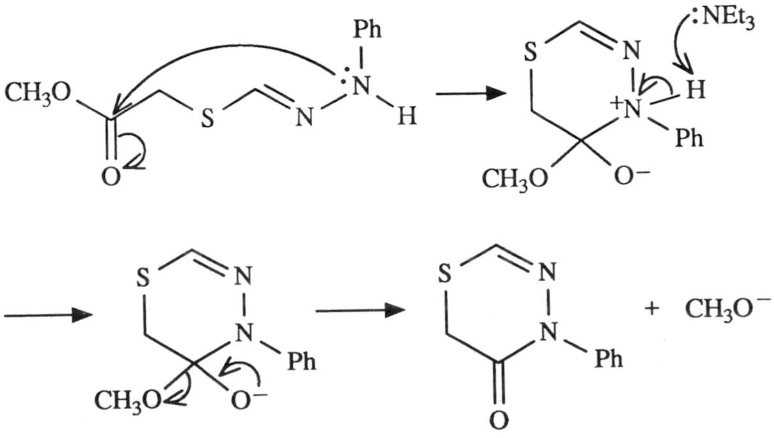

The other products, methoxide ion and triethylammonium ion, would equilibrate to give the weakest acid and weakest base (see Table 1-2).

$$CH_3O^- \quad + \quad H\overset{+}{N}Et_3 \quad \rightleftharpoons \quad CH_3OH \quad + \quad NEt_3$$

Example 2-14. Apply Helpful Hint 2-13 to the following reaction:

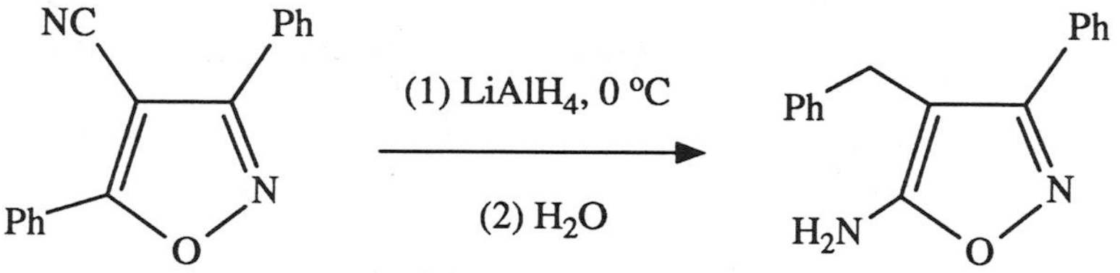

First, consecutively number the atoms of the starting material. In this example, numbering of the atoms in the product is obvious, because of the location of the phenyl groups and nitrogens:

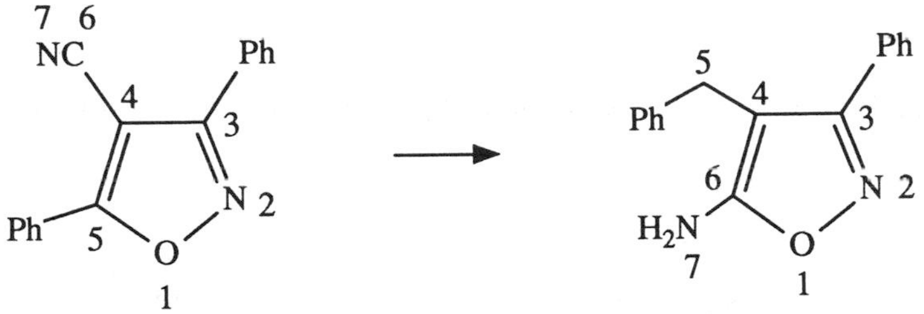

Without having to write any mechanistic steps, the numbering scheme allows us to decide that the C-5, O-1 bond breaks and a new bond forms between O-1 and C-6. This numbering scheme gives the least amount of rearrangement of the atoms when going from starting material to product. This information is invaluable when writing a mechanism for this reaction.

This example is derived fom Alberola, A.; Gonzalez, A. M.; Laguna, M. A.; Pulido, F. J. *J. Org. Chem.* **1984**, *49*, 3423–3424; a mechanism is suggested in this paper.

Helpful Hint 2-14. A paraphrase of the philosophical principle, commonly known as Ockham's razor, can often be applied to writing organic reaction mechanisms: When there is more than one possible mechanistic scheme, the simplest is often the best.

Problem 2-4. For each of the following transformations, number all relevant non-hydrogen atoms in the starting materials, and number the same atoms in the product.

a.

(See Problem 3-19 b for further exploration of this problem.)

b.

(See Example 4-7 for further exploration of this problem.)

Problem 2-5. In each of the following problems, an overall

*reaction is given, followed by a mechanism. In each case, there is
a more reasonable mechanism which can be written for the reaction; write one. For each, state why your mechanism is better
than the one written here and, if applicable, give the related
Helpful Hint number from this chapter and explain its relationship to the problem.*

a.

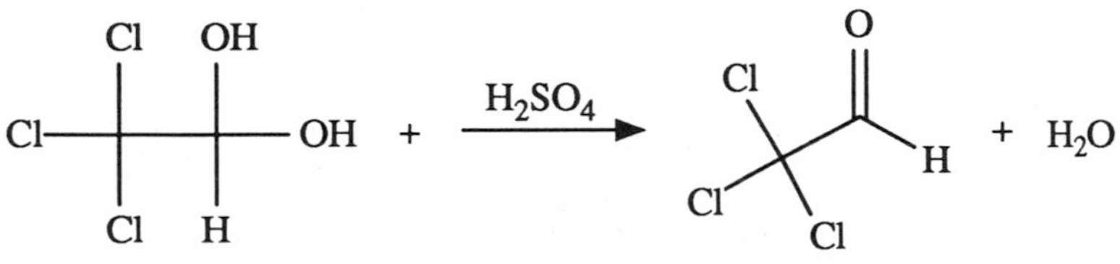

2-4

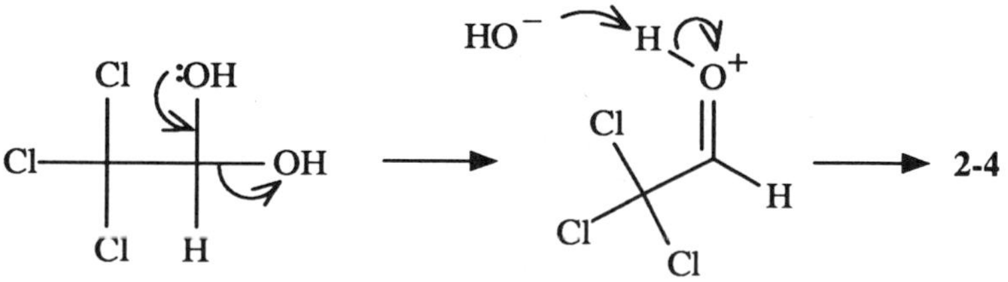

2-4

b.

2-5

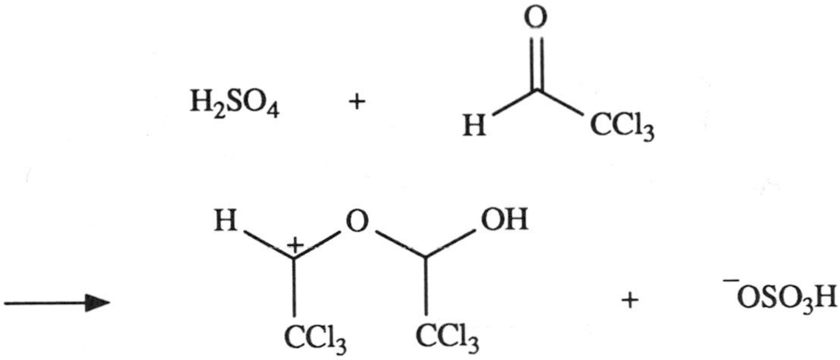

2-5

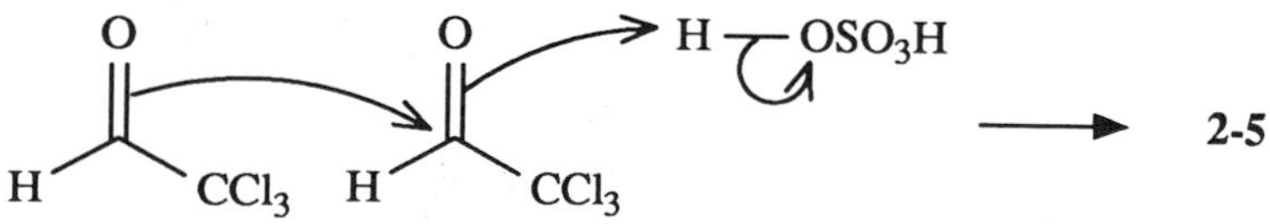

c.

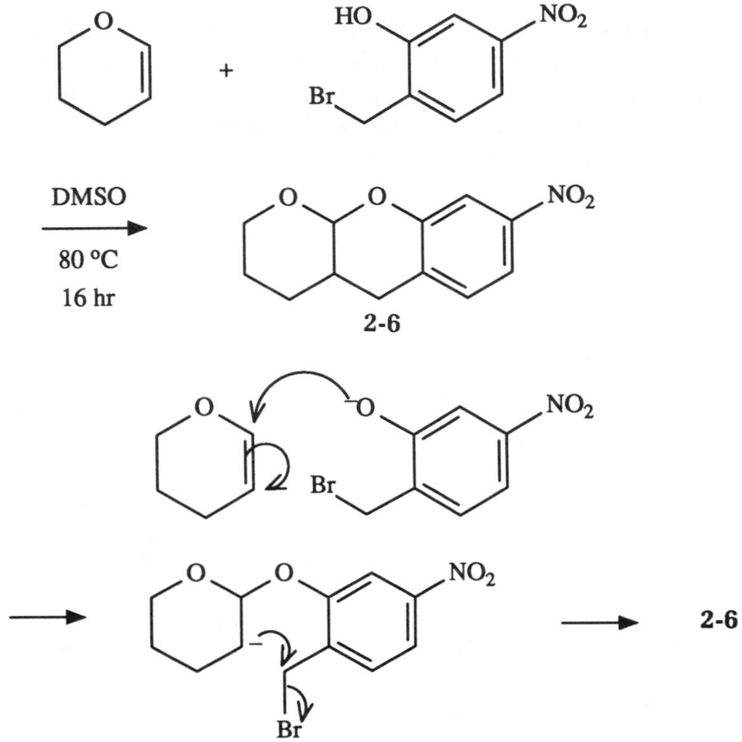

Kirby, A. J.; Martin, J. J. *J. Chem. Soc. Perkin Trans. II* **1983**, 1627–1632.

d.

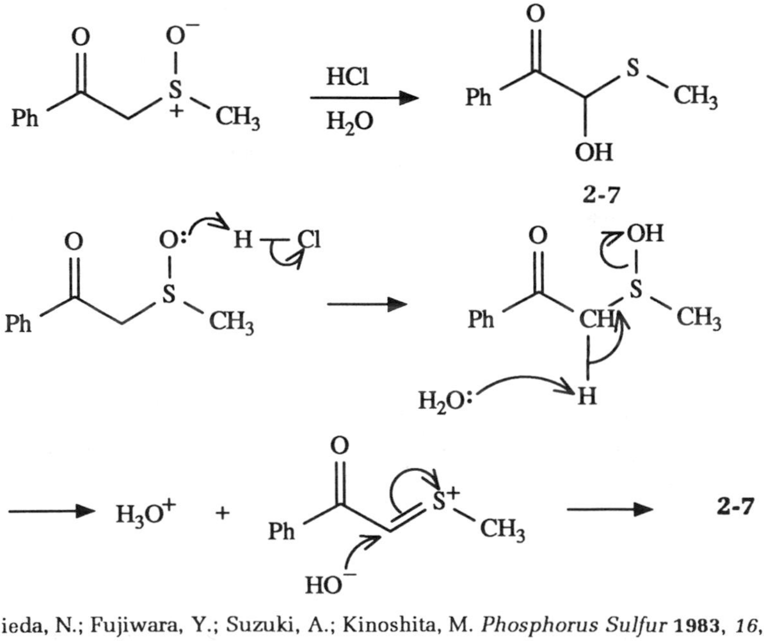

Kunieda, N.; Fujiwara, Y.; Suzuki, A.; Kinoshita, M. *Phosphorus Sulfur* **1983**, *16*, 223–232.

Answers to Problems

2-1.a.

2-1.b.

2-1.c.

2-2.a.

Caution: There is some question about the mechanism for this rearrangement. See the paper cited for detail.

2-2.b.

2-2.c.   removal leads to product

most acidic

Utilizing Table 1-2, there are several ways to compare the acidity of the protons in the boxes. The pK_a of the amide protons in acetamide is 25.5. The pK_a of the amide proton in this compound should be somewhat greater, because the second amido nitrogen will reduce resonance stabilization of the anion formed when the amide proton is removed. The precedent for the other boxed protons is the pK_a of the methylene protons of methyl cyanoacetate, 12.8. Because the carbonyl group in the given compound is an amide rather than an ester, the pK_a of its protons will be somewhat higher, but not nearly as high as the value for the amide proton.

Here is another possible comparison. The pK_a of the protons on the α carbon of ethyl acetate is 30, whereas the pK_a of those of methyl cyanoacetate is 12.8. Therefore, a cyano group lowers the pK_a of the α carbon protons by 30–12.8=17 units. The pK_a of the α carbon protons of an N,N disubstituted amide, N,N-diethylacetamide is 34.5. Assuming that the cyano group enhances the acidity of the methylene protons in the given compound by the same amount, their pK_a would be 34.5–17=16.5. Thus, the estimated pK_a of the methylene protons is at least 9 pK units lower than that of the amide proton, a factor of one billion.

A less quantitative approach is to rationalize the stabilities of the anions that result when a proton is removed from various regions of the molecule. Removal of hydrogen from nitrogen gives an anion stabilized by two resonance forms, one with negative character on nitrogen and one with negative character on oxygen:

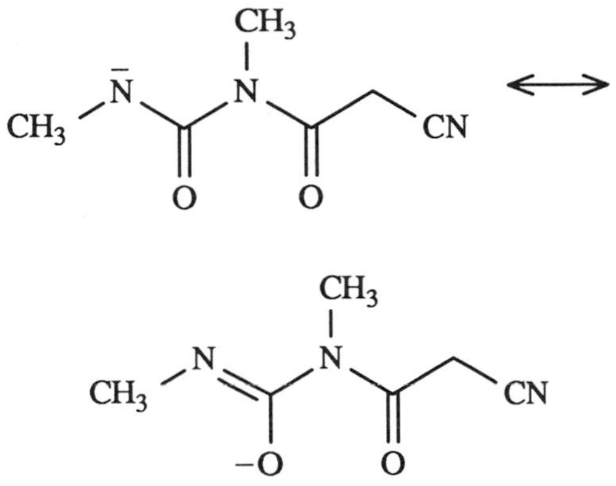

On the other hand, removal of hydrogen from carbon gives three resonance forms, with negative charges on nitrogen, oxygen, or carbon:

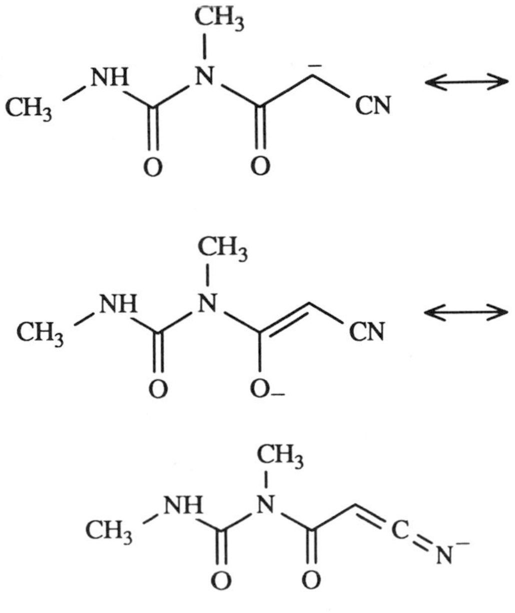

Thus, the second anion should be more stable.

Finally, removal of a proton from either of the methyl groups in the compound produces a localized anion, which would be much less stable than either of the other two possibilities.

The mechanism for the reaction could be written as follows. First the proton is removed by base.

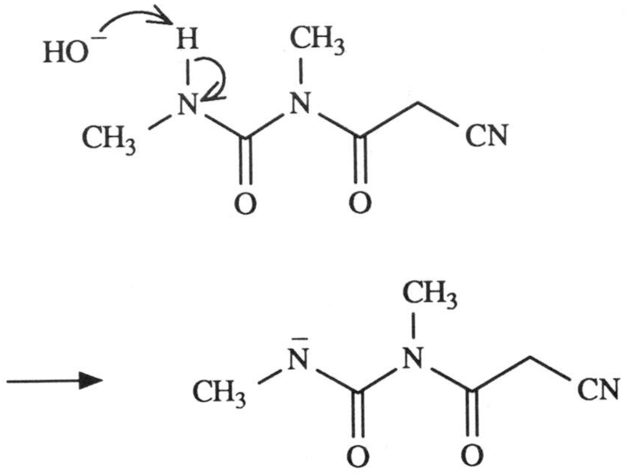

2-8

The resulting anion, **2-8**, can be written in a conformation which emphasizes the cyclization which then takes place.

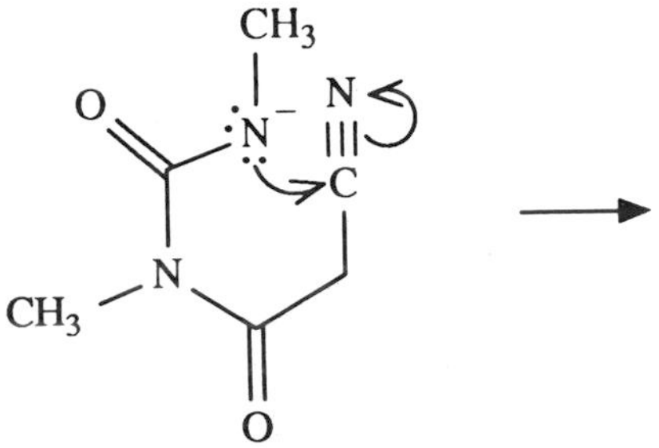

2-8

Now a tautomerization takes place to give the final product.

Product

2-2.d.

2-3.a.

Table 1-2 indicates that carbonyl oxygens are more basic than the oxygen of an alcohol. The protonated carbonyl is stabilized by resonance while the protonated alcohol is not.

The mechanism proposed in the cited paper suggests that the starting material is also dehydrated before cyclization occurs.

2-3.b.

2-3.c.

Table 1-2 is of help here. The pK_a of protonated acetophenone is −4.3, while the pK_a of protonated dimethyl sulfoxide is −1.5. Thus, dimethyl sulfoxide is almost 10^3 time more basic than acetophenone. These compounds are excellent models for the two functional groups in the compound given.

It is very difficult to decide which oxygen is more basic, based only on the composition of the molecule. On the one hand, since sulfur is slightly more electronegative than carbon (see Table 1-1), one might predict that the carbonyl oxygen is more basic. Furthermore, protonation of the carbonyl group leads to a cation which is stabilized by resonance delocalization into the aromatic ring:

On the other hand, the π bond between C and O should be stronger than the π overlap between S and O because C and O both use 2p orbitals to form π bonds while a π bond between S and O would involve 3p-2p overlap which is not as effective. This should make the oxygen of the sulfoxide functional group more negative. Additionally, sulfur stabilizes charges more effectively than do carbon and oxygen because of its higher polarizability.

Without calculations and/or hard data like that from Table 1-2, the choice of the most basic atom is difficult to make.

If hard data were not available, the qualitative arguments would suggest that either oxygen is fairly basic. Realizing that a mechanism will not always proceed through protonation of the most basic atom, it would be best to try writing mechanisms of reactions in acid for this compound by protonation of either of the two oxygens.

2-4.a. Start numbering the product at the carboethoxy group which has not been altered by the reaction. Once you notice that the keto group of the ketoester reactant is also unchanged, the remaining atoms easily fall into place. For the isocyanate, N-6 is obvious, and C-7 and O-8 follow.

2-4.b. The first thing to notice is that the position of the C-O bond, relative to the methyl group, has not changed. Thus, an initial try at numbering the product would leave those atoms in the same relative positions as in the starting material. This gives the following numbers:

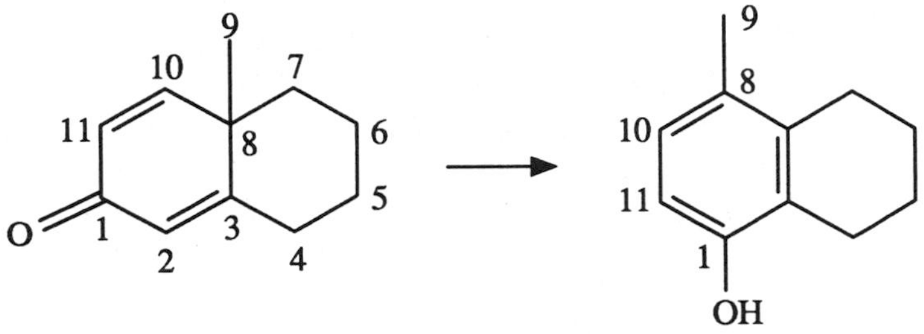

A strong possibility is that the left-hand ring of the product contains the same carbons that this ring contained in the starting material. Minimizing the changes in bonding gives the following:

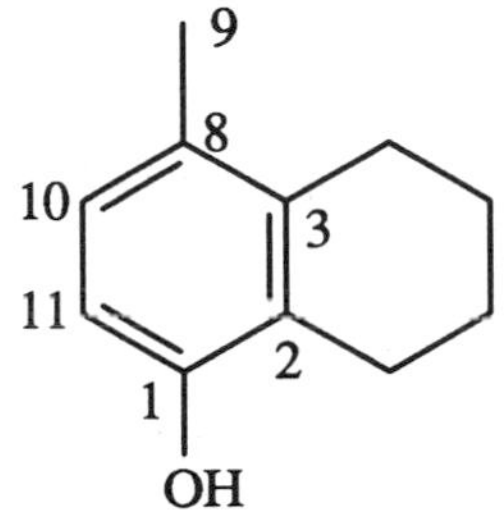

This leaves two possible numbering schemes for the right-hand ring:

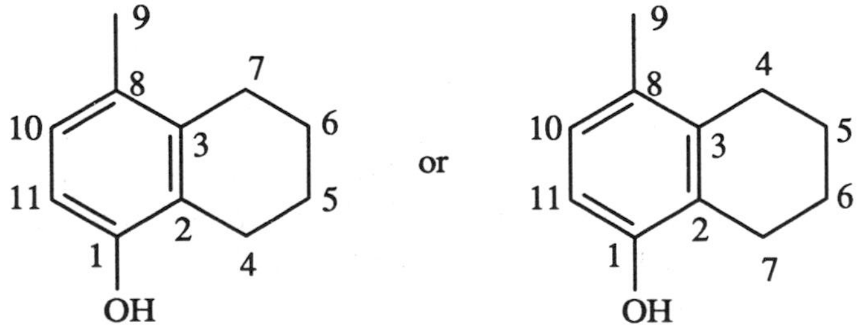

The second structure involves less rearrangement, but consideration of Example 4-7 indicates that, since a symmetrical intermediate is produced, either numbering scheme outlines a possible bond reorganization to form the product.

2-5.a.   Helpful Hint 2-5. This reaction is taking place in a strongly acidic medium. Therefore, strong bases like ⁻OH will be in such low concentration that they cannot be effective reagents in the reaction. A better mechanism would be to protonate the oxygen of one of the hydroxyl groups to convert it into a better (and neutral) leaving group, the water molecule.

2-5.b.   Helpful Hint 2-8. Break up the trimolecular step into a protonation of the carbonyl group, converting it to a better electrophile, followed by nucleophilic reaction with the oxygen of another molecule of starting material.

The nucleophilic addition of the carbonyl compound to another protonated molecule can be written either as shown or by using the π bond as the nucleophile instead of a lone pair on oxygen. This second kind of addition leads to the second resonance form shown instead of the first one.

These two representations for this step are equivalent.

The oxygen of one starting aldehyde molecule must be protonated before nucleophilic addition of the carbonyl oxygen of the second. Otherwise, a very basic anion is formed in a very acidic medium, the same problem discussed in part a.

2-5.c.   There are two major problems with the mechanism shown. Data from Table 1-2 indicate that the substituted phenol would not be significantly ionized in DMSO: protonated dimethyl sulfoxide has a pK_a of -1.5, whereas *m*-nitrophenol has a pK_a of 8.3. Also, the intermediate carbanion, resulting from the addition, is very unstable; it is not stabilized by resonance in any way. The following mechanism avoids these problems by forming an intermediate cation, **2-9**, which is stabilized by resonance, and leaving removal of the proton to the last step.

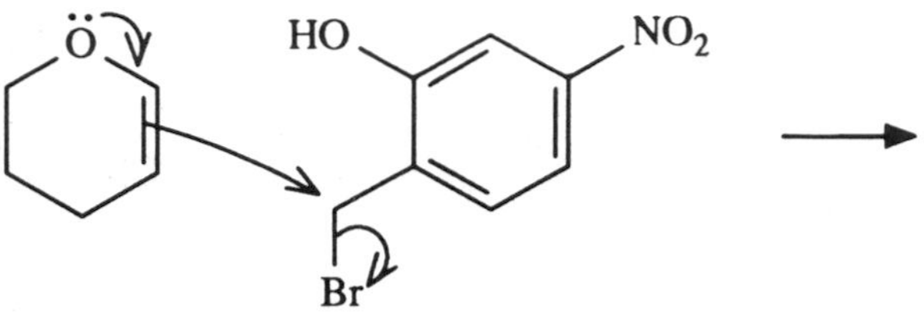

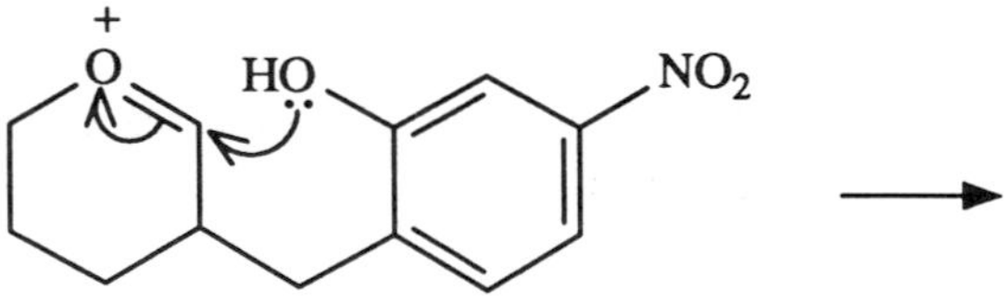

2-9

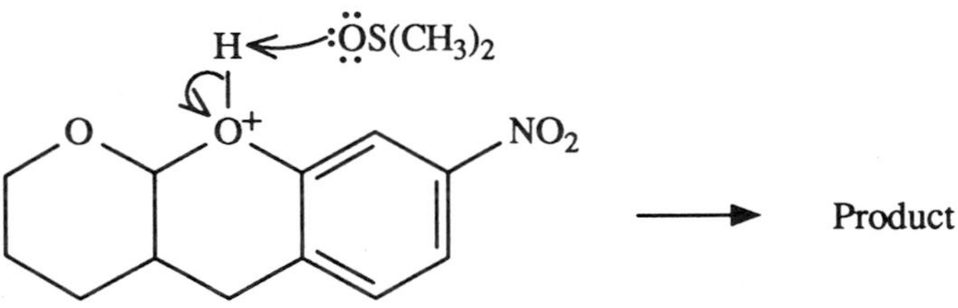

Product

Intermediate **2-9** could also be written as the other resonance form, **2-10**, which contributes to the stabilization of the positive charge:

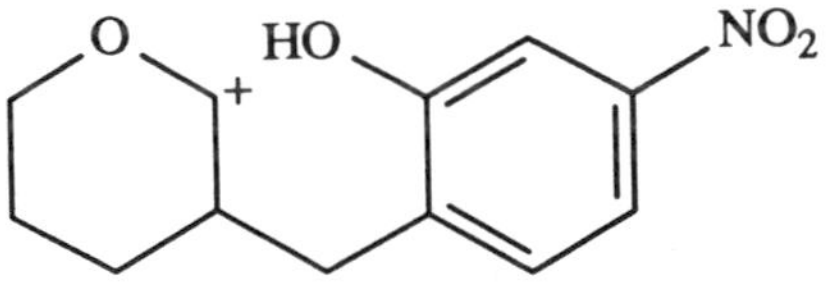

2-10

It is often useful to draw resonance forms for proposed intermediates, because their existence is an indication of stability which represents a driving force for the reaction. Removal of a proton from the OH group of the resonance hybrid of **2-9** and **2-10** is unlikely, because bromide ion is an even weaker base than DMSO: the pK_a of HBr is −9. Therefore, it is expected that ring closure, by nucleophilic reaction with the cation, takes place before removal of the proton.

2-5.d.   Helpful Hint 2-5. The given mechanism has a sequence which is unlikely in strong acid; that is, loss of hydroxide ion, and subsequent use of hydroxide ion as a nucleophile to form the product. In the following preferred mechanism, the oxygen is protonated first so that water can act as the leaving group and later as the nucleophile in the addition to intermediate **2-11**. This mechanism also uses the hydronium ion as the acid, since hydrochloric acid is completely ionized in aqueous solution. It is always a good idea to keep track of lone pairs of electrons on sulfur and oxygen by drawing them in. Also keep track of formal charges, on the atoms of interest, as the mechanism proceeds; in this case, sulfur and oxygen.

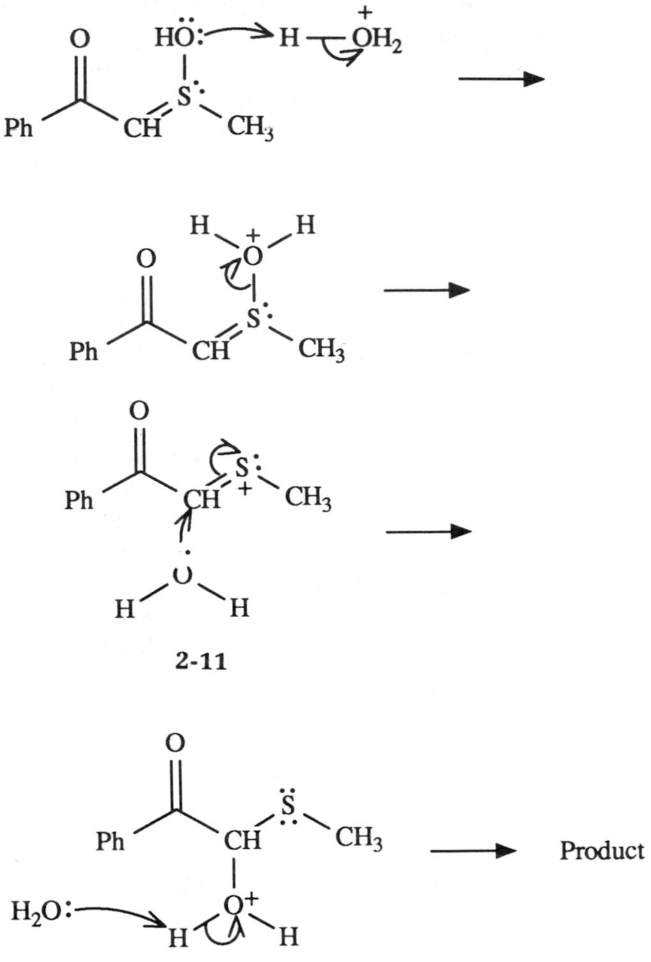

2-11

Formation of **2-12** (by doubly protonating the sulfoxide oxygen of starting material), from which simultaneous elimination of water and loss of a proton from the methylene group might be written, would not be as good a step as those shown above because the development of two adjacent positive centers is destabilizing.

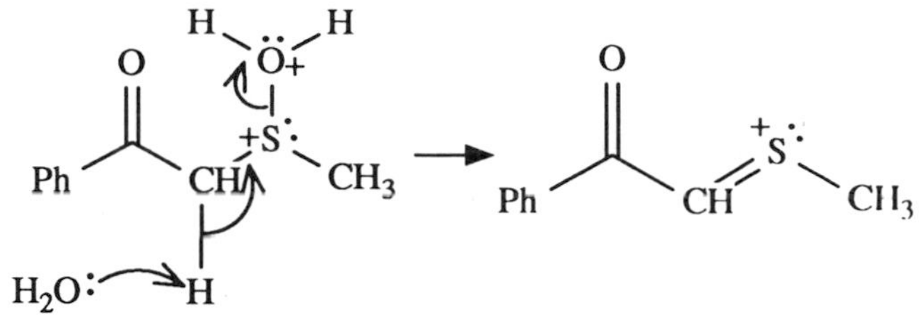

2-12

Removal of a proton always requires reaction with a base. In this reaction step, water could act as the base. Thus, the following representation is not strictly correct but, as we stated, is allowed by some instructors.

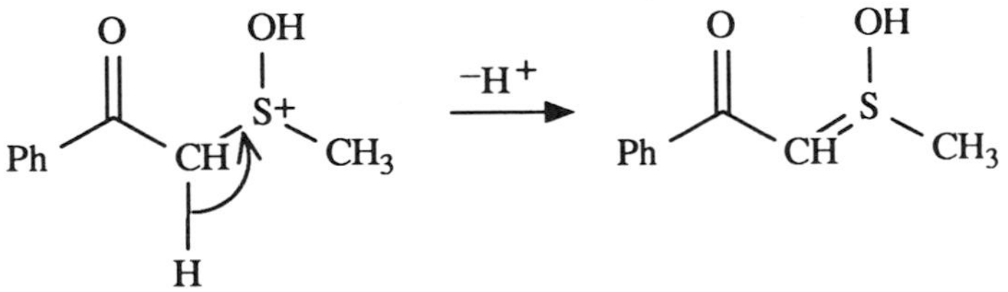

Showing **2-13** for removal of a proton is not accurate on two counts. First, the arrow indicates electron movement in the wrong direction; this would produce H⁻ instead of H⁺. (See Chapter 3 for further discussion of hydride loss.)' Second, proton removal always requires reaction with a base, even if it is a weak base.

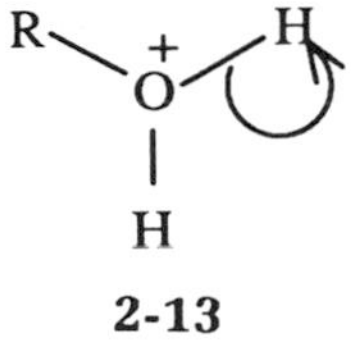

2-13

The following equation is an example of a [1,3] sigmatropic intramolecular shift of hydrogen. Chapter 6 will discuss why this type of tautomeric reaction is an unlikely process.

Chapter 3

Reactions of Nucleophiles and Bases

This chapter includes examples of aliphatic nucleophilic substitution at both sp^3 and sp^2 centers, aromatic nucleophilic substitution, E-2 elimination, nucleophilic addition to carbonyl compounds, 1,4-addition to α,β-unsaturated carbonyl compounds, and rearrangements promoted by base.

3-1 Nucleophilic Substitution

3-1-1 THE S_N2 REACTION

The S_N2 reaction is a concerted bimolecular nucleophilic substitution at carbon. It involves an electrophilic carbon, a leaving group and a nucleophile. The positive character of the carbon is supplied by the electron-withdrawing leaving group. The carbon undergoing reaction may also be attached to other electron-withdrawing substituents which enhance the reaction.

Example 3-1. The S$_N$2 reaction: a concerted process.

The electrons of the nucleophile interact with carbon at the same time that the leaving group takes both of the electrons in the bond between carbon and the leaving group. This particular example involves both a good leaving group and a good nucleophile.

$$^-CN \quad PhCH_2 \frown OSO_2CF_3 \longrightarrow PhCH_2CN \ + \ ^-OSO_2CF_3$$

Nucleophiles are discussed in Section 1-6. Table 3-1 lists typical leaving groups and gives a qualitative assessment of their effectiveness. Usually, the less basic the substituent, the more easily it will act as a leaving group. Relative leaving group abilities also depend upon the solvent and nucleophilicity of the nucleophile, but this does not significantly affect the mechanisms derived in this book.

TABLE 3-1

Leaving Group Abilities

Excellent

N_2, $^-OSO_2CF_3$ (triflate), nosylate, brosylate, tosylate, $^-OSO_2CH_3$ (mesylate)

Good	Fair	Poor
I^-, Br^-, Cl^-, SR_2	OH_2, NH_3, $^-OCOCH_3$	F^-, ^-OH, ^-OR

Very Poor

$^-NH_2$, ^-NHR, $^-NR_2$, R^-, H^-, Ar^-

Helpful Hint 3-1. Except in some reactions of organometallics, hydride, H⁻, almost never acts as a leaving group.

The other poor and very poor leaving groups also very rarely act as leaving groups in the S_N2 reaction. In other reactions, such as the nucleophilic substitution of derivatives of carboxylic acids, the negatively charged oxygen- and nitrogen-containing species are more common leaving groups. In elimination reactions, hydroxide acts as a leaving group only when there is considerable driving force for the reaction, such as formation of a double bond stabilized by resonance.

Problem 3-1. Explain why the following mechanistic step in the equilibrium between a protonated and an unprotonated alcohol is a poor one.

Example 3-2. The S_N2 reaction of an alcohol requires prior protonation.

The alcohol oxygen is protonated before substitution takes place. Thus, the leaving group is a water molecule, a fair leaving group, rather than the hydroxide ion, a poor leaving group.

A poorer mechanistic step would show hydroxide as the leaving group:

The S_N2 reaction occurs only at sp^3-hybridized carbons. The relative reactivities of carbons in the S_N2 reaction are $CH_3 > 1° > 2° >> 3°$, due to steric effects. Methyl, $1°$ carbons, and $1°$ and $2°$ carbons which are also allylic, benzylic, or alpha to a carbonyl group are especially reactive. For example, α-halocarbonyl compounds show augmented reactivity in the S_N2 reaction. (See Example 3-18.)

Example 3-3. Stereochemistry of the S_N2 reaction.

The S_N2 reaction always produces a 100% inversion of the configuration. Thus, when this mechanism is depicted accurately with arrows, the nucleophile is shown approaching the electrophilic carbon on the side opposite the leaving group (a 180° angle between the line of approach of the nucleophile and the bond to the leaving group).

Problem 3-2. For the following synthesis,

show how the alkylation of the phenolic oxygen (introduction of the benzyl group onto the oxygen) could be explained more accurately than by the following step and subsequent deprotonation of the phenolic oxygen.

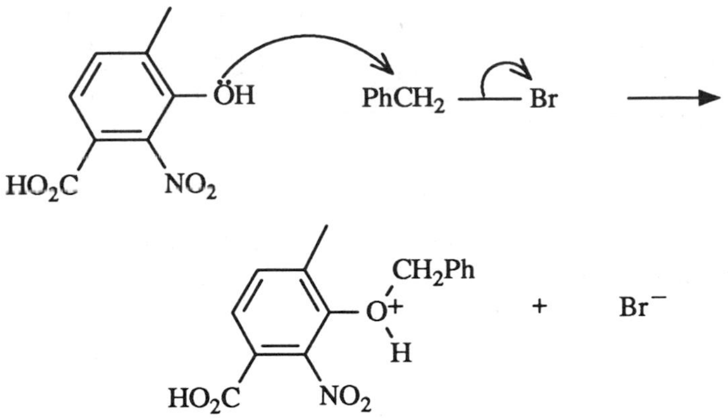

Pena, M. R.; Stille, J. K. *J. Am. Chem. Soc.* **1989**, *111*, 5417–5424.

Problem 3-3. Pick out the electrophile, nucleophile and leaving group in each of the following reactions, and write a mechanism for the formation of products.

a.

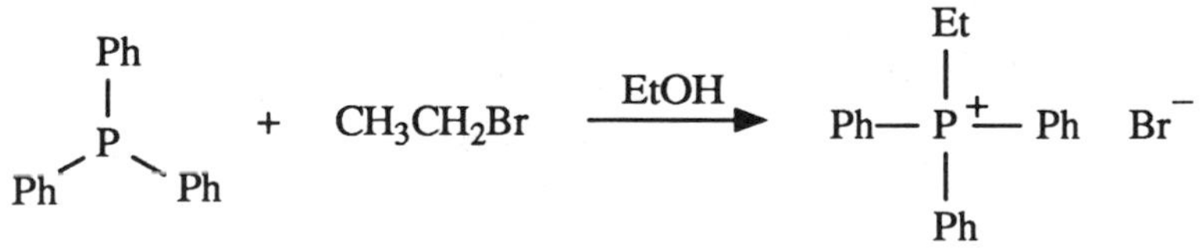

b.

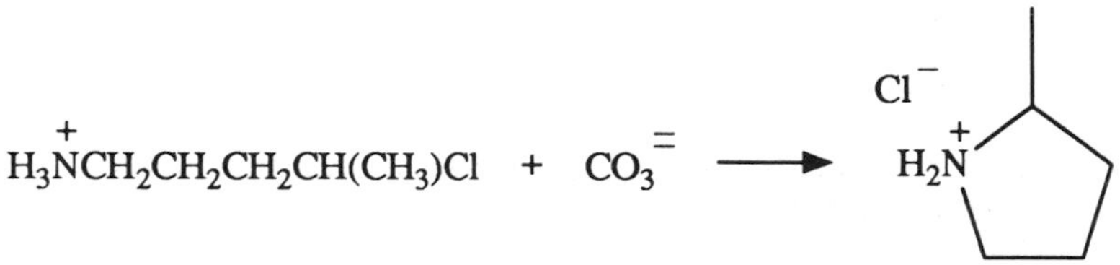

$$H_3\overset{+}{N}CH_2CH_2CH_2CH(CH_3)Cl \;+\; CO_3^{=} \longrightarrow$$

c.

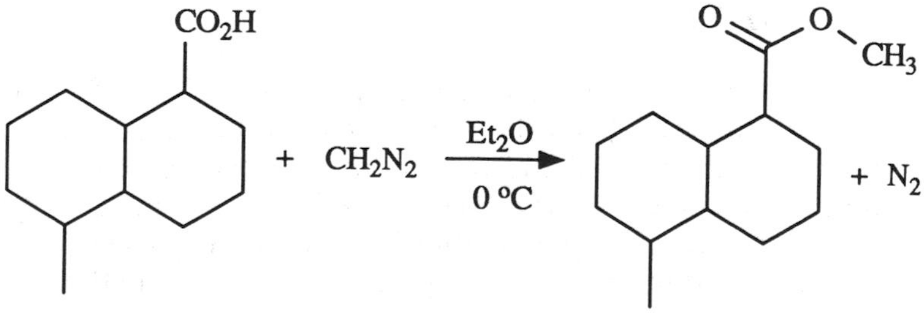

d.

Problem 3-4. Write step-by-step mechanisms for the following transformations:

a.

b.

Wenkert, E.; Arrhenius, T. S.; Bookser, B.; Guo, M.; Mancini, P. *J. Org. Chem.* **1990**, *55*, 1185–1193.

3-1-2 NUCLEOPHILIC SUBSTITUTION AT AN ALIPHATIC sp^2 CARBON

The familiar substitution reactions of derivatives of carboxylic acids with basic reagents illustrate nucleophilic substitution at aliphatic sp^2 carbons. (Substitution reactions of carboxylic acids, and their derivatives, with acidic reagents will be covered in Chapter 4.) The mechanisms of these reactions involve two steps: (1) addition of the nucleophile to the carbonyl group and (2) elimination of some other group attached to that carbon. Common examples include the basic hydrolysis and aminolysis of acid chlorides, anhydrides, esters and amides.

Example 3-4. Mechanism for hydrolysis of an ester in base.

Unlike the one-step S_N2 reaction, the hydrolysis of esters in base is a two-step process. The net result is substitution, but the first step is a nucleophilic addition to the carbonyl group, the carbon of which rehybridizes to sp^3, and the second step is an elimination in which the carbonyl carbon is rehybridized back to sp^2.

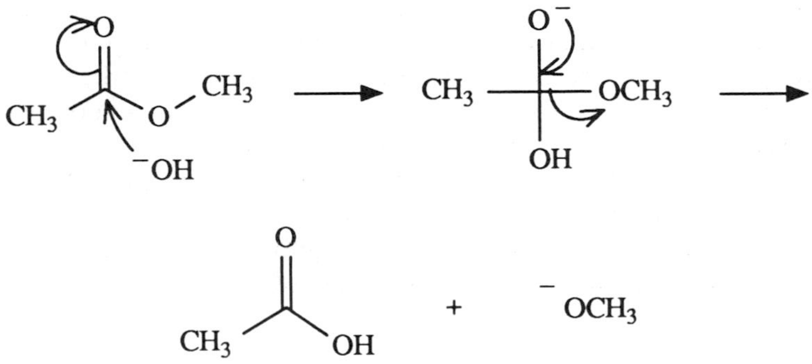

This is followed by removal of a proton from the acid, by the methoxide ion, to yield methanol and the carboxylate ion:

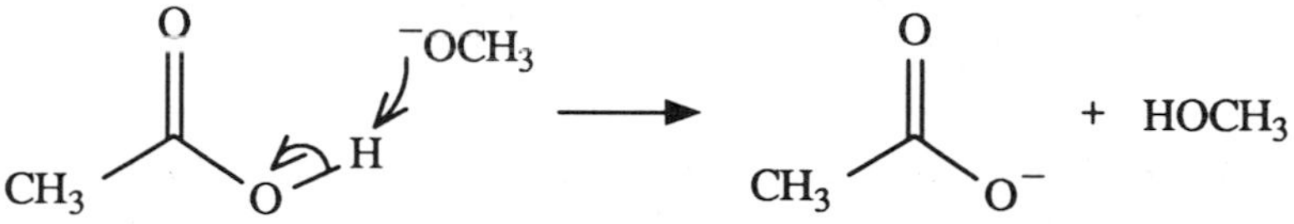

Another possible mechanism for this hydrolysis is an S_N2 reaction at the alkyl carbon of the ester:

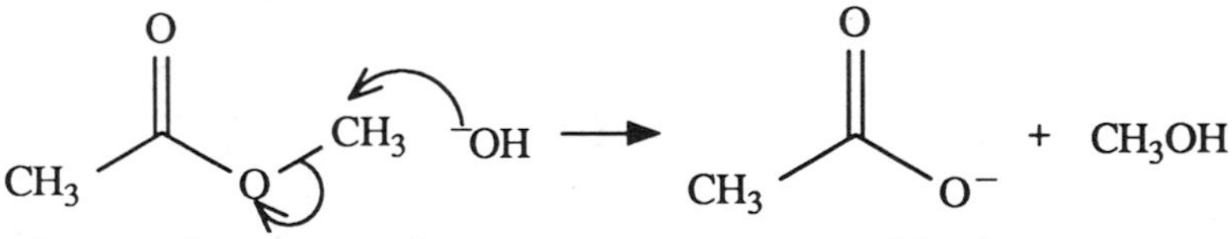

This single-step mechanism appears reasonable, because carboxylate is a fair leaving group and hydroxide is a very good nucleophile. However, labelling studies rule out this mechanism under common reaction conditions. The two-step mechanism must be favored, because the higher mobility of the π electrons of the carbonyl group makes the carbonyl carbon especially electrophilic.

Helpful Hint 3-2. Direct nucleophilic substitution at an sp^2-hybridized center is not likely under common reaction conditions. Thus, nucleophilic substitution reactions at such centers are usually broken up into two steps. (For exceptions to this hint, see Dietze, P.; Jencks, W. P. *J. Am. Chem. Soc.* **1989**, *111*, 5880–6 and references cited therein.)

There are several reasons why direct substitutions at sp^2-hybridized centers occur less readily than at sp^3 centers. First, because there is more s-character in the bond to the leaving group, this bond is stronger than the corresponding bond to an sp^3-hybridized carbon. Second, the greater mobility of the π electrons at an sp^2 center increases the likelihood that the interaction will cause electron displacement. Third, because of the planar configuration of the substituents around an sp^2 center, there is strong steric interference to the approach of a nucleophile to the side opposite the leaving group. Thus the mechanism for basic hydrolysis of an ester would **not be written** as below:

Problem 3-5. What mechanism, which is more likely than the one shown, might be written for the following transformation?

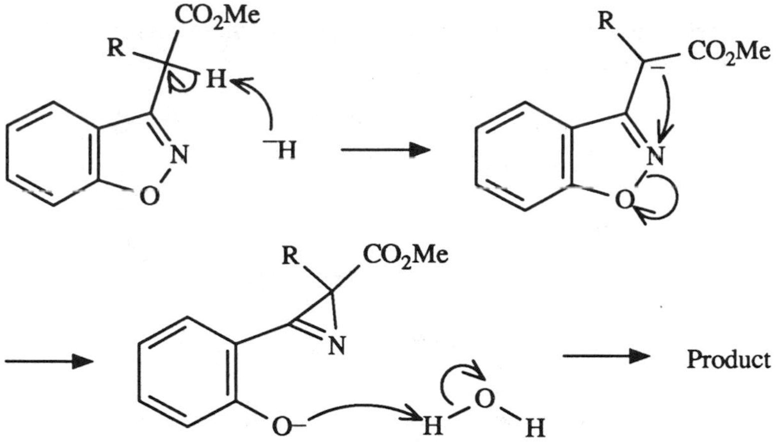

Problem 3-6. Write step-by-step mechanisms for the following transformations:

a.

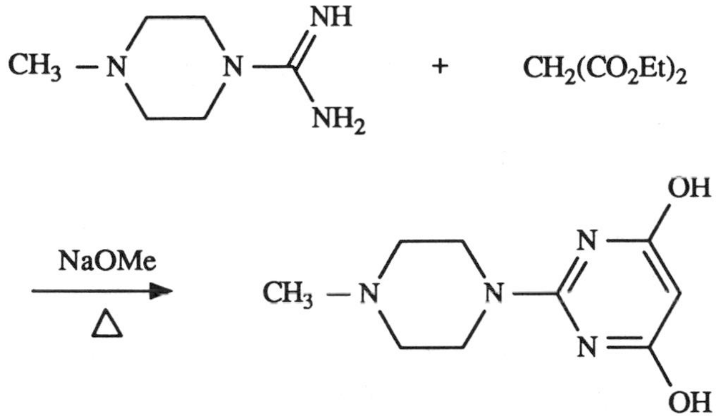

Gueremy, C,.; Audiau, F.; Renault, C.; Benavides, J.; Uzan, A.; Le Fur, G. *J. Med. Chem.* **1986**, *29*, 1394–1398.

b.

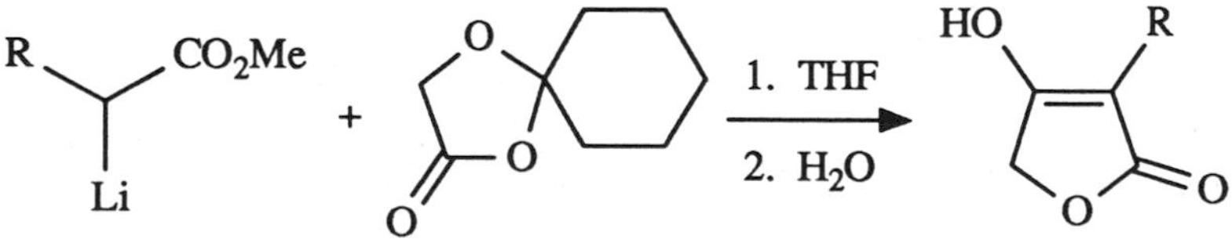

Ramage, R.; Griffiths, G. J.; Shutt, F. E.; Sweeney, J. N. A. *J. Chem. Soc. Perkin I* **1984**, 1539–1545.

c. Critically evaluate the following partial mechanism for the reaction given in part a:

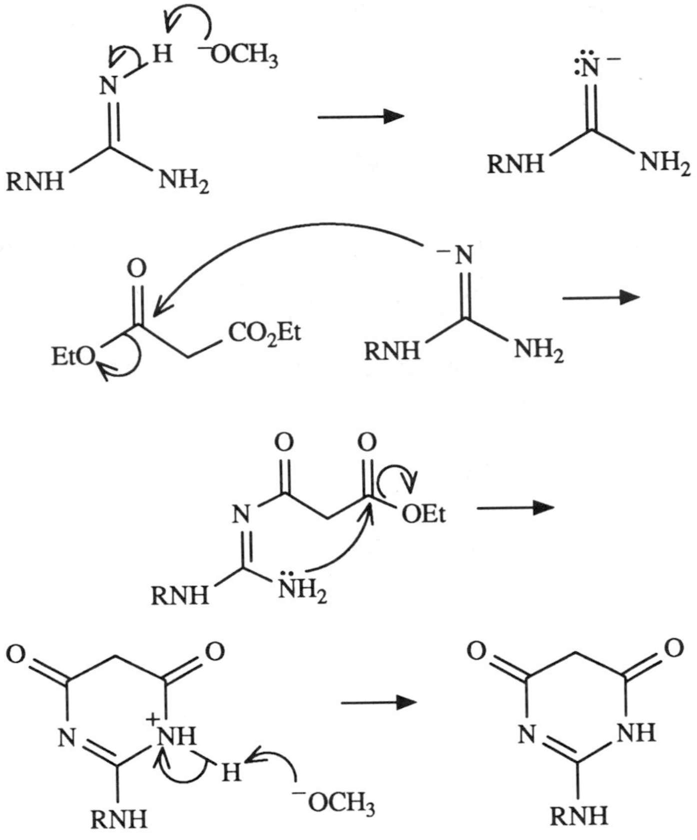

3-1-3 NUCLEOPHILIC SUBSTITUTION AT AROMATIC CARBONS

Both mechanisms for nucleophilic aromatic substitution occur in two important steps. In one mechanism, an addition is followed by an elimination. In the other mechanism, an elimination is followed by an addition.

Example 3-5. Relative reactivity in the addition-elimination mechanism.

When X=halogen, the observed relative reactivities of the starting materials are F>Cl>Br>I. This indicates that the first step is rate-determining. If the second step were rate-determining, the relative reactivities would be reversed, because the relative abilities of the leaving groups are $I^->Br^->Cl^->F^-$. The relative reactivities are as observed, because the greater the electron-withdrawing power of the halogen (see Table 1-1) the more it increases the electrophilicity of the aromatic ring, making it more reactive to nucleophiles.

The addition-elimination mechanism generally requires a ring activated by electron-withdrawing groups. These groups are especially effective at stabilizing the negative charge in the ring when they are located at the positions *ortho* and/or *para* to the eventual leaving group.

Problem 3-7. By drawing the appropriate resonance forms, show that the negative charge in the intermediate anion in Example 3-5 is stabilized by extensive electron delocalization.

Problem 3-8. Write a step-by-step mechanism for the following transformation.

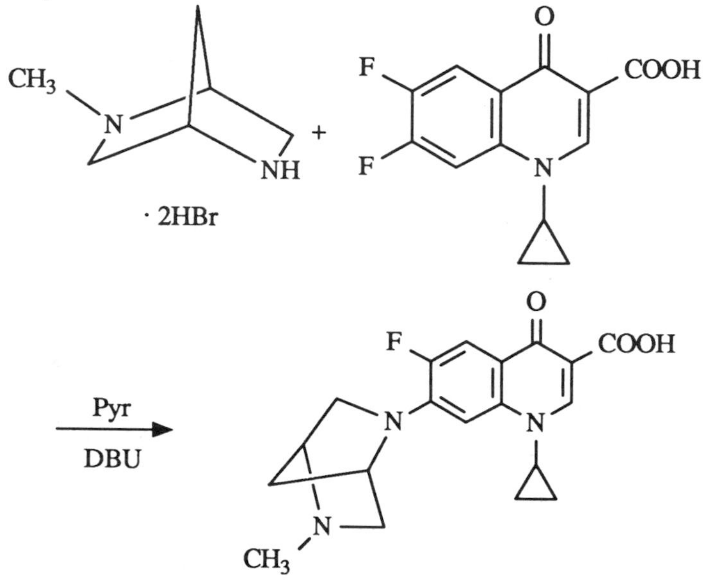

Braish, T. F.; Fox, D. E. *J. Org. Chem.* **1990**, *55*, 1684–1687. [This is the last step
  in the synthesis of danofloxicin, an antibacterial. Pyr (or Py) is a common
  acronym for pyridine; DBU is 1,8-diazabicyclo[5.4.0]undec-7-ene. A good
  reference for translation of acronyms is Daub, G. H.; Leon, A. A.; Silverman,
  I. R.; Daub, G. W.; Walker, S. B. *Aldrichimica Acta* **1984**, *17*, 13–23.]

In reactions which proceed by the elimination-addition mechanism, often called the aryne mechanism, the bases used are commonly stronger than those used in reactions proceeding by the addition-elimination mechanism. Also in this reaction, the aromatic
ring does not need to be activated by electron-withdrawing substituents, although a reasonable leaving group must be present.

**Example 3-6. An elimination-addition mechanism—aryne
intermediate.**

Write a mechanism for the following reaction:

The intermediate, with a triple bond, is called benzyne; for substituted aromatic compounds, this type of intermediate is called an aryne. In benzyne, the ends of the triple bond are equivalent, and either can react with a nucleophile.

This is not a normal triple bond. The six-membered ring does not allow the normal linear configuration of two sp-hybridized carbon atoms and their substituents. Thus, the carbons remain sp^2-hybridized, and the triple bond contains the σ-bond, the π-bond, and a third bond formed by overlap of the sp^2-hybridized orbitals which formerly bonded with the bromine and hydrogen atoms. This third bond is in the plane of the benzene ring and contains two electrons.

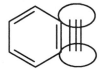

The rate determining step can be either proton removal or departure of the leaving group, depending on the acidity of the proton and the ability of the leaving group. In many cases, the relative rates are so close that the reaction cannot be distinguished from a concerted process.

Problem 3-9. In Example 3-6, if the carbon bound to the bromine in bromobenzene is enriched with ^{13}C, where is this label found in the product aniline?

Problem 3-10. Write a step-by-step mechanism for the following transformation:

Bunnett, J. F.; Skorcz, J. A. *J. Org. Chem.* **1962**, *27*, 3836–3843.

3-2 Other Elimination Reactions

Other important elimination reactions are the E-2 and the Ei, intramolecular. The E-2 reaction is a concerted process, with a bimolecular rate-determining step. In this case, "concerted" means that bonding of the base with a proton, formation of a double bond, and departure of the leaving group all occur in one step. The stereochemistry is usually **anti,** (but in some cases is) **syn**. The term *"anti"* means that the proton and leaving group depart from opposite sides of the bond which then becomes double. That is, the dihedral angle (measured at this bond) between their planes of departure is 180°. If they depart from the same side (the dihedral angle is 0°), the stereochemistry of the elimination is called *syn*.

Example 3-7. An *anti* E-2 elimination.

The dihedral angle between the proton and bromide is 180°, so this is an *anti* elimination.

Example 3-8. In an acid-catalyzed elimination of water from an alcohol, water is the leaving group.

The mechanism for an elimination step in the acid-catalyzed aldol condensation would be written as follows:

The following would be less likely for the formation of an α-β-unsaturated aldehyde in acid (see Helpful Hint 2-5).

Example 3-9. Under some conditions, hydroxide can act as a leaving group.

A 3-hydroxyaldehyde (or ketone) will undergo elimination under basic conditions if the double bond being formed is especially stable, e. g., conjugated with an aromatic system. Such eliminations can occur under the reaction conditions of the base-promoted aldol condensation. An example is the formation of 3-phenylbut-2-enal by an E-2 elimination from 3-hydroxy-3-phenylbutanal.

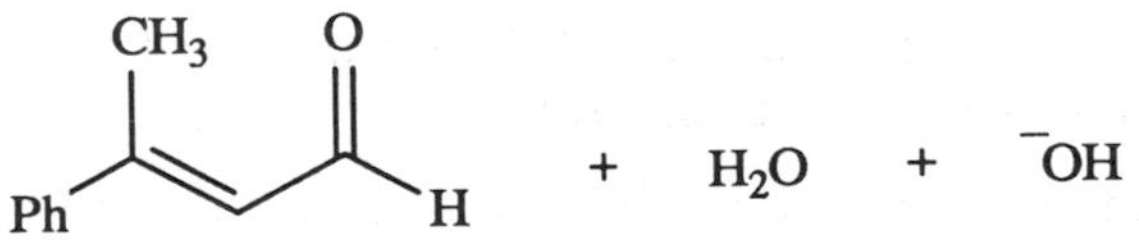

In another type of elimination reaction, called Ei or intramolecular, the base, which removes the proton, is another part of the same
molecule. Such eliminations from amine oxides, or sulfoxides, have
5-membered ring transition states. These transition states are more
stable with *syn* than with *anti* orientations of proton and leaving
group, producing a very high *syn* stereoselectivity.

**Example 3-10. An Ei reaction: pyrolytic elimination from a
sulfoxide.**

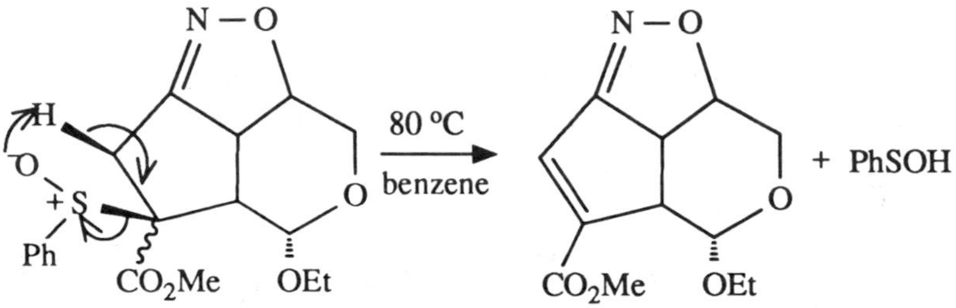

Curran, D. P.; Jacobs, P. B.; Elliott, R. L.; Kim, B. H. *J. Am. Chem. Soc.* **1987,** *109,*
5280–82.

*Problem 3-11. Write a mechanism for the following reaction.
What is the other product?*

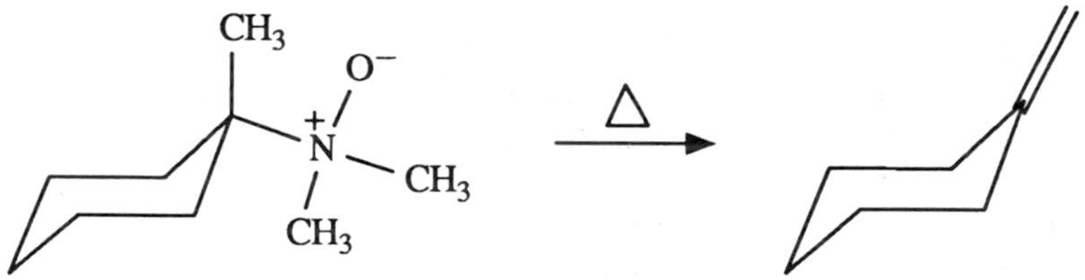

3-3 Nucleophilic Addition to Carbonyl Compounds

Nucleophilic addition to the carbonyl groups of aldehydes and ketones occurs readily. Although such addition reactions are often in

equilibria which favor the starting materials, they are productive when the initial adduct is removed by further reaction.

3-3-1 ADDITION OF GRIGNARD AND ORGANOLITHIUM RE-AGENTS

Reactions of either Grignard or organolithium reagents with most aldehydes, ketones, or esters produce alcohols. Reactions of organolithium reagents with carboxylic acids, or of Grignard reagents with nitriles produce ketones.

Example 3-11. The Grignard reaction of phenylmagnesium bromide with benzophenone.

The electron pair in the carbon magnesium bond of phenylmagnesium bromide is the nucleophile and the carbonyl carbon of the ketone is the electrophile. Also, magnesium is an electrophile and the carbonyl oxygen is a nucleophile, so the salt of an alcohol is the product of the reaction. The alcohol itself is generated by an acidic workup. Lithium reagents behave similarly.

3-1

Example 3-12. The stability of the intermediate determines what product is produced.

The intermediate formed by addition of a Grignard reagent to a nitrile, or by addition of a lithium reagent to a carboxylic acid salt, is stable, and subsequent workup produces a ketone. On the other hand, the intermediate formed by addition of a Grignard reagent to an ester is unstable and decomposes to a ketone, under reaction conditions. Because ketones are more reactive than esters in this reaction, another equivalent of Grignard reagent adds, and the final product is a 3° alcohol.

A. Grignard addition to a nitrile.

3-2

The protonation of the intermediate, **3-2**, to give **3-3**, is similar to the protonation of **3-1** in the previous example.

3-3   **3-4**

The mechanism for hydrolysis of **3-3** to **3-4** is closely related to the reverse of the mechanism for formation of a hydrazone, Example 3-13. (See the answer to Problem 3-14 c.)

B. Addition of an organolithium reagent to a carboxylic acid.

The reaction requires two moles of organolithium reagent per mole of acid. The first mole of organolithium reagent neutralizes the carboxylic acid, giving a salt.

The second mole adds to the carbonyl group giving a dilithium salt, **3-5**, which is stable under the reaction conditions. Sequential hydrolysis of each O-Li group in acid, during workup, gives a dihydroxy compound, **3-6**, which is the hydrate of a ketone. This loses water to give the more stable ketone. The mechanism of this last step is similar to the mechanism given in Example 4-12.

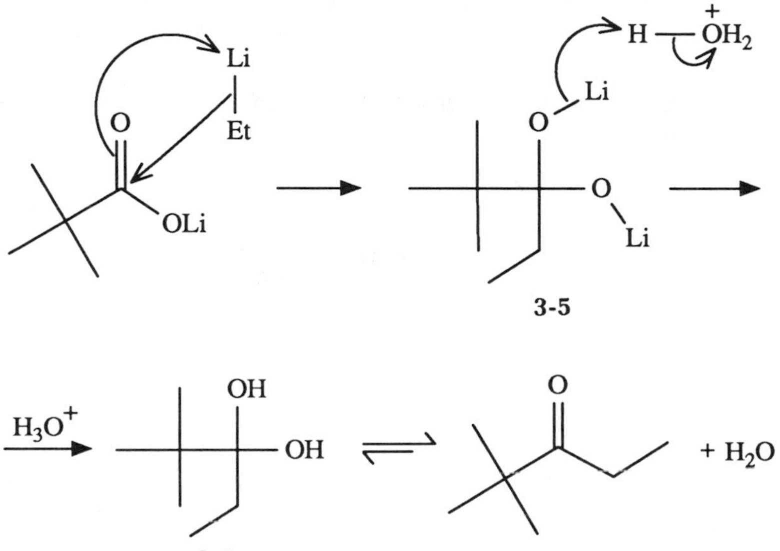

3-5

3-6

C. Grignard reaction of an ester.

Esters react with two moles of Grignard reagent to give the salt of an alcohol. For example, reaction of ethyl benzoate with two moles of phenylmagnesium bromide gives a salt of triphenylmethanol. The first addition gives an intermediate, **3-7**, which is unstable under the reaction conditions.

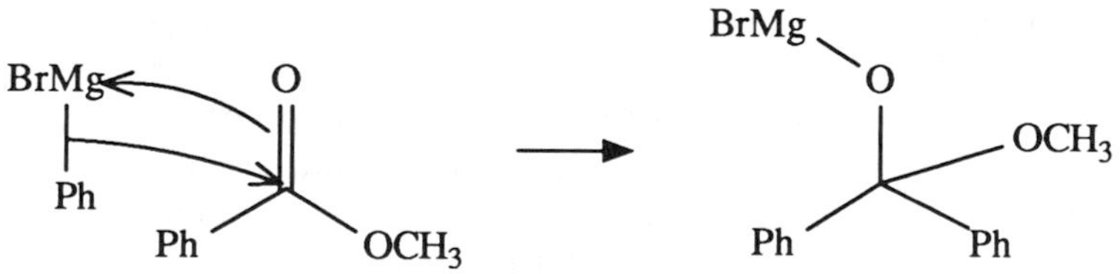

3-7

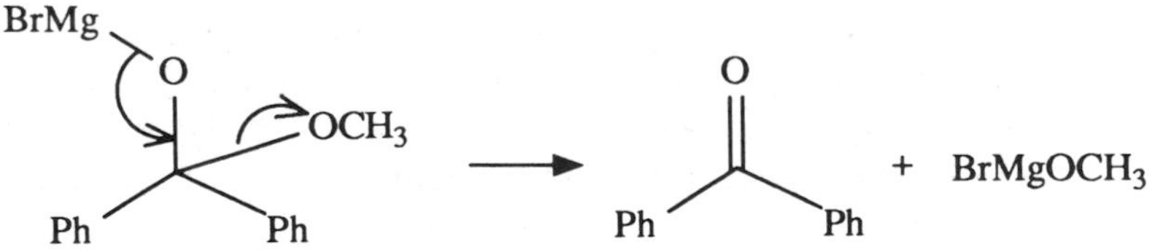

3-7

Another molecule of phenylmagnesium bromide now reacts with benzophenone; this reaction is shown in Example 3-11.

Problem 3-12. Write step-by-step mechanisms for the following reactions:

a.

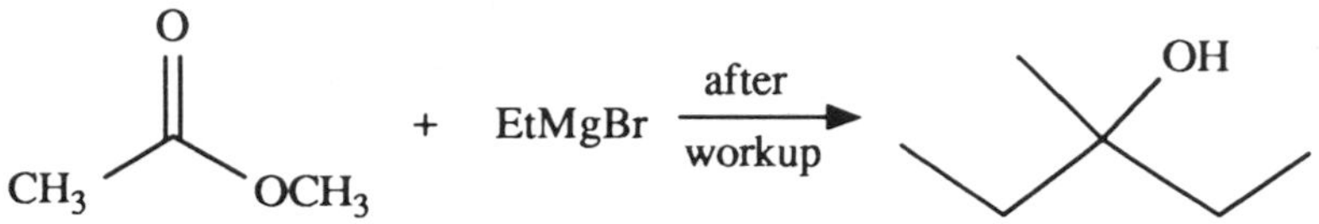

b.

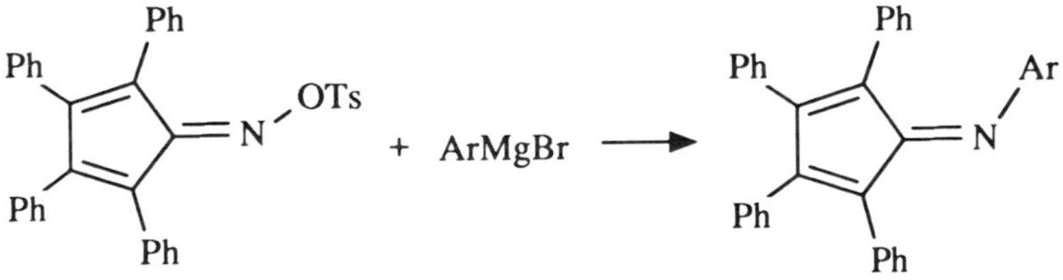

Hagopian, R. A.; Therien, M. J.; Murdoch, J. R. *J. Am. Chem. Soc.* **1984**, *106*, 5753–5754.

3-3-2 REACTION OF NITROGEN-CONTAINING NUCLEOPHILES WITH ALDEHYDES AND KETONES

A number of reactions of nitrogen-containing nucleophiles with aldehydes and ketones involve addition of the nitrogen to the carbon of the carbonyl group, followed by elimination of water to produce a double bond. Common examples are: reactions of 1° amines to produce substituted imines, reactions of 2° amines to produce enamines, reactions of hydrazine or substituted hydrazines to produce hydrazones, reactions of semicarbazide to give semicarbazones, and reactions of hydroxylamine to produce oximes. These reactions are usually run with an acid catalyst.

In the synthesis of imines and enamines by this method, the water produced in the reaction must be removed azeotropically to drive the reaction to the right. Example 3-12 A illustrates the fact that, in aqueous acid, ketone rather than imine is favored.

Example 3-13. Mechanism for formation of a hydrazone.

The first step in a mechanism for the following synthesis of a phenylhydrazone is an equilibrium protonation of the carbonyl oxygen.

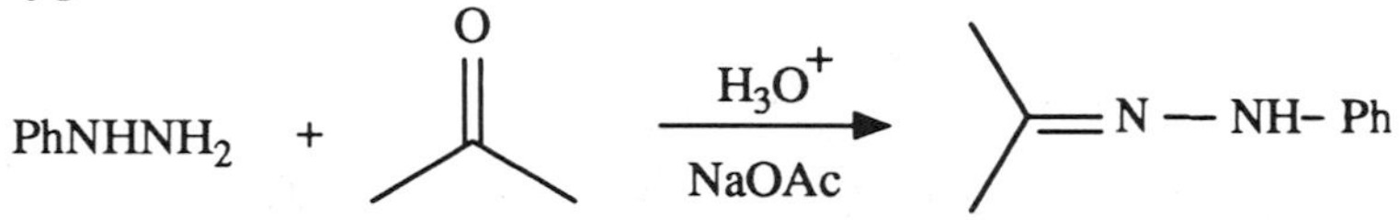

The protonated carbonyl group is then more susceptible to reaction with a nucleophile than the neutral compound: note that the protonated carbonyl group is a resonance hybrid:

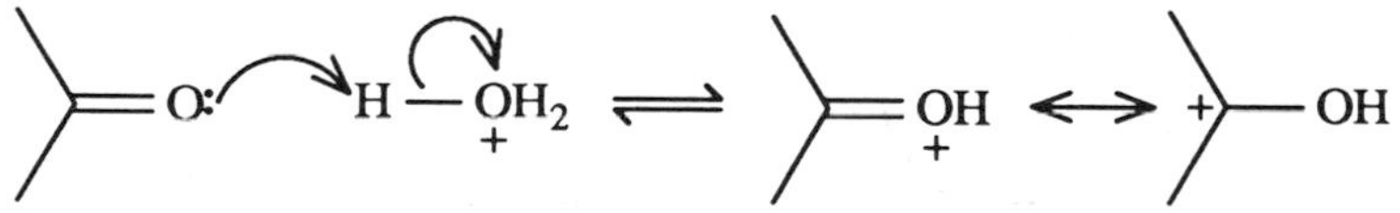

The more nucleophilic nitrogen of the hydrazine reacts at the electrophilic carbon of the carbonyl group. Loss of a proton, facilitated by base, is followed by acid-catalyzed elimination of water.

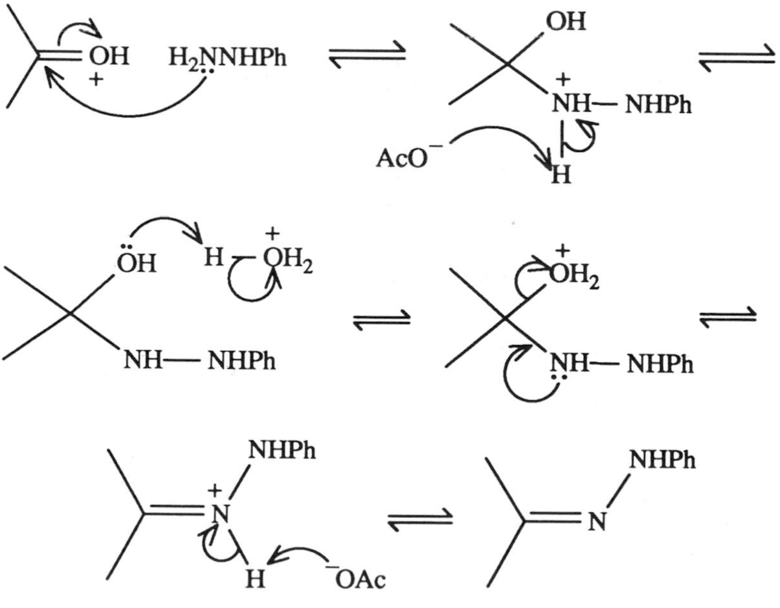

Problem 3-13. Explain why the nitrogen in phenylhydrazine which acts as the nucleophile is not the nitrogen with the phenyl substituent.

Problem 3-14. Write step-by-step mechanisms for the following transformations:

a.

b.

Hagopian, R. A.; Therian, J. J.; Murdoch, J. R. *J. Am. Chem. Soc.* **1984**, *106*, 5753–5754.

c.

3-3-3 THE ALDOL CONDENSATION

The aldol condensation reaction involves the formation of an anion on a carbon α to an aldehyde or ketone carbonyl group, followed by nucleophilic attack of that anion on the carbonyl group of another molecule. The reaction may involve a self-reaction of an aldehyde or ketone or may involve formation of the anion of one compound and attack at the carbonyl of a different compound. The latter is called a mixed aldol condensation.

Example 3-14. Condensation of acetophenone and benzaldehyde: nucleophilic addition of an anion to a carbonyl group followed by an elimination.

This reaction is a mixed aldol condensation of an aldehyde and a ketone.

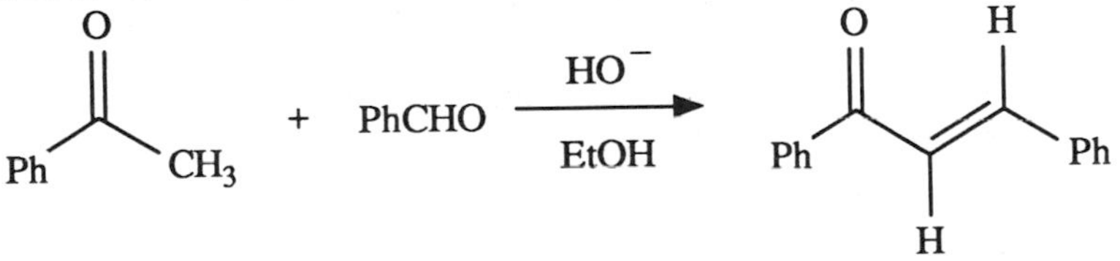

Consider a step-by-step mechanism for this process. The first step will be removal of a proton from the carbon α to the carbonyl group of the ketone to give a resonance-stabilized anion. (Note that removal of the proton directly attached to the aldehyde carbonyl does not give a resonance-stabilized anion, and there are no hydrogens on the carbon α to the aldehyde carbonyl.)

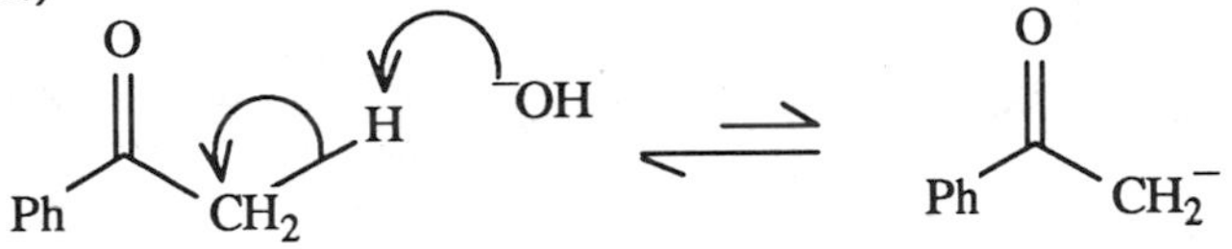

The equilibrium in this reaction favors starting material; in Problem 1-11 b, the equilibrium constant for this reaction was calculated to be approximately 10^{-9}. Nonetheless, the reaction continues to a stable product. Next, the carbonyl group of the aldehyde undergoes nucleophilic addition by the enolate anion to give **3-8**:

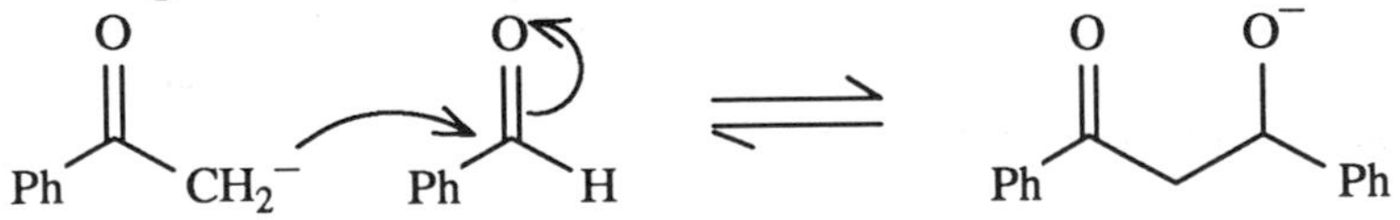

3-8

Why is there preferential attack at the aldehyde carbonyl? First, when the carbonyl group of a ketone reacts, it forms a sp^3-hybridized carbon less stable than that formed from an aldehyde because of the steric bulk of the two carbon substituents. (The aldehyde carbonyl has only one carbon substituent.) Second, the two carbon substituents stabilize the positive character of the carbonyl carbon of the ketone, making it less reactive.

Anion **3-8** can remove a proton from the solvent which is often ethanol.

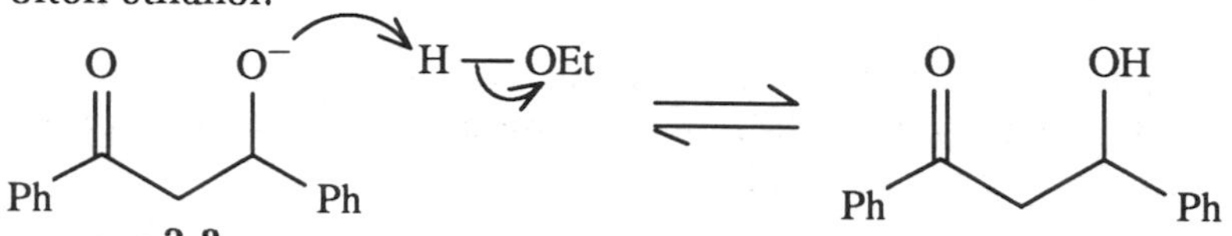

Finally, a base-promoted E-2 elimination of water occurs to give the product. This elimination is driven energetically by the formation of a double bond, which is stabilized by conjugation with both a phenyl group and a carbonyl group.

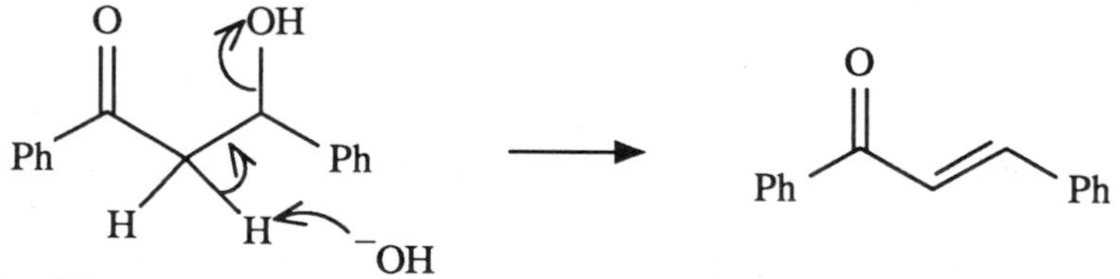

Problem 3-15. Write step-by-step mechanisms for the following reactions:

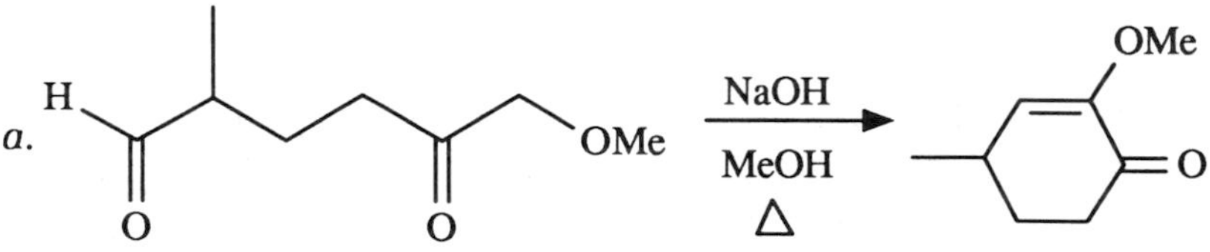

a.

Gadwood, R. C.; Lett, R. M.; Wissinger, J. E. *J. Am. Chem. Soc.* **1984**, *106*, 3869–3870.

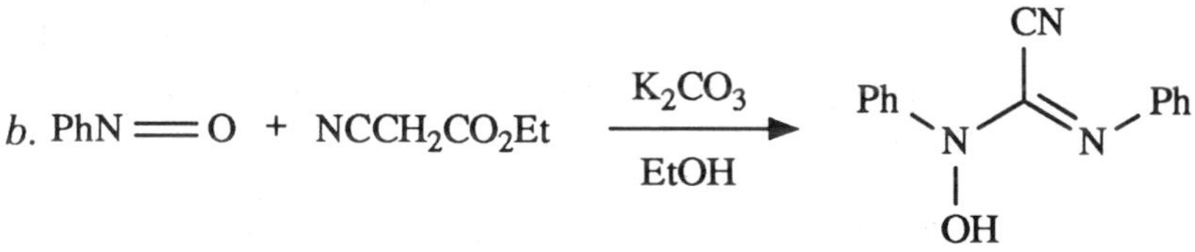

b. PhN══O + $NCCH_2CO_2Et$

Bell, F. *J. Chem. Soc.* **1957**, 516–518.

3-3-4 THE MICHAEL AND 1,4-ADDITION REACTIONS

Strictly speaking, the Michael reaction is addition of a carbon nucleophile to the β-position of an α-β-unsaturated carbonyl compound or its equivalent. It may also be called a 1,4-addition reaction: the carbonyl oxygen is counted as 1 and the β-carbon as 4. The conjugation of the π bond with the carbonyl group imparts positive character to the β-position, making it susceptible to reaction with a nucleophile. The product of this reaction, an enolate ion, is also stabilized by resonance.

When nucleophiles, other than carbon, add to α-β-unsaturated carbonyl compounds, the process is called a 1,4-addition.

Example 3-15. A typical Michael reaction.

This example shows addition of a fairly stable carbanion (stabilized by two adjacent carbonyl groups) to an α,β-unsaturated ester.

The initially formed adduct is also an enolate ion, stabilized by resonance.

This anion can remove a proton from the solvent to give the neutral product.

There are many examples in the literature where the Michael reaction is followed by subsequent steps.

Example 3-16. 1,4-Addition followed by subsequent reaction.

The overall reaction is:

Abell, C.; Bush, B. D.; Staunton, J. *J. Chem. Soc. Chem. Commun.* **1986**, 15–17.

Analysis of the starting material indicates an acidic phenolic hydroxyl, a thioester susceptible to base-promoted hydrolysis and an α-β, doubly unsaturated ketone which could undergo 1-4 addition followed by subsequent reaction. From the structure of the product, it is clear that both the thioester and the α-β-unsaturated ketone undergo reaction. Since hydroxide is the base, a proton will be readily removed from the phenolic hydroxyl group, forming **3-9**.

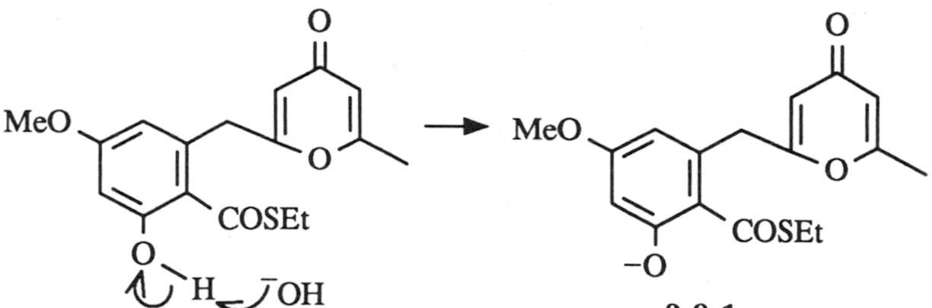

Whether the ester or one of the positions β to the carbonyl group reacts first, cannot be ascertained from the data given. Thus, both possibilities will be discussed.

Mechanism 1

Drawing another resonance form of **3-9**, **3-9-2**, shows that the oxygen of the thioester has negative character and could act as an intramolecular nucleophile:

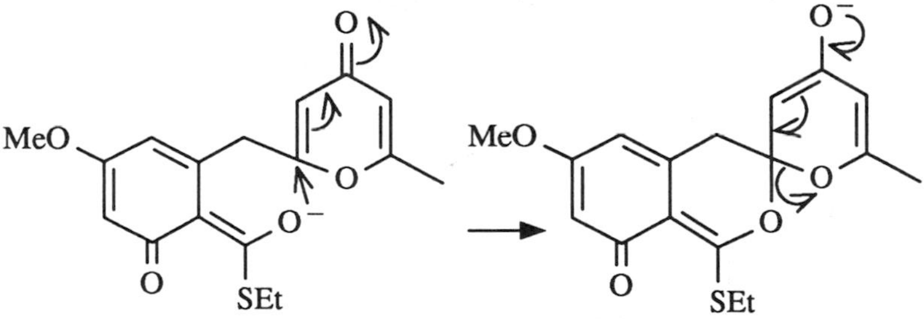

Comparison of **3-10** with the product reveals a central ring with the same atomic skeleton as the product but a right-hand ring which does not. Thus the latter opens to an enolate ion, **3-11**.

This can remove a proton from solvent to give **3-12** which can undergo addition of hydroxide to give a resonance stabilized anion, **3-13**.

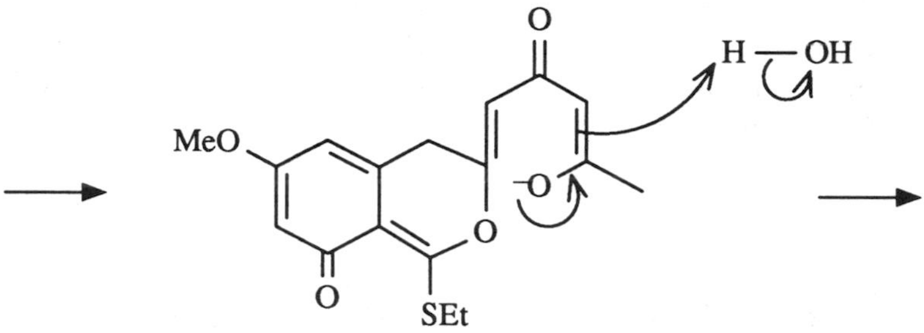

3-11

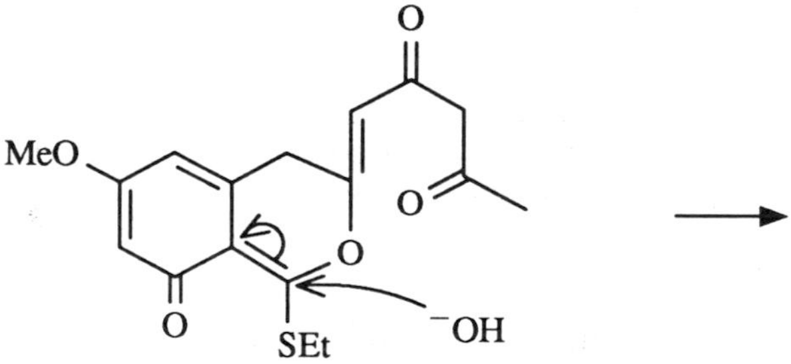

3-12

Ion **3-13** can lose ethyl thiolate giving **3-14**.

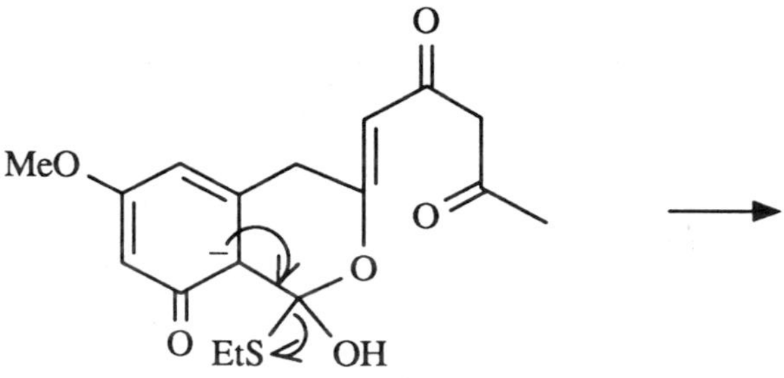

3-13

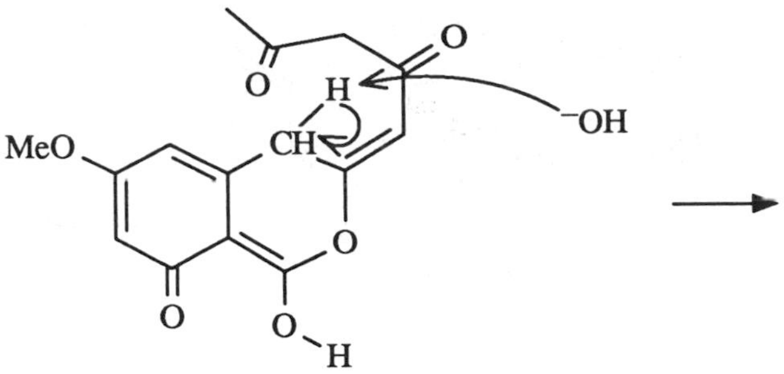

3-14

Intermediate **3-14** contains a number of acidic protons. Removal of some of these would give anions which probably react to give side products, and removal of others may result in unproductive equilibria. Removal of the proton shown, gives a resonance stabilized anion, **3-15**, which can react with the terminal carbonyl group of the side chain to form the third ring.

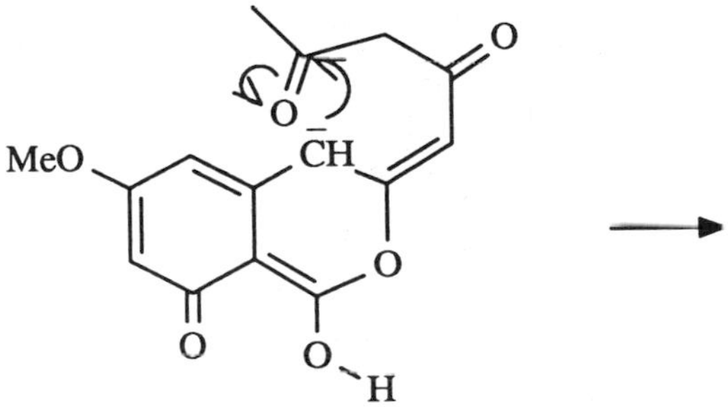

3-15

The intermediate **3-16** removes a proton from solvent and undergoes elimination of water to give **3-18**.

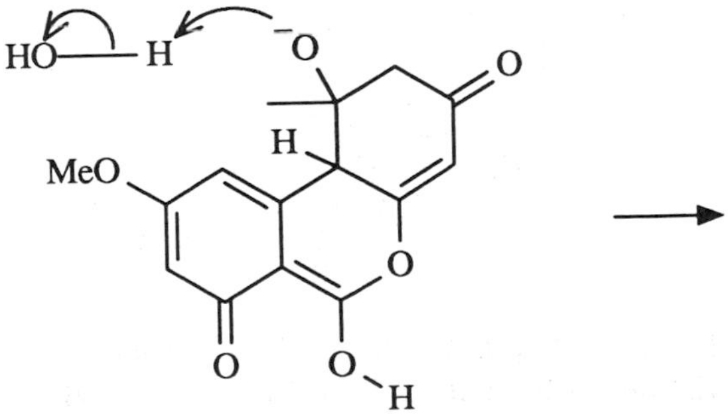

3-16

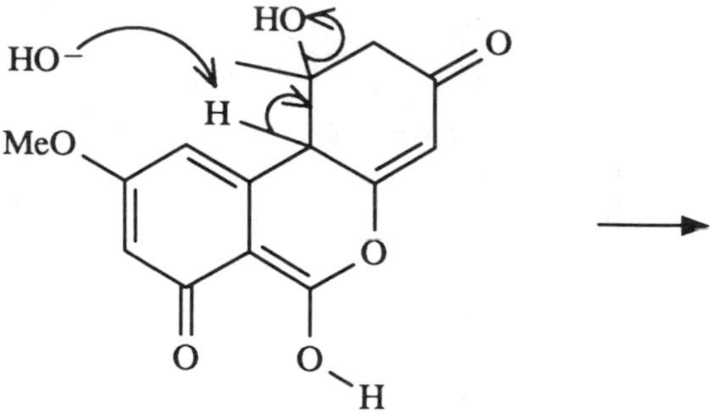

3-17

Removal of a proton from the right-hand ring of **3-18** gives a phenolate ion, **3-19**.

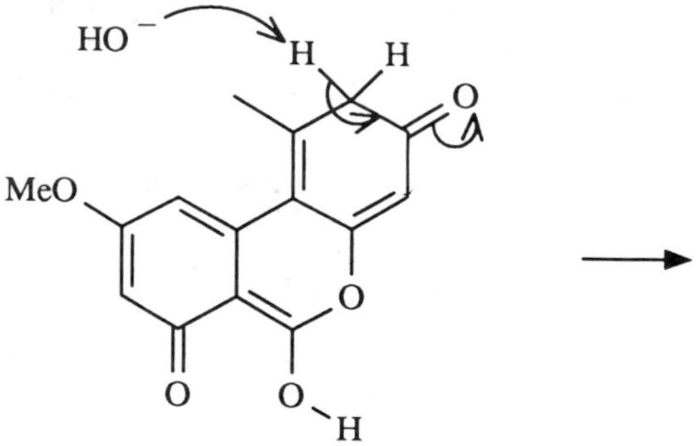

3-18

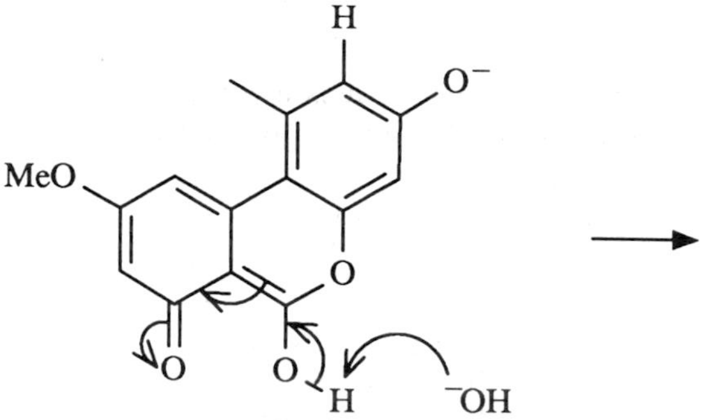

3-19

Removal of another proton from **3-19** gives a diphenolate, **3-20**, which will be protonated to give the product upon workup in aqueous acid.

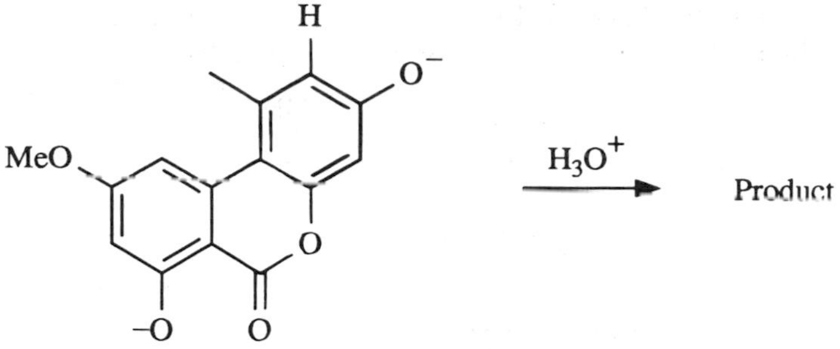

3-20

In this reaction as in many others, the exact timing of steps, especially proton transfers, is difficult to anticipate. For example, the proton shown being removed from **3-19** may actually be removed in an earlier step.

Mechanism 2

In this mechanism formation of the right-hand ring occurs before formation of the middle ring. After the formation of **3-9**, 1,4-addition of hydroxide ion to the α,β-unsaturated ketone occurs.

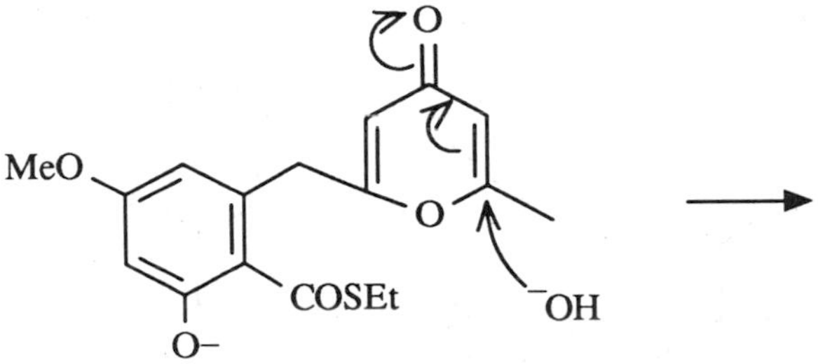

3-9

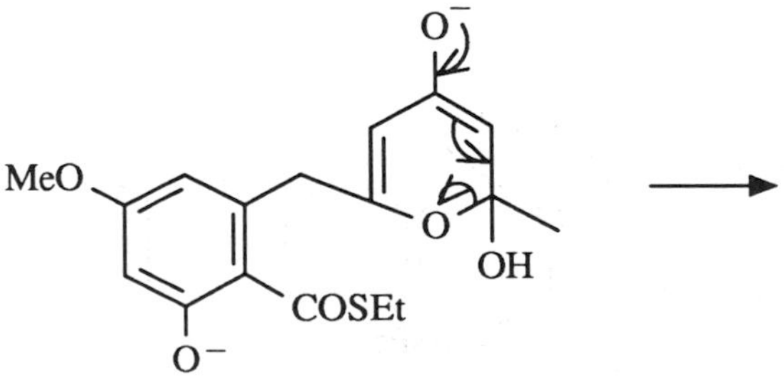

3-21

Ring opening of **3-21** follows to give a new enolate, **3-22**, which can be protonated at carbon by water to give **3-23**.

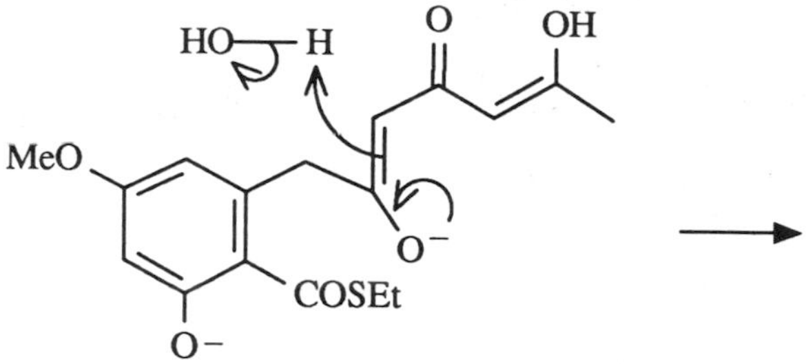

3-22

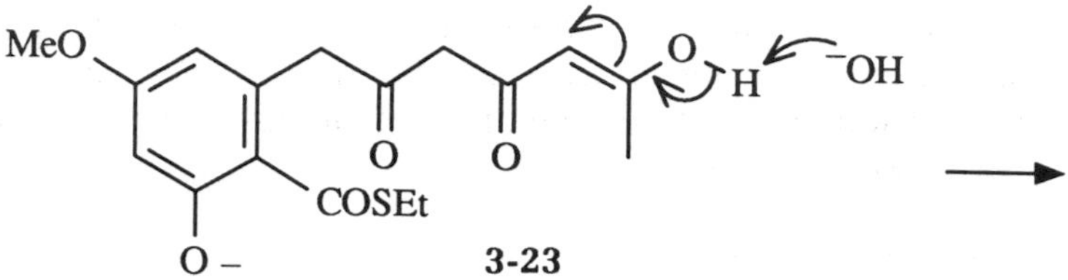

3-23

Base-promoted tautomerization of the enol in **3-23** gives **3-24**. The latter can also be formed by a series of steps which start with attack of hydroxide on the other carbon β to the ketone in the right-hand ring of **3-9**. These steps follow a course analogous to the one depicted.

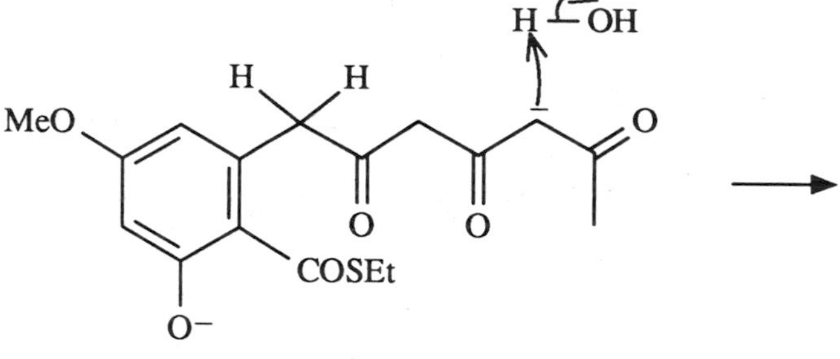

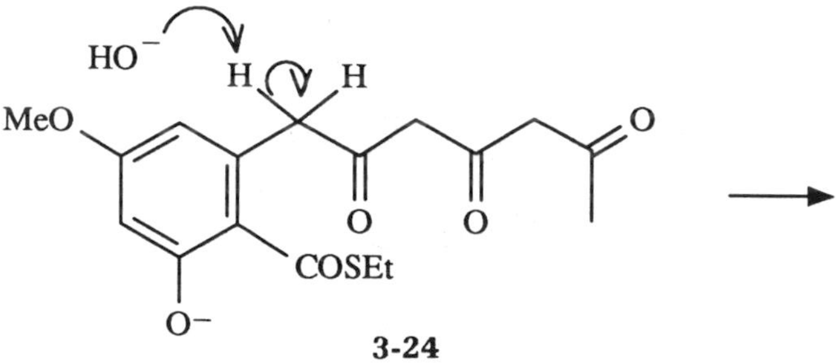

3-24

The mechanism then continues with an intramolecular aldol condensation.

The resulting β-hydroxy ketone, **3-25**, can undergo elimination of water, giving **3-26**. Base-promoted tautomerization of one of the protons in the box in **3-26** gives the enol, and removal of the proton in the circle then gives the phenolate ion, **3-27**.

3-25

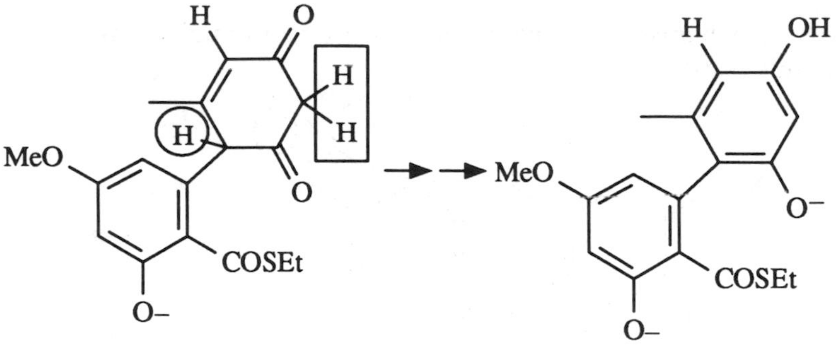

3-26   **3-27**

This nucleophilic phenolate adds to the carbon of the thioester; then ethylthiolate is eliminated.

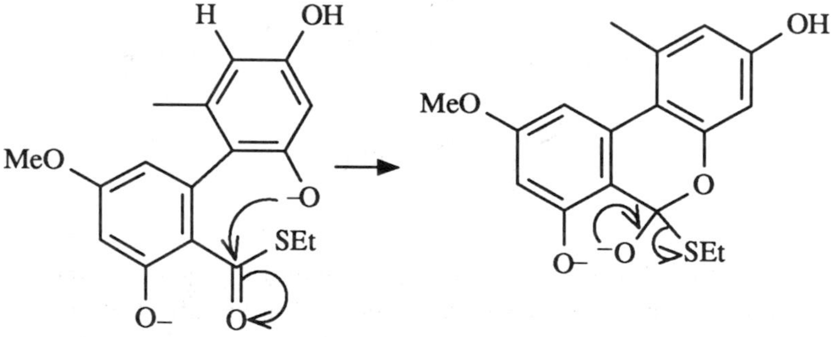

3-28

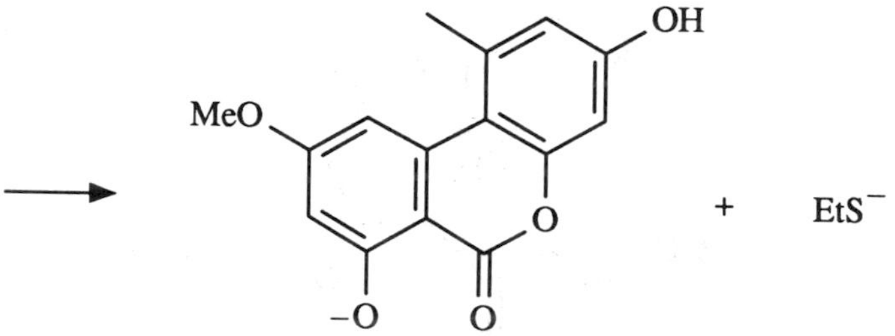

3-29

Once the phenolate ion **3-27** has reacted, the phenol in the right-hand ring of either **3-28** or **3-29** will react with hydroxide to give a new phenolate in this ring. One of these possibilities is represented below. (The phenolate ion in the right-hand ring of **3-27** reduces the acidity of the other phenolic group in that ring. Thus, we anticipate that the proton of the second phenolic group is removed in a later step.)

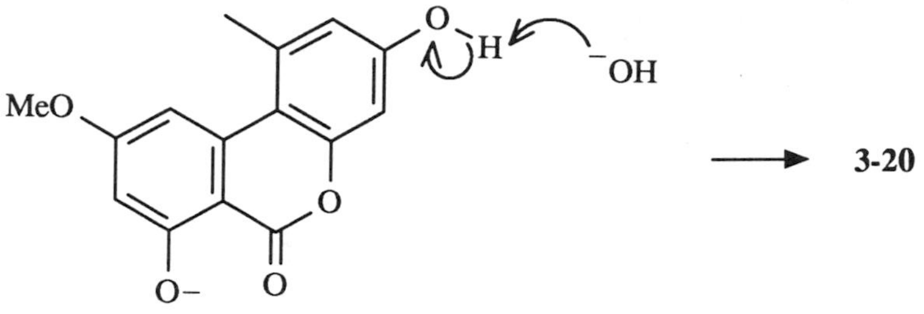

3-29

Both mechanisms for the reaction seem reasonable. The authors of the paper cited showed that **3-30** also cyclizes to the product in excellent yield. Note that **3-30** is the phenol corresponding to the intermediate phenolate **3-24** of mechanism 2. This evidence does not prove that **3-24** is an intermediate in the reaction, but does support it as a viable possibility.

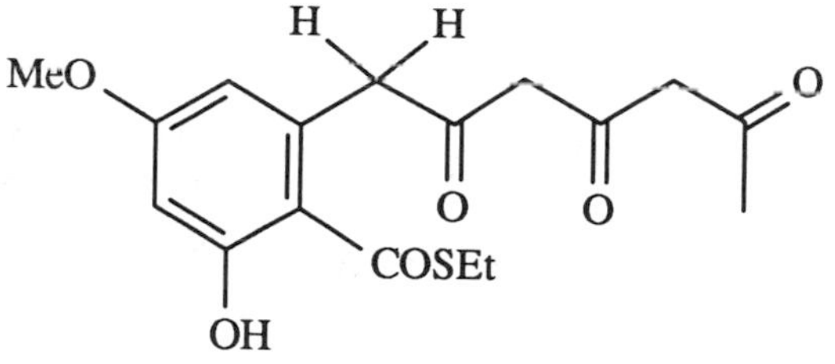

3-30

Problem 3-16. Why is the following mechanistic step unlikely? How would you change the mechanism to make it more reasonable?

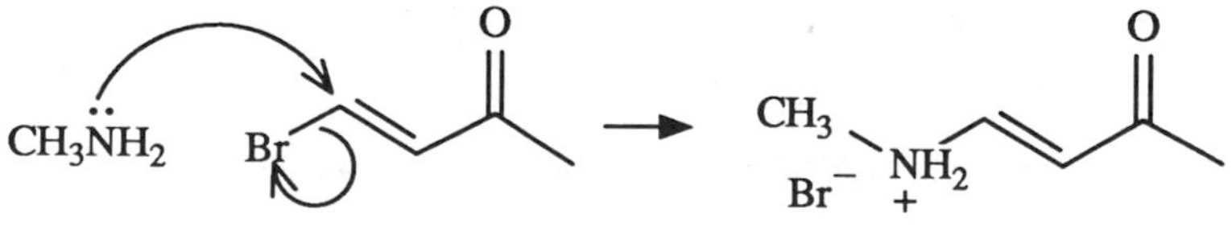

Problem 3-17. Write step-by-step mechanisms for the following transformations.

a.

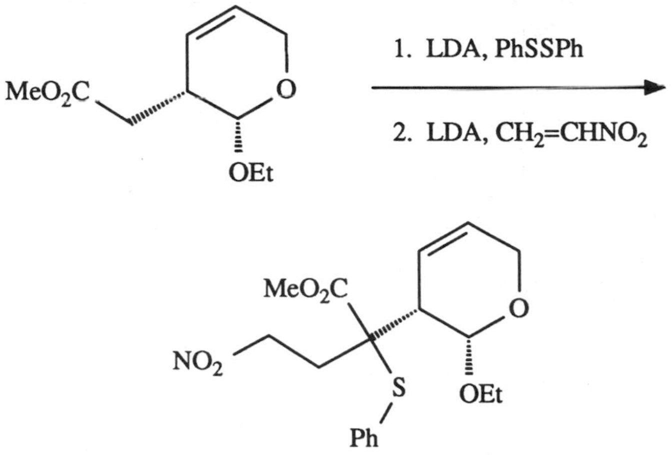

1. LDA, PhSSPh

2. LDA, CH_2=$CHNO_2$

Curran, D. P.; Jacobs, P. B.; Elliott, R. L.; Kim, B. H. *J. Am. Chem. Soc.* **1987**, *109*, 5280–5282.

LDA, lithium diisopropylamide, is a strong base toward the proton but not a good nucleophile, because of steric inhibition by the two isopropyl groups directly attached to the nitrogen anion.

b.

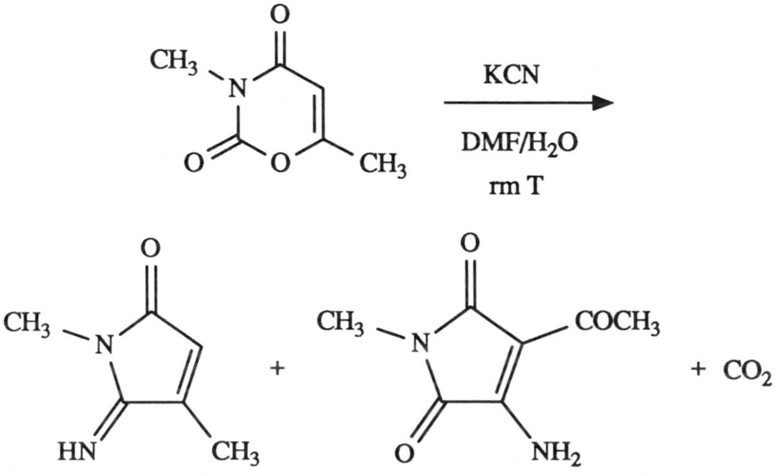

KCN

DMF/H_2O

rm T

+

+ CO_2

Yogo, M.; Hirota, K.; Maki, Y. *J. Chem. Soc. Perkin I* **1984**, 2097–2102.

3-4 Base-Promoted Rearrangements

3-4-1 THE FAVORSKII REARRANGEMENT

A typical Favorskii rearrangement involves reaction of an α-halo ketone with a base to give an ester or carboxylic acid, as in the following example:

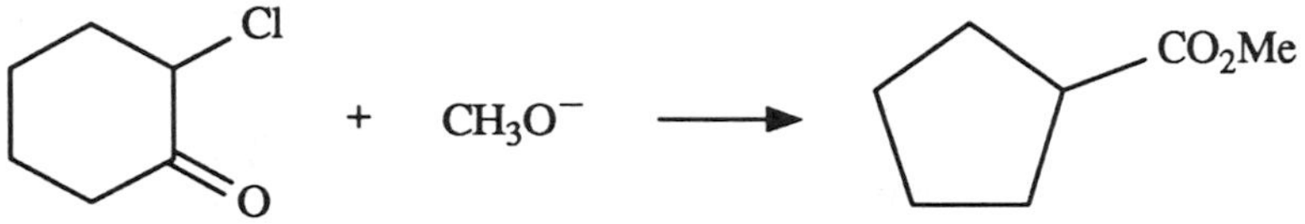

Labelling studies have shown that the two α carbons, in the starting ketone, become equivalent during the course of the reaction. This means that a symmetrical intermediate must be formed. One possible mechanism, which is consistent with this result, follows:

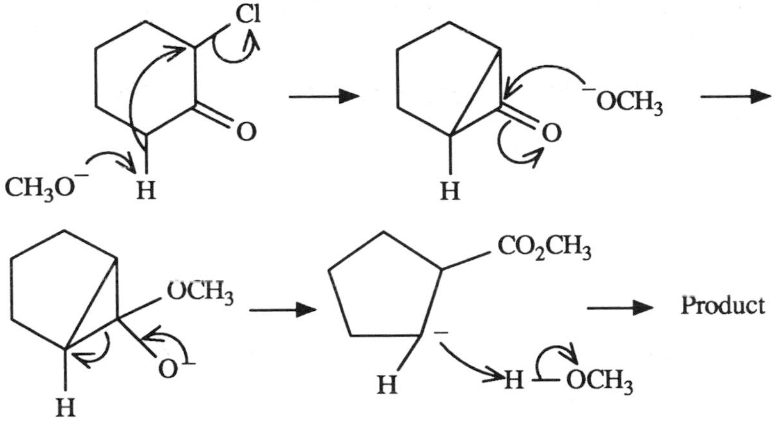

3-4-2 THE BENZILIC ACID REARRANGEMENT

This is a rearrangement of an α-diketone, in base, to give an α-hydroxycarboxylic acid. The reaction gets its name from the reaction of benzil to give benzilic acid:

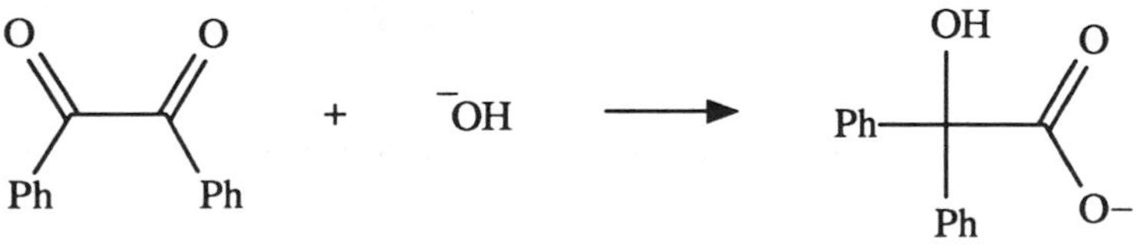

The mechanism involves nucleophilic addition of the base to one carbonyl group, followed by transfer of the substituent on that carbon to the adjacent carbon:

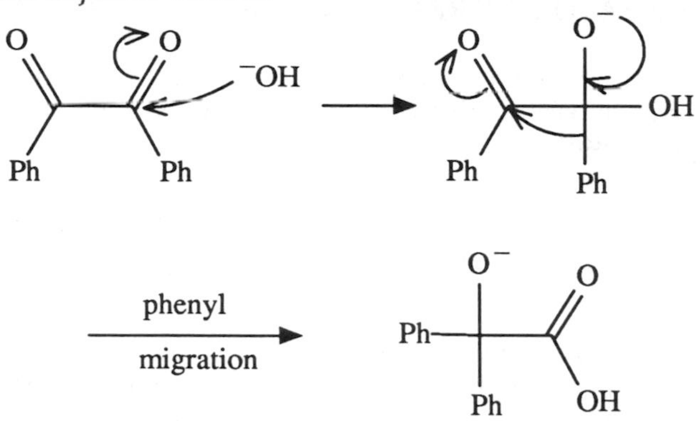

The final steps, under the reaction conditions, are protonation of the alkoxide and deprotonation of the carboxylic acid to give the corresponding carboxylate salt.

Problem 3-18. Write step-by-step mechanisms for the following transformations:

a.

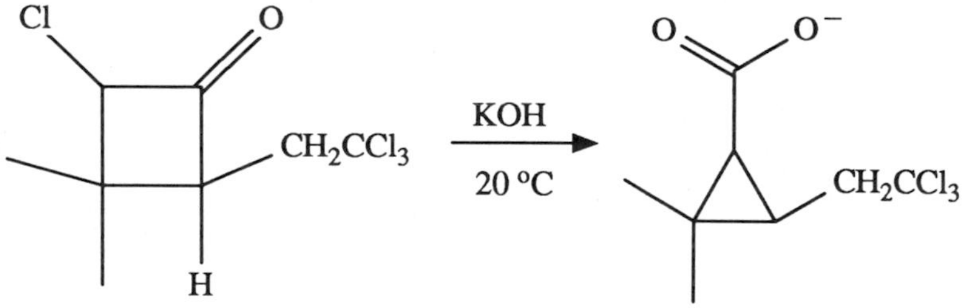

Martin, P.; Greuter, H.; Bellus, D. *J. Am. Chem. Soc.* **1979,** *101,* 5853-5854.

b.

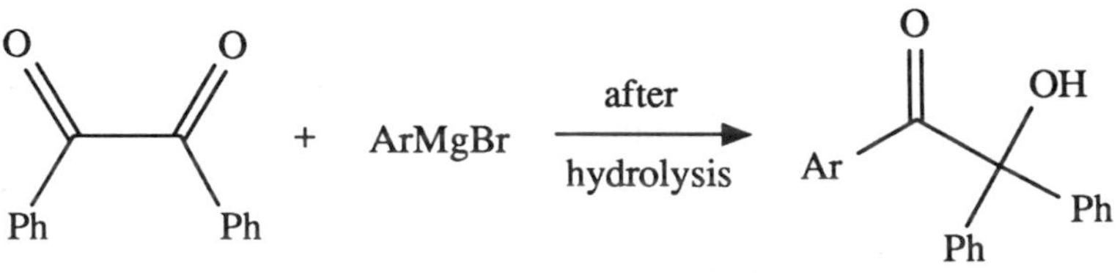

March, J. *Advanced Organic Chemistry,* 3rd ed., **1985,** New York: Wiley, p. 970.

3-5 Reaction Mechanisms in Basic Media

Example 3-17. Nucleophilic addition followed by rearrangement.

Write a mechanism for the following transformation:

First, number the atoms in starting material and product, to ascertain how the atoms have been reorganized.

Numbering indicates that C-1 has become attached to C-3, and that the methyl group on S-2 must come from methyl iodide. Focussing attention on positions 1 and 3 reveals that: (1) the protons at position 1 are acidic because they are benzylic; that is, if a proton is removed from this position, the resulting anion is stabilized by resonance. (2) position 3, a thiocarbonyl carbon, is an electrophile and should react with nucleophiles. The first step of the reaction might then be as follows:

The next step would be nucleophilic reaction of the carbanion with the electrophilic carbon of the thiocarbonyl group; this reaction joins carbons 1 and 3, as was predicted from the numbering scheme.

The resulting three-membered ring intermediate (thiirane) is not stable under the reaction conditions. How do we know this? There is no three-membered ring in the product! The ring strain in a three-membered ring and the negatively charged sulfur facilitate the ring opening. There are three possible bonds that could be broken when the electron pair on the thiolate makes a π bond with the carbon to which it is attached. Each possibility gives an anion whose stability can be approximated by comparing the relative strengths of the corresponding acids (formed when each anion is protonated). These values can be approximated by choosing compounds from Table 1-2 with structures as close as possible to the structural features of interest. Breaking the C-N bond would give the dimethylamide ion (the pK_a's of aniline and diisopropylamine are 31 and 36, respectively); breaking the C-C bond would give back the starting material (the pK_a of toluene is 43); and breaking the C-S bond would give a new thiolate ion (the pK_a of ethanethiol is 11). Thus, based on the thermodynamic stability of the product, the C-S bond of the ring would break. In fact, breaking the C-S bond leads to an anion which is structurally related to the final product of the reaction. However, keep in mind that although thermodynamics is often helpful, it does not always predict the outcome of a reaction.

3-31

The original transformation is completed by an S_N2 reaction of the thiolate ion, **3-31**, with methyl iodide.

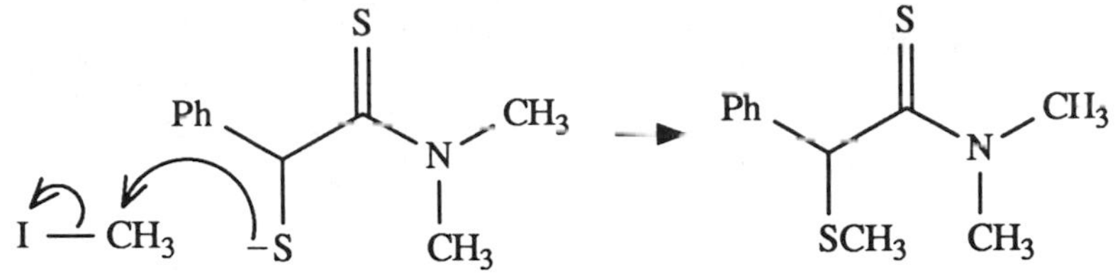

Example 3-18. A combination of proton exchange, nucleophilic addition and nucleophilic substitution.

Write a mechanism for the following transformation:

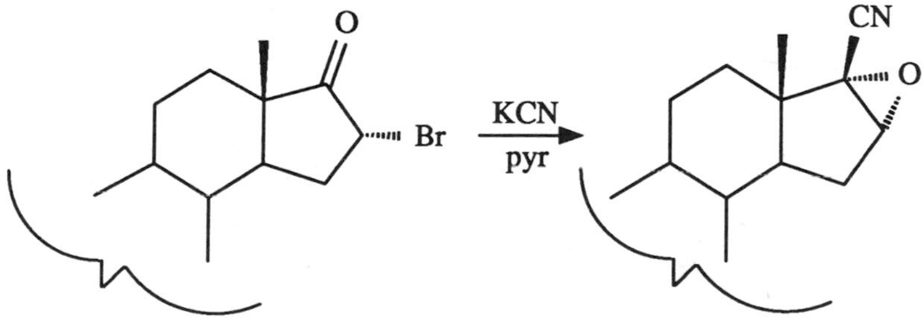

The rings shown in the substrate are the C and D rings of a steroid molecule. The brackets indicate that A and B rings are attached; but, for simplicity, we leave them out while writing the mechanisms below.

First, consider the reaction medium. In pyridine, cyanide is very basic and is also an excellent nucleophile. Since nucleophilic substitution at a position α to a carbonyl is facile, one possible step is nucleophilic substitution of Br by CN. This would be an S_N2 reaction with 100% inversion.

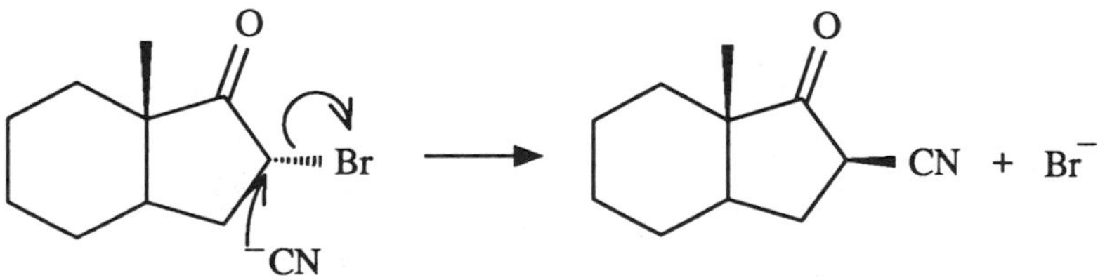

However, there is no reasonable pathway from the product of this reaction to the final product.

Another possible step is nucleophilic reaction of cyanide at the electrophilic carbonyl carbon:

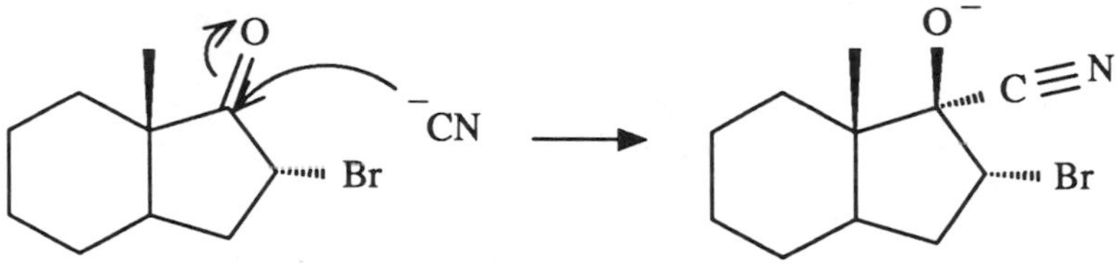

The cyanide approach has been directed so that the alkoxide produced is *anti* to the halide; this is the position most suitable for an S_N2-like reaction to give an epoxide:

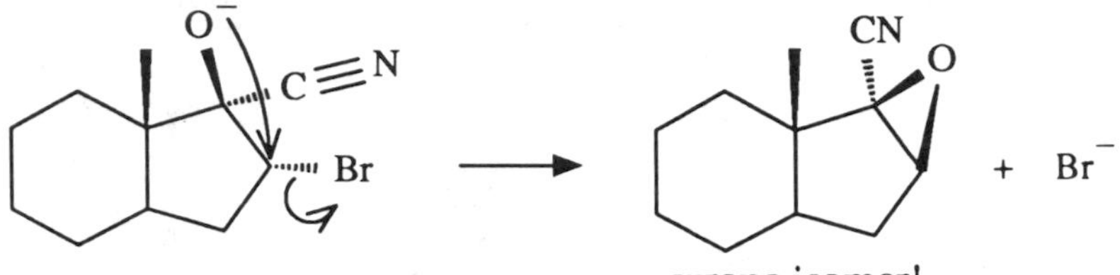

wrong isomer!

However, this reaction leads to the wrong stereochemistry for the product.

A third possible mechanism can explain the stereochemical result: a proton is removed from, and then returned to the α carbon, such that the starting material is "epimerized" before cyanide reacts. (Epimerization is a change of stereochemistry at one carbon atom.)

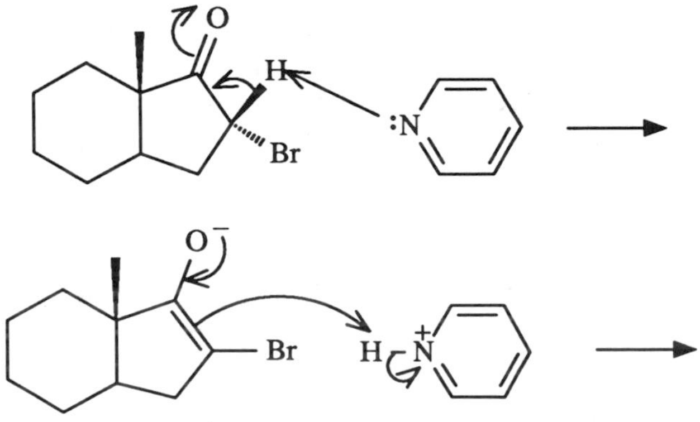

It is of interest that the bromo compound isomeric to the starting material, and the starting material are in equilibrium under
the reaction conditions, and that both starting materials lead to
roughly 75% yield of the same product.

Numazawa, M.; Satoh, M.; Satoh, S.; Nagaoka, M.; Osawa, Y. *J. Org. Chem.* **1986**,
51, 1360–1362.

*Problem 3-19. Write reasonable step-by-step mechanisms for the
following transformations:*

a.

Note: the sulfone is first treated with excess butyllithium to form
a dianion, and then the α-chlorocarbonyl compound is added.

Eisch, J. J.; Dua, S. K.; Behrooz, M. *J. Org. Chem.* **1985**, *50*, 3674–3676.

b.

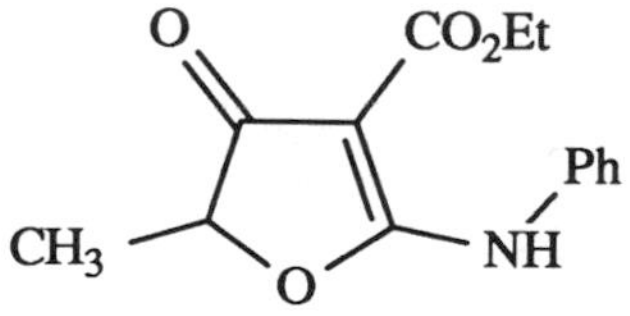

Mack, R. A.; Zazulak, W. I.; Radov, L. A.; Baer, J. E.; Stewart, J. D.; Elzer, P. H.; Kinsolving, C. R.; Georgiev, V. S. *J. Med. Chem.* **1988**, *31*, 1910–1918.

c.

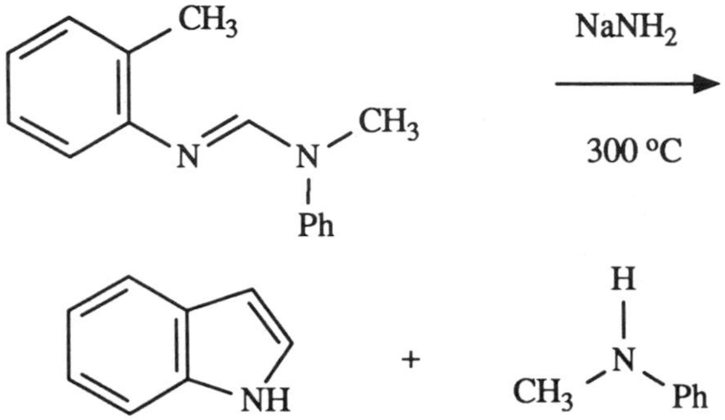

Lorenz, R. R.; Tullar, B. F.; Koelsch, C. F.; Archer, S. *J. Org. Chem.* **1965**, *30*, 2531–33.

d.

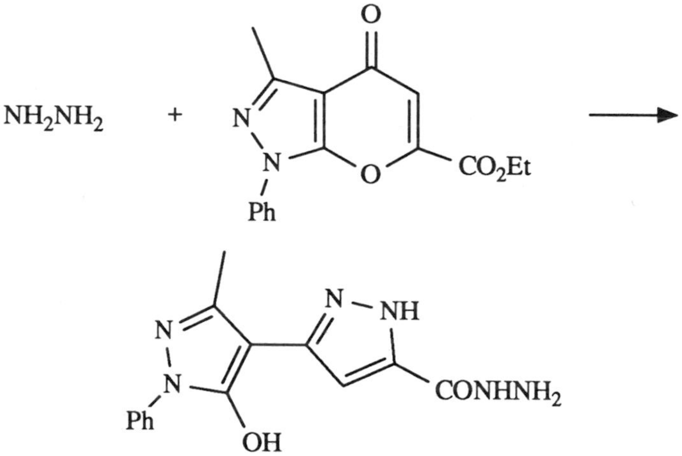

Khan, M. A.; Cosenza, A. G. *Afinidad* **1988**, *45*, 173–4; *Chem. Abs.* **1988**, *109*, 128893.

e.

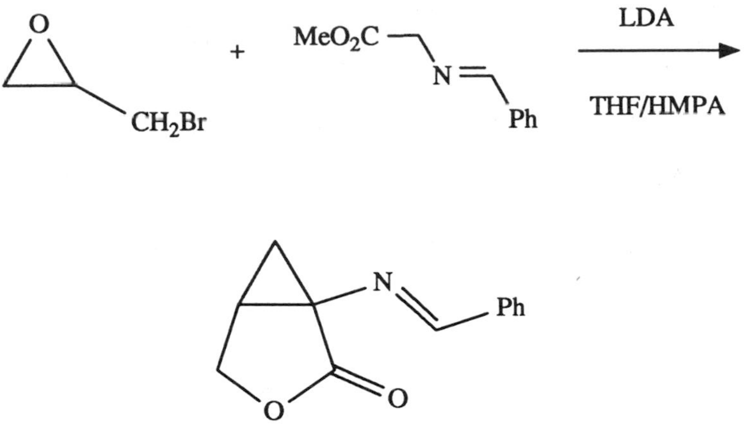

Bland, J.; Shah, A.; Bortolussi, A.; Stammer, C. H. *J. Org. Chem.* **1988**, *53*, 992–995.

Problem 3-20. Consider the following reaction:

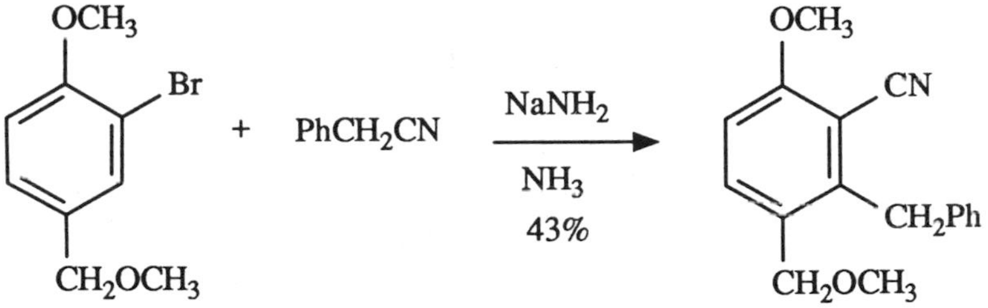

Two mechanisms for the reaction are written below. Both proceed through formation of the anion of phenylacetonitrile:

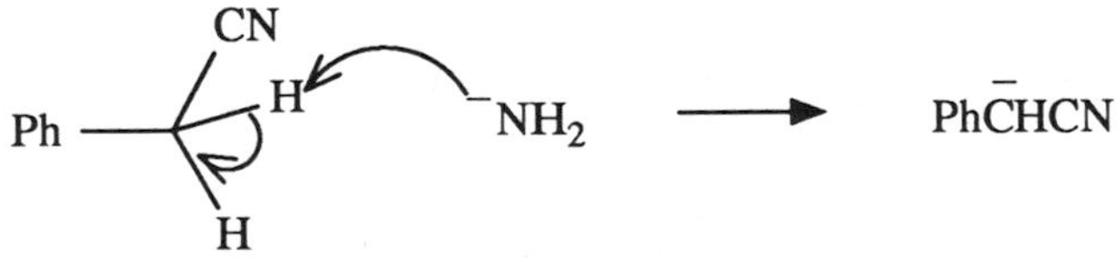

Decide which is the better mechanism, and discuss the reasons for your choice.

Mechanism 1

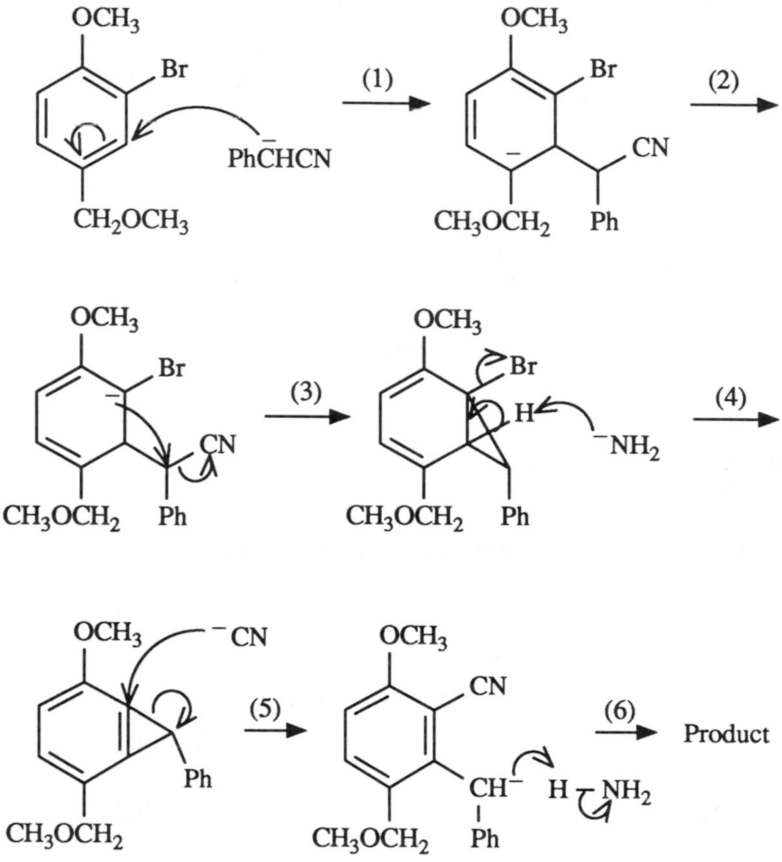

Mechanism 2

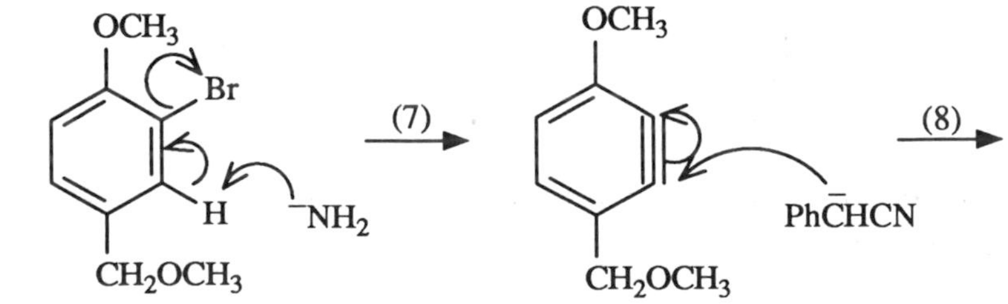

Khanapure, S. P.; Crenshaw, L; Reddy, R. T.; Biehl, E. R. *J. Org. Chem.* **1988**, *53*, 4915–4919.

Problem 3-21. The following esterification is an example of the Mitsunobu reaction. Notice that the reaction goes with inversion of the configuration of the alcohol.

Mitsunobu, O.; Eguchi, M. *Bull. Chem. Soc. J.* **1971**, *44*, 3427–3430. For a review article on the versatility of the reaction, see Mitsunobu, O. *Synthesis*, **1981**, 1–28.

The reaction mechanism is believed to proceed according to the following outline: Triphenylphosphine reacts with the diethyl azodicarboxylate to give an intermediate, which is then protonated by the carboxylic acid to form a neutral salt. This salt then reacts with the alcohol to form dicarboethoxyhydrazine and a new salt, which reacts further to give triphenylphosphine oxide and the ester. Using this outline as a guide, write a mechanism for the reaction.

Answers to Problems

3-1. First, although the atoms are balanced on both sides of the equation, the charges are not. A proton, H^+, should be shown on the right side of the equation, rather than a hydride, H^- (see Helpful Hints 2-2 and 2-3). Also, as stated in Helpful Hint 3-1, hydride is an extremely poor leaving group. The curved arrow, on the left side of the equation, is pointed in the wrong direction. This should be apparent from the fact that the positively charged oxygen atom is far more electronegative than hydrogen. Thus, the arrow should point toward oxygen, not away from it.

Loss of a proton is often written in mechanisms simply as $-H^+$. We will not do this, since protons are always solvated in solution. Therefore, in this book, with the exception of Chapter 7, the loss of a proton will always be shown as assisted by a base. However, the base need not be a strong base. Thus, for the transformation indicated in the problem, the following could be written:

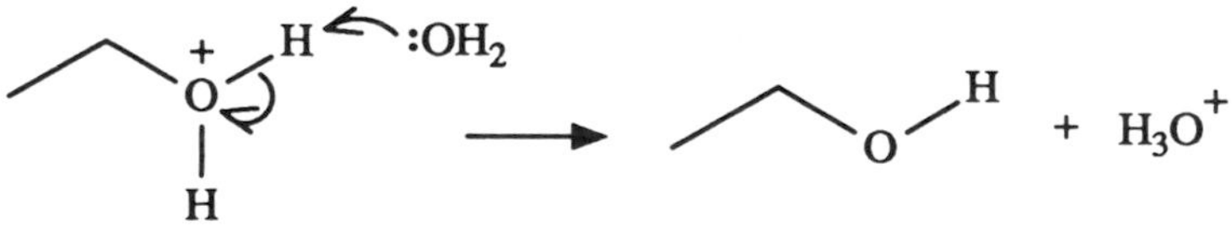

3-2. According to Table 1-2, the pK_a of bicarbonate is 10.2 while the pK_a of phenol is 10.0. However, the combined effect of the substituents would make this phenol considerably more acidic than a pK_a of 10.0 would indicate. In carbonate solution, it will be almost entirely converted to the corresponding phenoxide ion, which will act as the nucleophile. (Clearly, the negatively charged ion is a much better nucleophile than neutral phenol.)

The carboxylic acid group in the starting material will also be converted to a salt in carbonate solution. While the phenoxide is a better nucleophile than the carboxylate ion, it may not react as rapidly as the carboxylate with benzyl bromide because it is much more sterically hindered (by the *ortho* methyl and nitro groups). S_N2 reactions are slowed considerably by steric hindrance.

3-3.a.   The methylene carbon of the ethyl bromide is the electrophile, the lone pair of electrons on the phosphorus of triphenylphosphine is the nucleophile, and the bromide ion is the leaving group.

The product salt is stable, and no further reaction takes place. If you wrote that ethoxide was formed and acted as a nucleophile to give further reaction, you neglected to consider the relative acidities of HBr and EtOH. (See Table 1-2.) The relative pK_a's indicate that the following reaction does not occur:

$$Br^- \ + \ EtOH \ \rightleftharpoons \ HBr \ + \ EtO^-$$

HBr has pK_a of -9.0 and EtOH has a pK_a of 16. Thus the K_a for this reaction is 10^{-25}!

     Prior ionization of ethyl bromide to the carbocation is unlikely, because the primary carbocation is very unstable.

3-3.b.   Base is present to neutralize the amine salt, giving free amine. The free amine is the nucleophile, the carbon bearing the chlorine is the electrophile, and chloride ion is the leaving group. The driving force for this intramolecular substitution reaction is greater than that of an intermolecular reaction, because of entropic considerations.

Under the basic reaction conditions, the salt shown and the neutral product would be in equilibrium.

3-3.c.   The overall reaction can be separated into two sequential substitution reactions. In the first, the proton of the carboxylic acid is the electrophile, the carbon of diazomethane is the nucleophile, and the carboxylate anion is the leaving group. In the second step the carboxylate ion is the nucleophile, the methyl group of the methyl diazonium ion is the electrophile, and nitrogen is the leaving group. Loss of the small stable nitrogen molecule provides a lot of driving force to the reaction. (See Helpful Hint 2-12.)

3-3.d.   In acid, the first step is protonation of the oxygen of the epoxide, to convert it into a better leaving group. This is followed by ring opening to the more stable carbocation (the one stabilized by conjugation with the phenyl group), followed by nucleophilic reaction of water at the positive carbon to give the product.

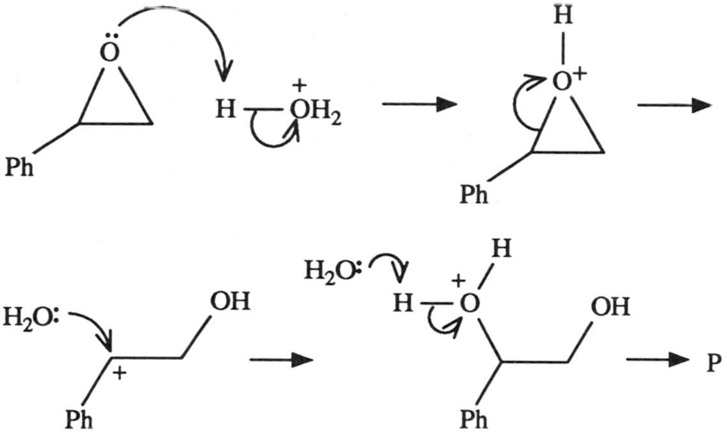

If this reaction were run in base, the following mechanism would apply:

This is one of the few examples where RO⁻ acts as a leaving group. The reason this reaction takes place is that it opens a highly strained three-membered ring. On the other hand, in water itself, the following mechanism is not valid, because water is not a good enough nucleophile to drive the reaction:

3-4.a. The carbon skeleton has rearranged in this transformation. Moreover, numbering corresponding atoms in product and starting material indicates that it is not the ethyl group, but the nitrogen, which moves. In other words, carbon 3 of the ethyl group is attached to carbon 2 in both the starting material and the product. This focuses attention on the nitrogen which is a nucleophile. Because chlorine is not present in the product, an intramolecular nucleophilic substitution is a likely possibility. Such an intramolecular S_N2 reaction is sometimes called neighboring group participation. This reaction gives a three-membered ring intermediate which can open in a new direction to give the product. In this second S_N2 reaction hydroxide is the nucleophile and the CH_2 group of the three-membered ring is the electrophile.

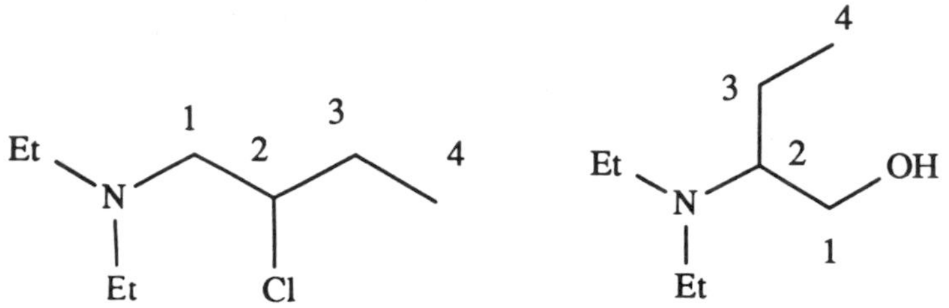

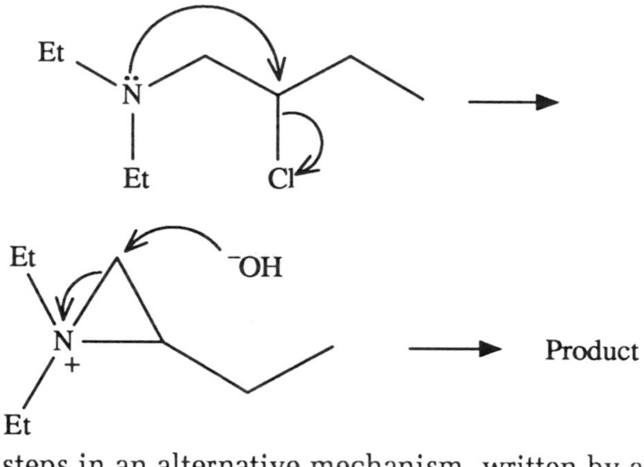

Some steps in an alternative mechanism, written by a student, are shown below:

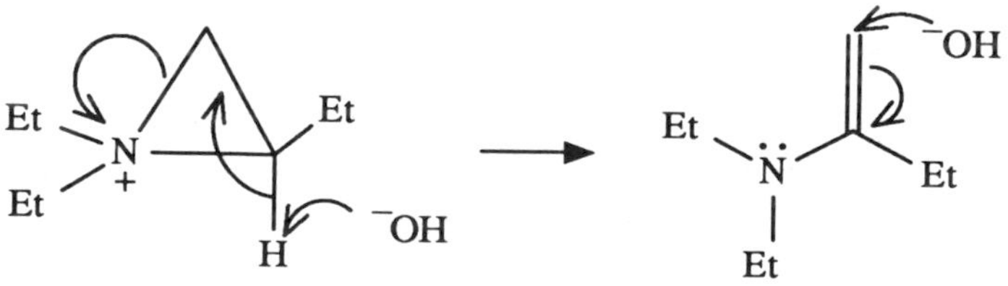

An unlikely step is the addition of hydroxide to the double bond. Double bonds of enamines, like this one, tend to be nucleophilic rather than electrophilic. That is, the resonance interaction of the lone pair of electrons on nitrogen with the double bond is more important than the inductive withdrawal of electrons by the nitrogen. Another way of looking at this is to realize that the final carbanion is not stabilized by resonance, and thus is not likely to be formed in this manner.

3-4.b.   The first step is removal of the most acidic proton, the one on the central carbon of the isopropyl group. The anion produced is stabilized by conjugation with both the carbonyl group and the new double bond to the isopropyl group. Removal of no other proton would produce a resonance-stabilized anion. The second step is nucleophilic reaction of the anion with the electrophilic carbon, the one activated by two bromines. Bromide ion acts as the leaving group.

3-5. The mechanism shown violates Helpful Hint 3-2, because the second step shows nucleophilic substitution at an sp^2-hybridized nitrogen occurring as a single concerted process. Change it to two steps; addition followed by elimination:

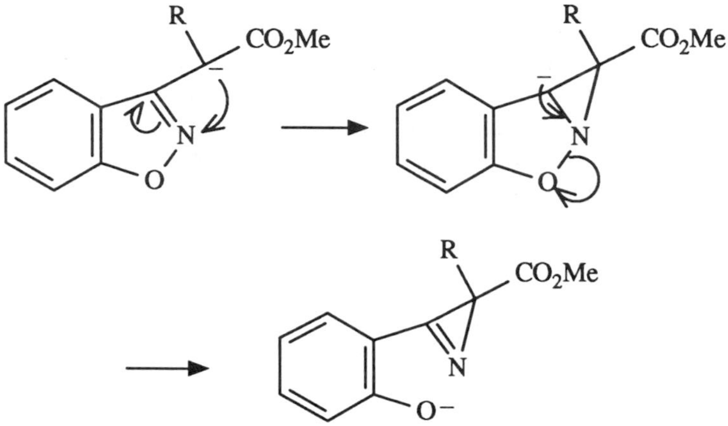

3-6.a. Because a strong base is present, the mechanism is not written with one of the neutral guanidino nitrogens acting as the nucleophile. Instead, the first step is removal of a proton from the guanidino NH_2 group. This gives a resonance stabilized anion, **3-32**. Removal of the proton from the imino nitrogen would produce a less stable anion, because it is not resonance stabilized. In the following, R = the N-methylpyrazine ring.

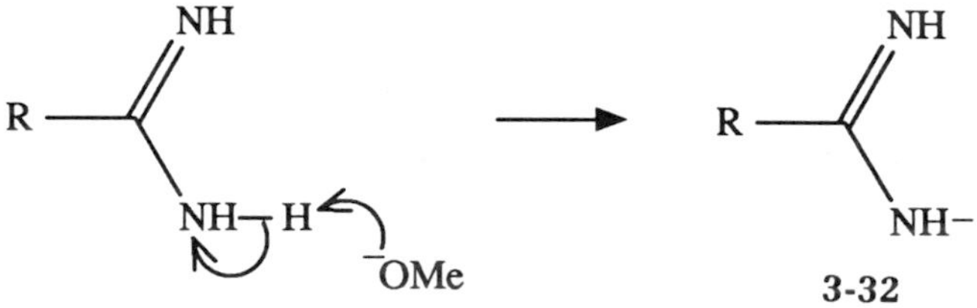

Another reasonable reaction which takes place under the reaction conditions, is removal of one of the protons on the carbon α to the carbonyl groups in diethyl malonate. However, this reaction does not lead to the product, and is an example of an unproductive reaction. Another unproductive reaction is ester interchange in the diethyl malonate; in this case, sodium methoxide would react with the ethyl ester to produce a methyl ester. However, this reaction is inconsequential, because methyl and ethyl esters have similar reactivity, and the alkyl oxygens with their substituents are lost in the course of the reaction.

The nucleophilic nitrogen of **3-32** adds to the carbonyl group; then ethoxide ion is lost.

Intermediate **3-33** reacts, by a route completely analogous to the previous steps, to give **3-35**, a tautomer of the product. That is, the proton removed from the amino nitrogen of **3-33** leads to a resonance stabilized anion, **3-34**. (The anion formed by removal of a proton from the imide nitrogen would not be resonance stabilized.) The nucleophilic anion, **3-34**, adds to the remaining ester carbonyl; then elimination of ethoxide gives **3-35**.

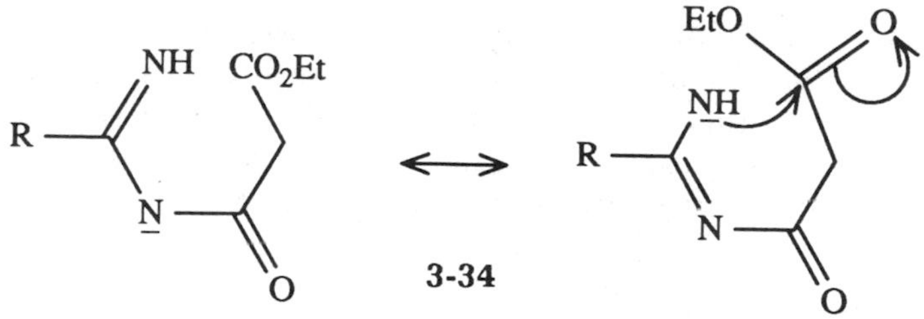

3-34

Intermediate **3-35** undergoes two tautomerizations to give the product. The first tautomerization involves removal of the proton on nitrogen, because this gives an anion which is considerably more delocalized than the anion which would be produced by removal of a proton on carbon. Either enolate oxygen of **3-36** can pick up a proton from either ethanol or methanol.

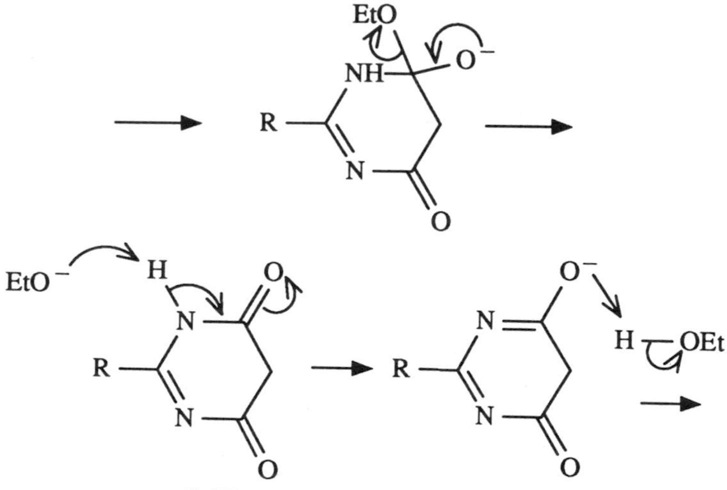

3-35   **3-36**

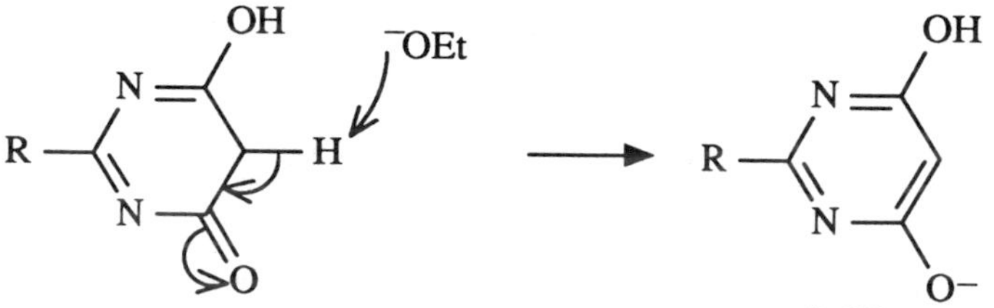

3-37

Finally, removal of a second acidic proton forms the product phenolate, **3-37**. In fact the reactions shown for this mechanism are reversible, and an important driving force for the reaction is production of the stable phenolate salt. The neutral product would be obtained by acidification of the reaction mixture upon workup.

Note: in a tautomerism like that shown for **3-36**, transfer of a proton is generally written as an intermolecular process, not an intramolecular process as pictured below:

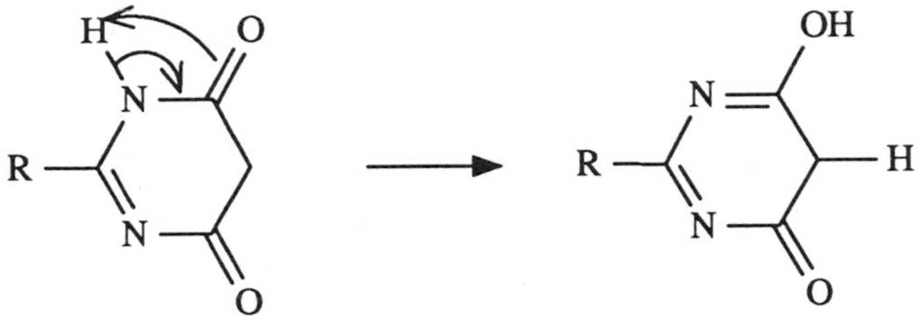

As we will see in Chapter 6, orbital symmetry and concomitant geometric considerations strongly support the intermolecular character of such proton transfers.

3-6.b.   Since the product contains no saturated carbon chain longer than one carbon, it is unlikely that the carbons of the 6-membered ring are part of the product. Furthermore, since the functionality attached to the 6-membered ring is an acetal, it is probable that cyclohexanone is the other product. Also, the methoxy group of the ester is not present in the product which suggests a nucleophilic substitution at the carbonyl of the ester.

Numbering starting material and product also aids in analysis. The relationship between the starting materials and the product is seen more clearly if a tautomer, **3-39**, of the product is depicted.

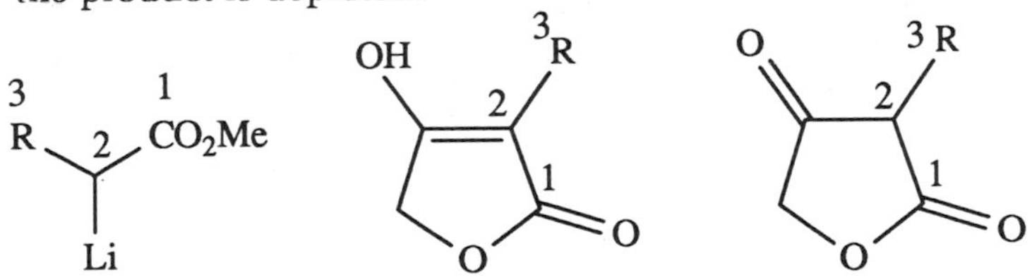

Comparison of **3-39** with **3-40**, the other starting material, shows that two carbons are introduced from **3-40**.

3-40   **3-39**

If the other C=O bond in **3-39** were derived from the lithium reagent, carbon 5 would no longer be bonded to carbon 4. It is difficult to write a mechanism that would facilitate this bonding change. Below two possibilities are represented: the introduction of the ring oxygen (O-6) in **3-39** from either of the ring oxygens in **3-40**.

The analysis by numbering the atoms; the presence of a nucleophile, the lithium reagent; and an electrophile, the ester carbonyl group of **3-40**, suggest nucleophilic substitution of the ester as a likely first step.

The elimination part of this initial nucleophilic substitution might be concerted with the loss of cyclohexanone, or might occur as a separate step.

Elimination by breaking a carbon-carbon bond instead of a carbon-oxygen bond, is less likely because the carbanion formed would be much less stable than the alkoxide ion.

The alkoxide ion, **3-41**, reacts intramolecularly with the ester to produce a lactone.

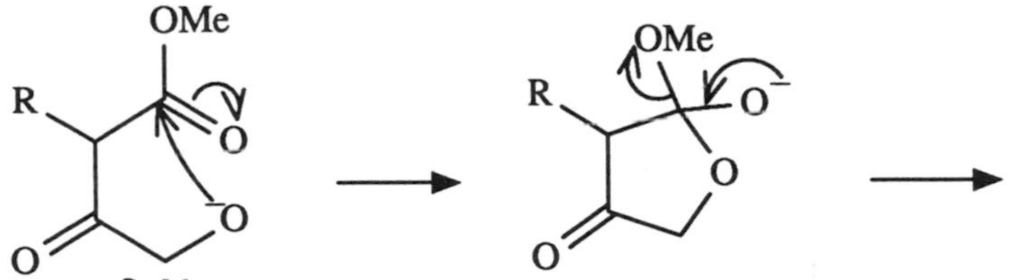

3-41

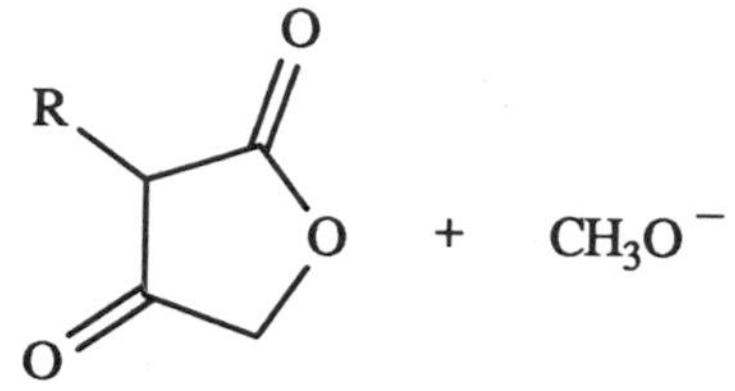

The remaining steps show tautomerization of the ketone.

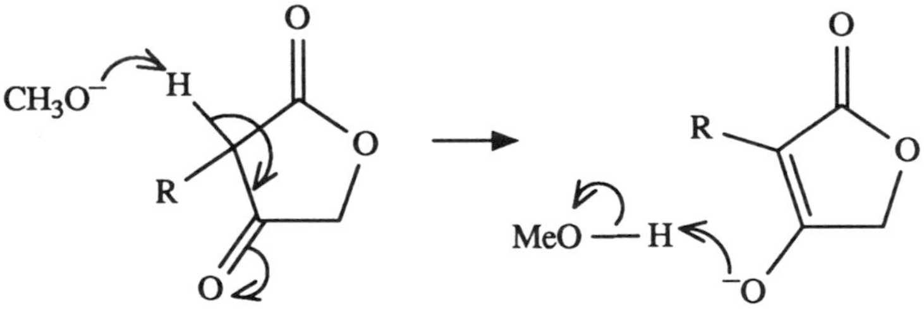

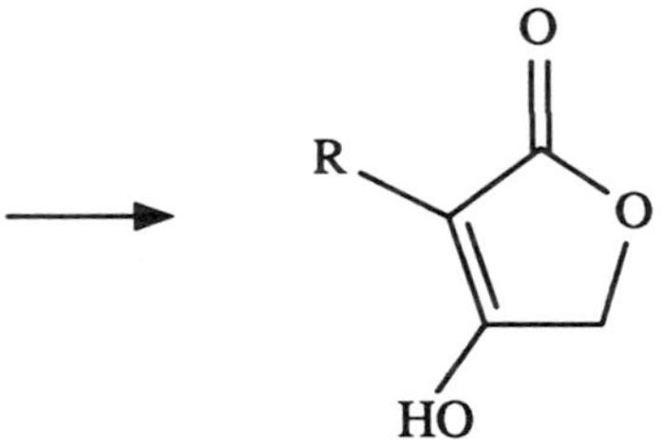

The authors of the cited paper write removal of a proton from **3-41** to give **3-42** prior to the ring closure step.

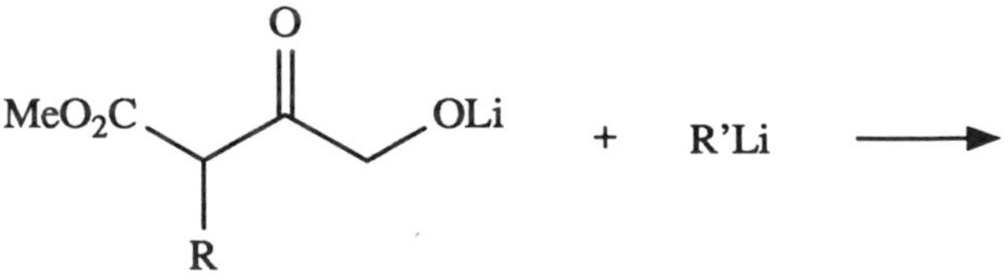

3-41

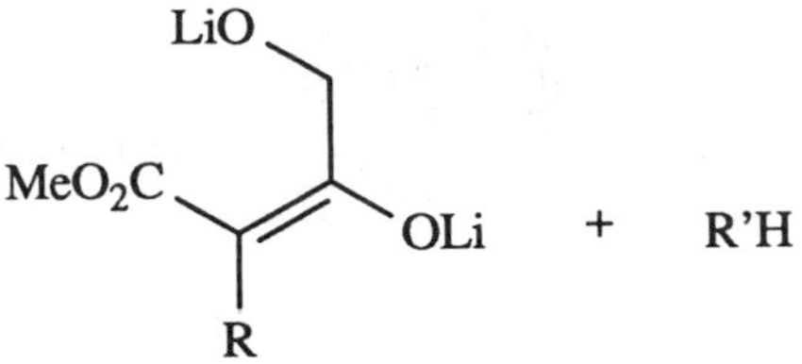

3-42

This proton, α to both the ester and ketone, is very acidic and easily removed. However, the ring closure reaction will be slowed because the alkoxide ion will be attacking a carbonyl which is already somewhat negative because of delocalization. (See **3-43**.)

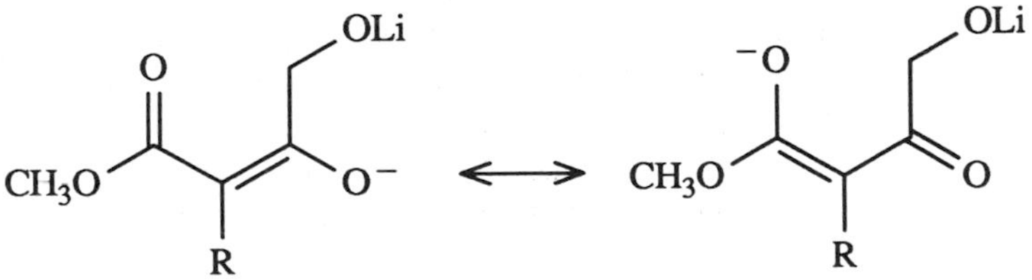

3-43

3-6.c.   The first step should be removal of a proton from an amino group, because this gives a resonance-stabilized anion. (Use the amino group which becomes substituted in the course of the reaction.)

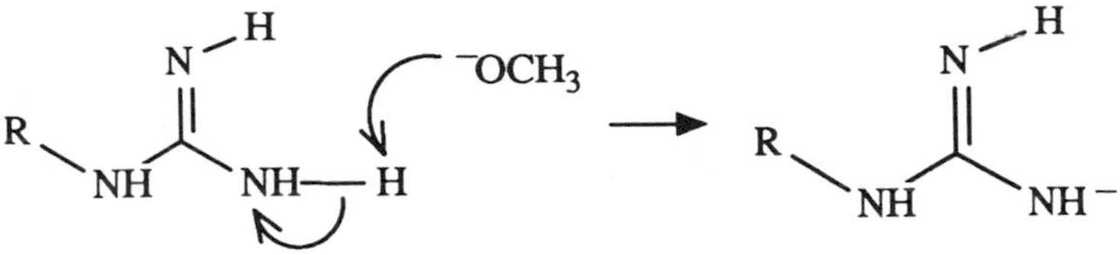

Substitution of the ethoxy group by the guanidino group should be a two step process: addition followed by elimination:

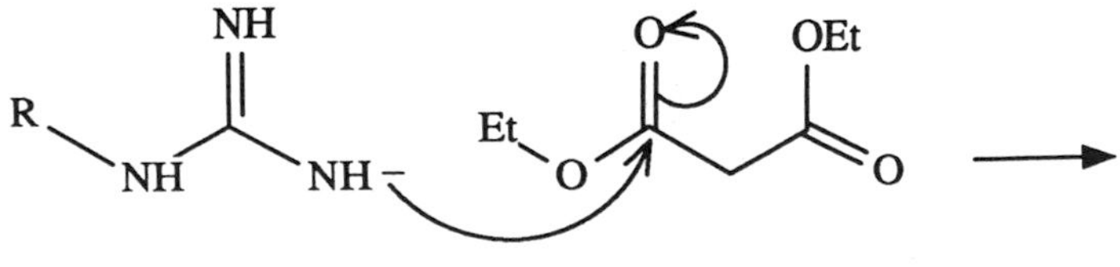

In the next step, because of the basicity of the medium and the acidic protons which are present, the amino proton is removed prior to nucleophilic substitution of the second ester carbonyl. Because of the resonance stabilization possible for the resulting anion, the amino proton, not the imino proton, reacts. Removal of a proton prior to cyclization also eliminates the need for the last step shown in the problem.

3-7.

Resonance forms, in which the nitro group is shown in its alternative forms,

do not add to the stability of the anion, relative to the stability of the starting material, since both the starting material and the intermediate have such resonance forms. Thus,
these forms are often omitted from answers to questions
like this one.

3-8.   The first steps, which are not shown, are the removal of the
protons from the protonated amine starting material. Those
steps would look similar to the final step of the mechanism
shown.

R = cyclopropyl

Substitution of the particular fluorine shown is favored, because the intermediate anion formed is stabilized by delocalization of the charge on the keto oxygen (resonance). Reaction at the carbon bearing the other fluorine would result in an intermediate in which the negative charge could not be delocalized onto this oxygen.

Initial formation of an aryne, followed by nucleophilic attack, is not a likely mechanism. The most important factors which detract from such a mechanism are: (1) the base is not a strong base and (2) the carbon-fluorine bond is very strong.

3-9. Since the aryne intermediate is symmetrical, half the ^{13}C label will be on the carbon bearing the amino group, and half will be on the carbon *ortho* to the amino group.

The two outer products at the base of the pyramid are, of course, the same: aniline with the label in the *ortho* position.

3-10. The most acidic proton in the molecule is on the carbon α to the nitrile. This proton is removed first. The second step is elimination of HCl to give an aryne. (As indicated in this chapter, this might also be a two-step process: removal of the proton, followed by loss of the chloride ion.) The aryne intermediate is usually written with a triple bond, and since the aromatic system is completely delocalized, it can be written as shown in **3-44**.

The anion in the side chain reacts as a nucleophile with the electrophilic aryne. The resulting anion, **3-45**, can remove a proton from ammonia to give **3-46**. Since the product has been achieved, we usually stop writing the reaction mechanism at this point. However, in the reaction mixture amide will remove a proton from the carbon α to the cyano group of **3-46**. Only during workup will the anion be protonated to give back **3-46**.

3-44

3-45

3-46

Some students, when answering this question, have used NH_4^+ as the reagent for protonation of **3-45**. However, since amide ion in ammonia is a very strongly basic medium, the concentration of ammonium ion would be essentially zero.

The representation, **3-47**, shown below, is equivalent to the usually written structure, **3-44**; but keep in mind that the extra π bond drawn, the one highlighted, is not an ordinary π bond. It is formed by the overlap of two sp^2 orbitals, one on each carbon. This bond is perpendicular to the other π bonds shown in **3-47**.

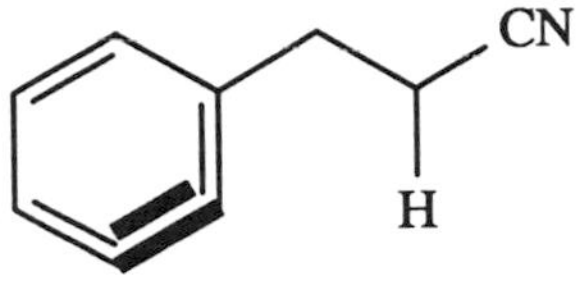

3-47

3-11.

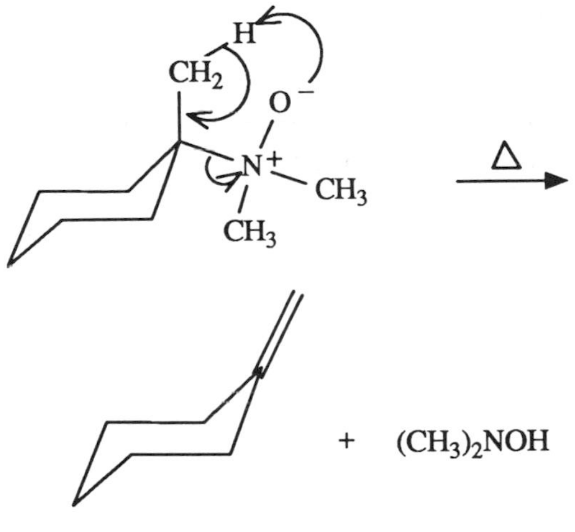

+ $(CH_3)_2NOH$

An interesting aspect of this elimination reaction is that it gives only the isomer with the exocyclic double bond. This results from the strict stereochemical requirements of the 5-membered ring transition state: all of the atoms must lie in the same plane. This rules out the alternate reaction, removal of a proton from a ring carbon, because too much distortion of the cyclohexane ring would be required.

3-12.a. This reaction is analogous to those of Example 3-12. However, we can also apply our knowledge that esters undergo nucleophilic substitution. Therefore, the initial reaction of the Grignard reagent will give a ketone:

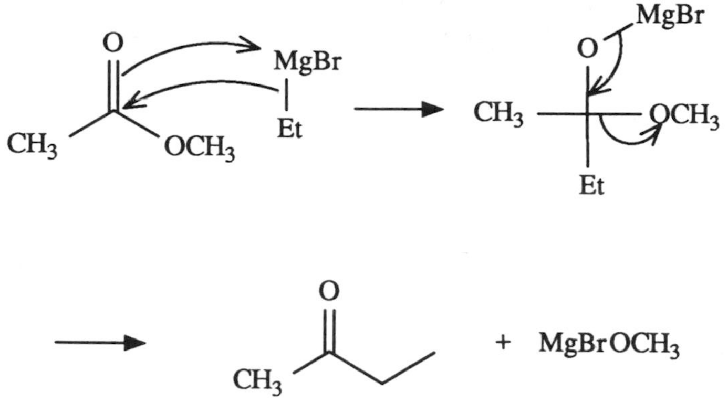

Then the ketone reacts with another mole of Grignard reagent to give the salt of an alcohol, which will be converted to the alcohol during acidic workup.

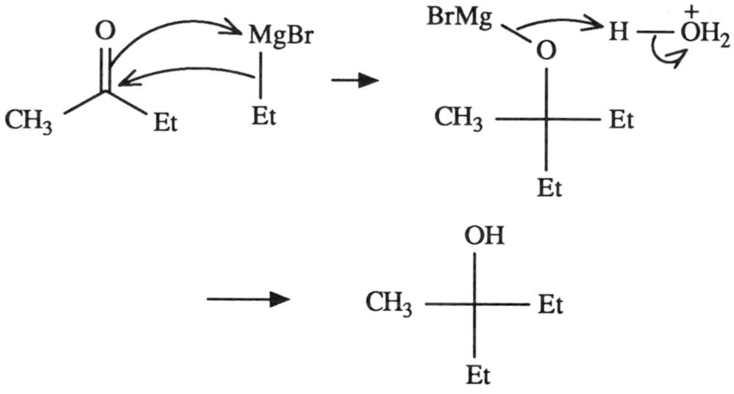

3-12.b. This reaction is an unusual addition to a carbonyl derivative. Normally, the nucleophile would react at the carbon atom of the C=N group, but because positive character of this carbon would impart antiaromatic character to the ring, it is less positive than usual. Conversely, to the extent that the carbon of the C-Mg bond of **3-48** is negative, this intermediate is stabilized, because the five-membered ring is an aromatic system.

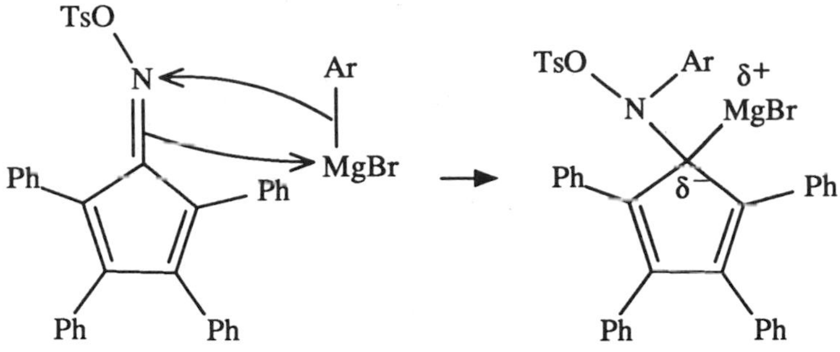

3-48

Subsequent elimination of the tosylate group gives the product:

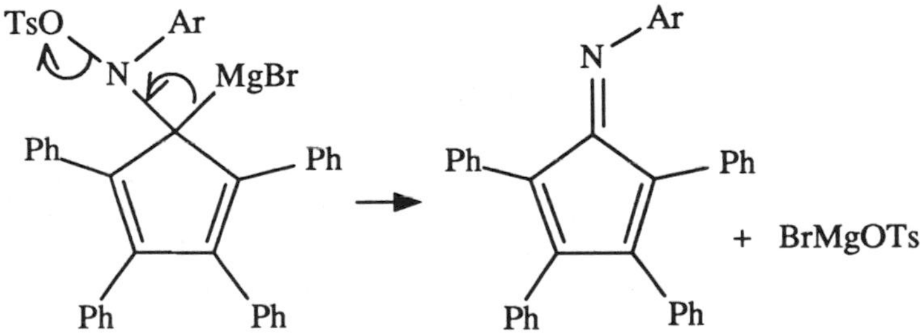

A less likely mechanism would be direct displacement of the tosylate anion by the Grignard reagent. That would be an S_N2 reaction at an sp^2 atom (see Helpful Hint 3-2).

3-13.　The nitrogen attached to the phenyl ring is less nucleophilic, because the lone pair of electrons is delocalized into the aromatic ring.

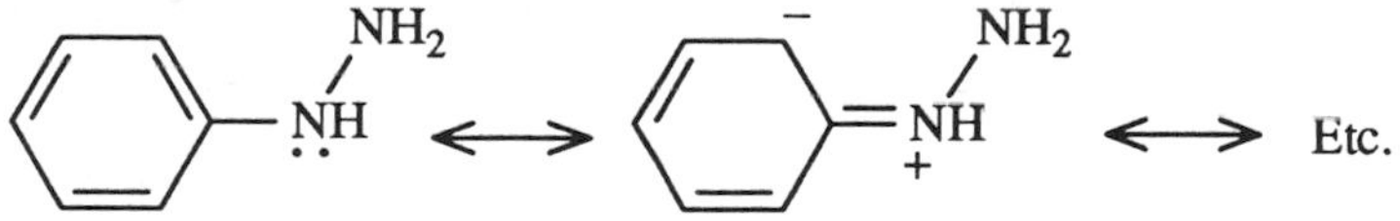

3-14.a.　This is another example of the reaction of an amine with a carbonyl compound, in the presence of an acid catalyst. The first steps are protonation of the carbonyl group, nucleophilic addition of the amine, and deprotonation of the nitrogen to give intermediate **3-49**.

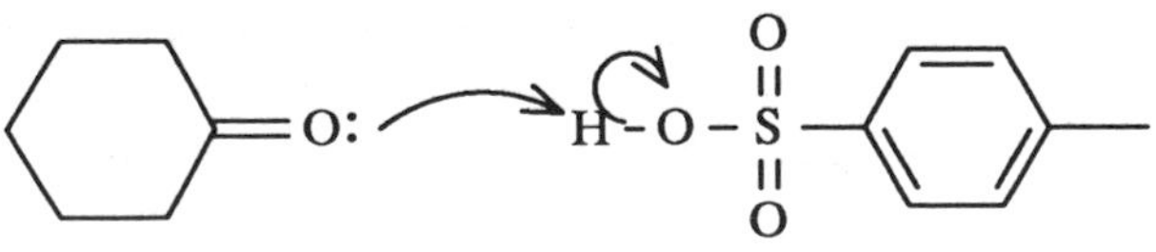

In **3-49**, unlike in the intermediate of Example 3-13, there is no proton on the nitrogen. Thus, it is not possible to form an imine via loss of a water molecule. Instead dehydration, *via* loss of a proton from carbon, gives the product which is called an enamine.

Notice that in the last step the catalyst is regenerated, and water is produced. The preferred direction of the equilibria in this problem is toward starting material. However, the enamine can be obtained in significant amount if the product water is removed from the reaction mixture as it is produced.

3-14.b. Analysis of the starting materials and products reveal that the =NPh group has been replaced by =NOH. An outline of steps can be formulated from this simple fact: (1) reaction of hydroxylamine at the carbon of the C=N (2) elimination of the NPh group. Details to be worked out include identifying the actual nucleophile and electrophile in (1) and the actual species eliminated in (2). These can be ascertained by considering the relative acidities and basicities of the species involved.

Because the nitrogen in hydroxylamine hydrochloride has no lone pairs of electrons, the salt cannot be the nucleophile. The pyridine in the reaction mixture can remove a proton from the nitrogen hydroxylamine hydrochloride. This equilibrium favors starting material (note the relative pK_a's below), but an excess of pyridine will release some hydroxylamine.

$$pK_a - 8.03 \qquad pK_a = 5.2$$

Moreover, because the salts of pyridine and hydroxylamine are sources of protons, it is unlikely that the anion derived from hydroxylamine, $^-$NHOH, or an anion like **3-50** will be formed as an intermediate. The anion in **3-50** can be compared to the anion formed when aniline ($pK_a = 30$) acts as an acid. Such a strong base will not be produced in significant amount in a medium in which there is protonated amine (see pK_a's above).

3-50

Thus, the nitrogen of the imine will be protonated prior to nucleophilic reaction at the carbon.

The electrophilic protonated imine reacts with nucleophilic hydroxylamine.

Before the phenyl-substituted nitrogen acts as a leaving group, it too is protonated to avoid the poor leaving group $PhNH^-$.

3-14.c. In Example 3-12a it was stated that the mechanism for this reaction was similar to the reverse for the formation of a hydrazone. In the first step the basic nitrogen of the imine is protonated. This converts the molecule into a better electrophile which adds water as a nucleophile at the positive carbon.

The resulting intermediate can be deprotonated at oxygen and protonated at nitrogen.

The result is to convert the nitrogen into a better leaving group.

The oxygen in the molecule provides driving force for the reaction by stabilizing the positive charge. Deprotonation gives the product.

3-15.a. While there are three carbons from which a proton could be removed to produce an enolate ion, only one of the possibilities leads to the product shown.

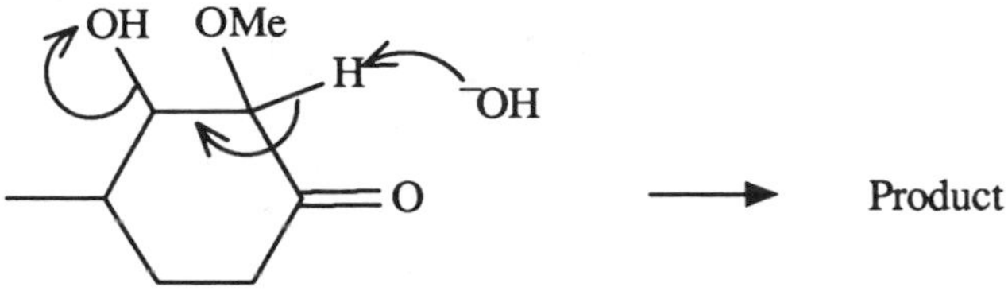

A common shortcut that students take is to write the following mechanistic step for loss of water:

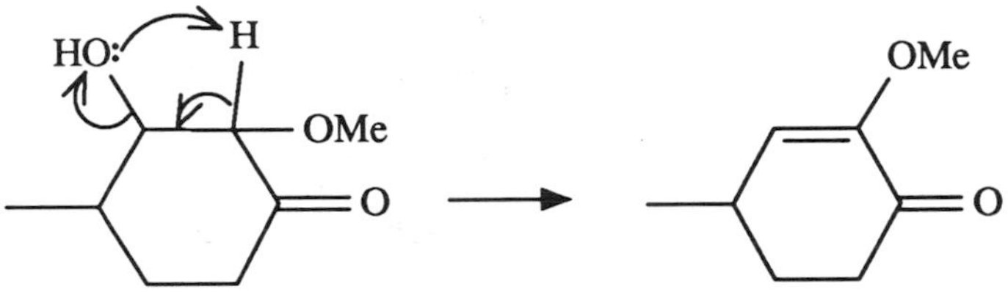

Since hydroxide ion is a much stronger base than the alcohol used as a base in this step, the previously written elimination is a better step.

3-15.b.   The first step of this reaction is removal of the very acidic proton α to both a cyano and a carbethoxy group. There are hydroxide ions present in 95% ethanolic solutions of carbonate, so either hydroxide ion or carbonate ion can be used as the base.

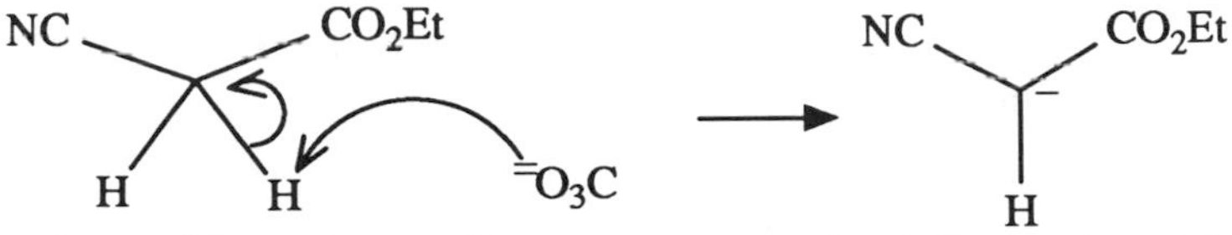

The resulting anion can then carry out a nucleophilic reaction with the electrophilic nitrogen of the nitroso group of nitrosobenzene.

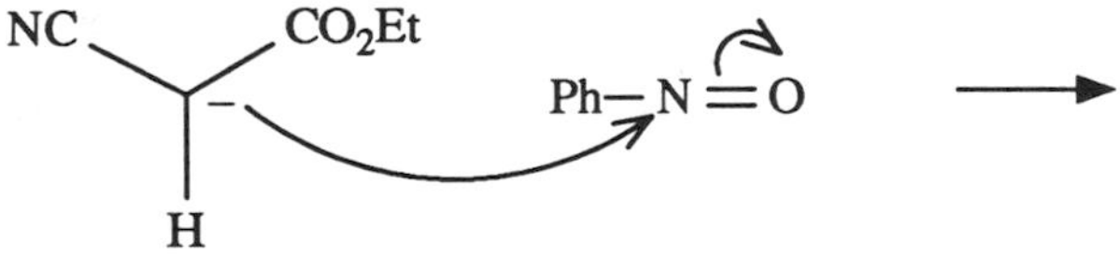

The resulting oxyanion can remove a proton from solvent.

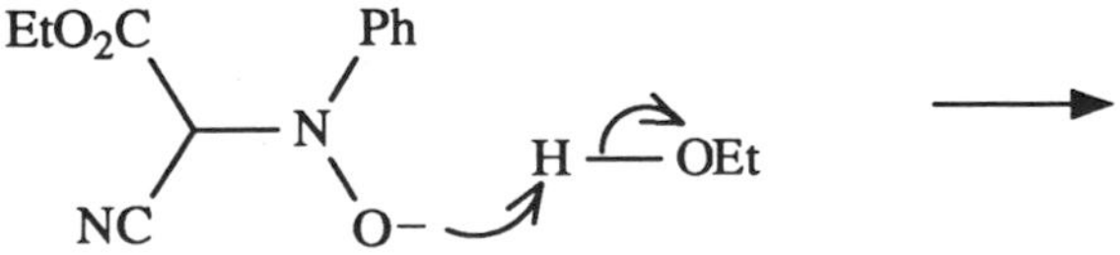

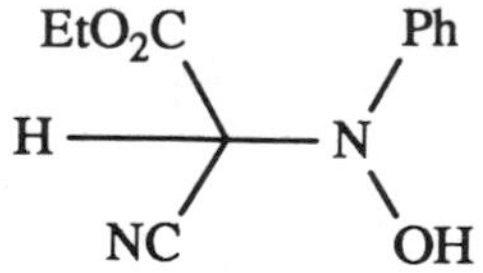

3-51

After the first addition, the reaction repeats itself with removal of the second α-proton in **3-51** and addition to a second molecule of nitrosobenzene.

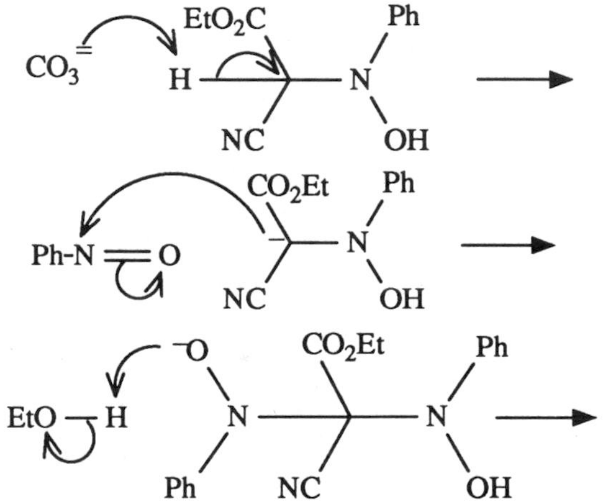

Finally, hydroxide can add to the carbethoxy group, and the intermediate undergoes elimination to give the products.

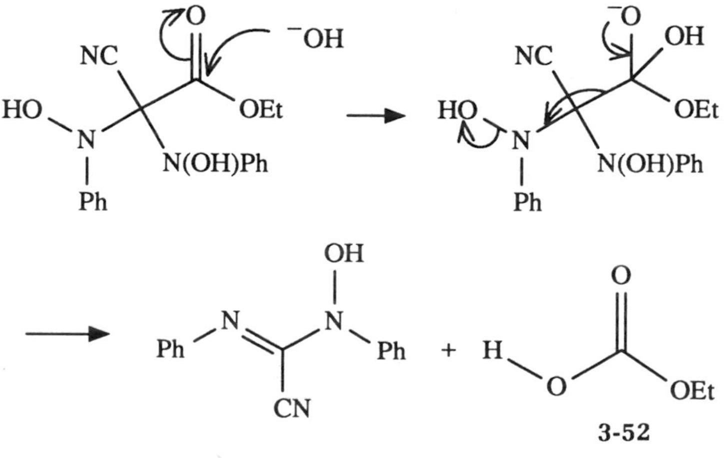

3-52

Product, **3-52**, a half ester of carbonic acid, is unstable and would decompose under the reaction conditions to CO_2 and HOEt.

Another possible mechanism for the final stages of the reaction involves an intramolecular nucleophilic reaction:

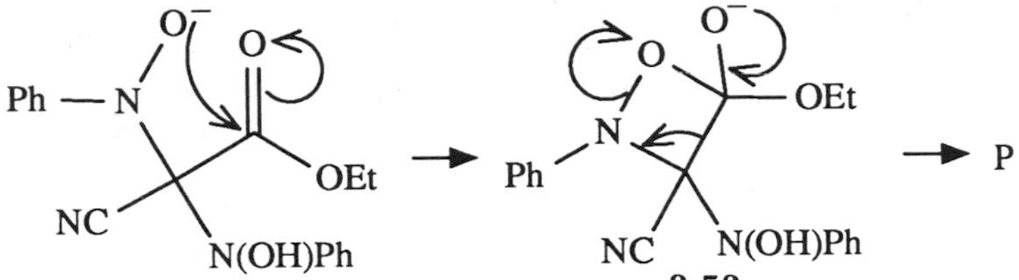

The fact that **3-53** is not necessary, does not eliminate it as a possible intermediate in the reaction.

A student wrote the following as a mechanism for the final elimination step:

If structures of the eliminated fragments had been drawn, this unlikely step might have been avoided. Besides hydroxide ion, the cation shown below is the other product.

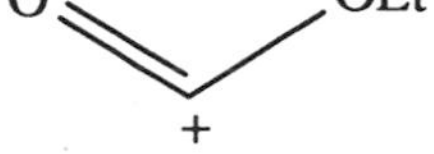

Formation of this cation, a very strong acid, would not be expected in a basic medium. (See Helpful Hint 2-5.)

3-16. This mechanism shows a direct nucleophilic substitution at an sp^2-hybridized carbon, which is unlikely. An alternative is addition of the amine to the α-β-unsaturated system, followed by elimination of bromide.

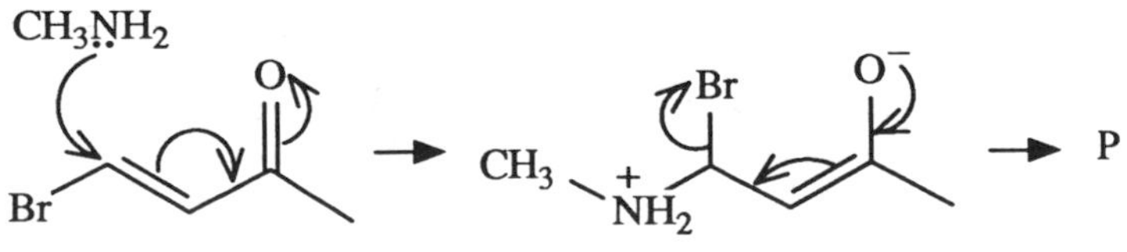

This is addition at the usual position for a 1,4-addition re-action, but the usual subsequent addition of a proton at the 1 position (oxygen of the carbonyl group) is replaced by elimination of bromide ion.

3-17.a.   Notice that both new groups in the molecule are attached to the carbon next to the carbonyl group. Thus, the first steps are removal of a proton from the carbon α to the car-bonyl group, to give an anion and a nucleophilic substitu-tion effected by that anion.

Then a second anion is formed which adds to the β carbon of the α-β-unsaturated nitro compound. The nitro group can stabilize the intermediate anion by resonance, analo-gous to a carbonyl group.

3-54

The diisopropylamine formed when LDA acts as a base is much less acidic than the proton α to the nitro group (see Table 1-2). Thus, protonation of the anion **3-54** must take place during workup.

3-17.b. Cyanide is a good nucleophile. A 1,4-addition to the α,β-unsaturated carbonyl puts the cyano group in the right place for subsequent reaction.

Formation of the other product involves addition to a car-bonyl group as the first step.

The final steps are a tautomerization.

3-18.a.

3-18.b.

3-19.a.   The reaction mechanism requires two separate nucleophilic steps; i.e. the mechanism requires two nucleophiles
not one. The reaction cannot be run with the α-chloro
compound in the presence of n-butyl lithium, because reactions between these reagents would give several important side products. Thus, the dilithium derivative, **3-55**, of
the starting sulfone is formed first, and then the chloro
compound is added.

There are two electrophilic positions in the chloro compound: the carbonyl carbon and the carbon α to it which
bears chloride as a leaving group. Nucleophilic reaction at
either position by **3-55** is a reasonable reaction, and we will
illustrate both possibilities. Reaction at the carbonyl gives
3-56 which can close to a cyclopropane.

The cyclopropane ring then opens to give the product.

3-57

The other mechanism involves **3-55** as the nucleophile in an S_N2 displacement at the highly reactive chloro-substituted carbon α to the carbonyl. The remaining anion, **3-58**, reacts with the carbonyl group to give **3-57**.

3-58

$\longrightarrow$ **3-57**

The authors of the cited paper favor the first mechanism based upon the product of the reaction of **3-55** with 1-chloro-2,3-epoxypropane.

Notice the resemblance of this reaction to Example 3-17. In the example, the benzylic carbon is inserted between the S and C=S. In this reaction the carbon introduced by the phenylsulfonyl anion is inserted between the benzoyl group and the α carbon of the starting carbonyl compound. In both cases, this "insertion" is effected by making a three-membered ring by forming one bond, then opening the three-membered ring by breaking a different bond.

3-19.b. This reaction is run in the presence of base. The most acidic hydrogen in the starting materials is on the carbon between the ester and ketone functional groups. If that proton is removed, the resulting anion can act as a nucleophile and add to the carbonyl group of the isocyanate. The oxyanion formed (stabilized by resonance with the nitrogen) can undergo an intramolecular nucleophilic substitution to produce the 5-membered ring. Base-catalyzed tautomerization gives the final product.

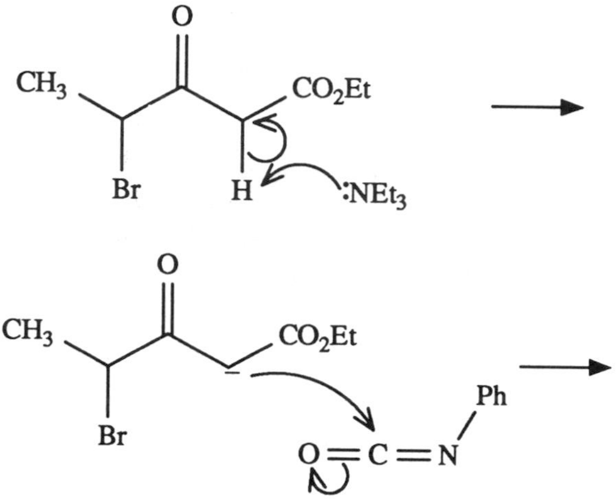

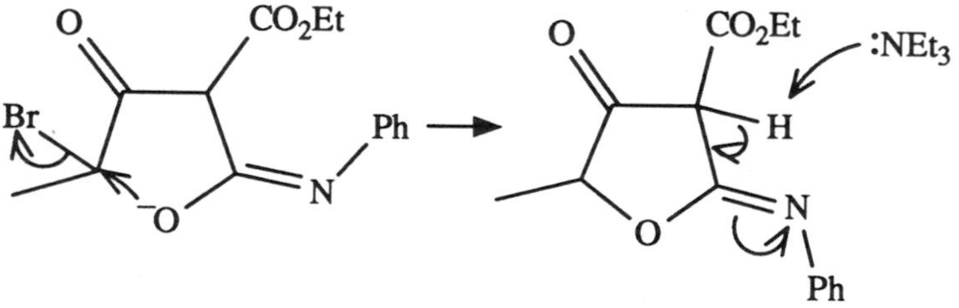

3-59

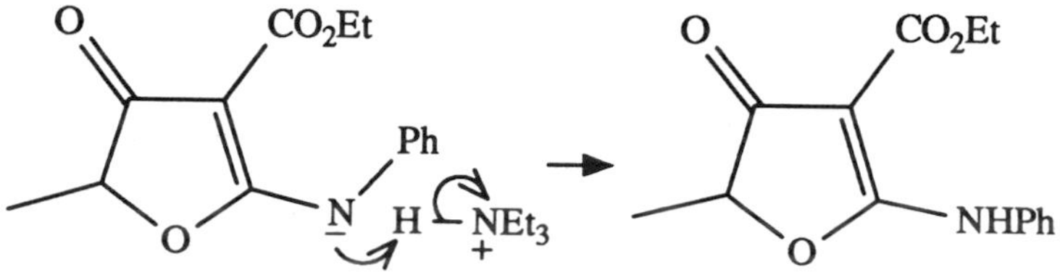

3-60

The anion, intermediate to the tautomers **3-59** and **3-60**, is a resonance hybrid:

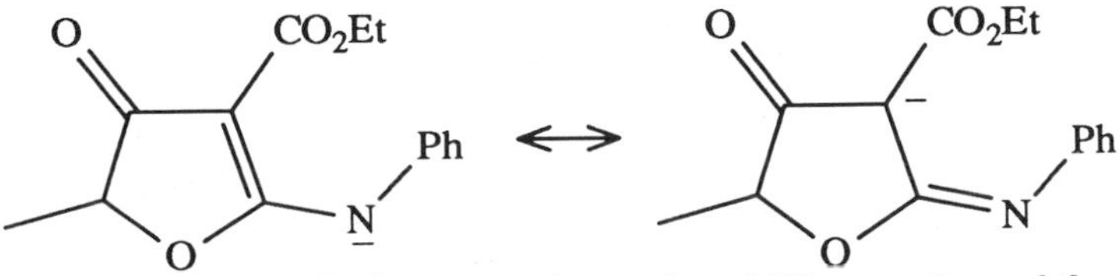

A much less likely first step is nucleophilic reaction of the carbonyl oxygen of the isocyanate with the carbon attached to bromine:

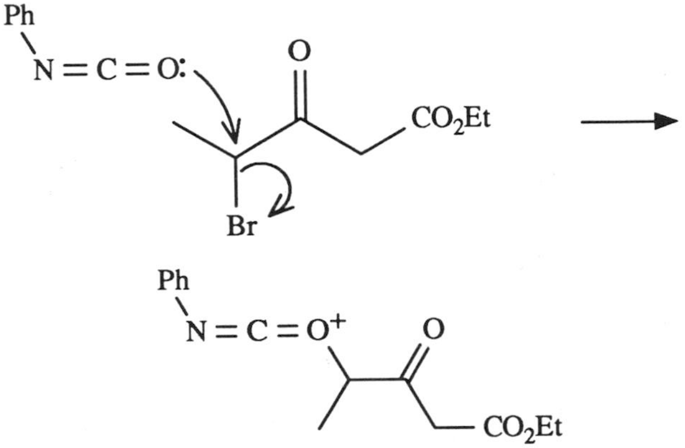

In general, although carbonyl groups will act as bases with strong acids, their nucleophilicity is quite low. We can get a rough idea of the basicity of the carbonyl group, relative to the acidity of the proton actually removed, from Table 1-2. The basicity of acetone is estimated from the pK_a of its conjugate acid, −2.85. The pK_a of the proton should be similar to that of ethyl acetoacetate, 11. Thus the ketoester is much more acidic than the carbonyl group is basic, and thus it is much more likely that the proton would be removed. Nonetheless, a positive feature of this mechanism is that there is a viable mechanism leading to product. (The next step would be removal of a proton on the carbon between the ketone and ester.)

3-19.c. In this reaction, the amide anion, a very strong base, removes a benzylic proton. Cyclization, followed by loss of methylphenylamide, and tautomerization leads to the product. Notice that nucleophilic reaction of the benzylic anion and loss of methylphenylamide are separate steps, in conformance with Helpful Hint 3-2.

The anion corresponding to the product is considerably more stable than either the amide ion or the phenylamide ion, so protonation of the final anion will take place during workup.

As in the previous problem, it is necessary to be able to distinguish between tautomers and resonance forms. The following two structures are tautomers. Thus a mechanism needs to be written for their interconversion.

3-19.d.   Note that two moles of hydrazine have reacted to give the product. It also appears that the introduction of each mole is independent of the other, and thus each mechanism can be shown separately. The hydrazinolysis of the ester is done first.

A mechanism for reaction with the other mole of hydrazine follows. The reaction sequence shown is preferred, because the anion produced by the first step, addition at the β position, is resonance-stabilized. The anion produced by reaction of hydrazine at the keto-carbonyl carbon would not be resonance stabilized.

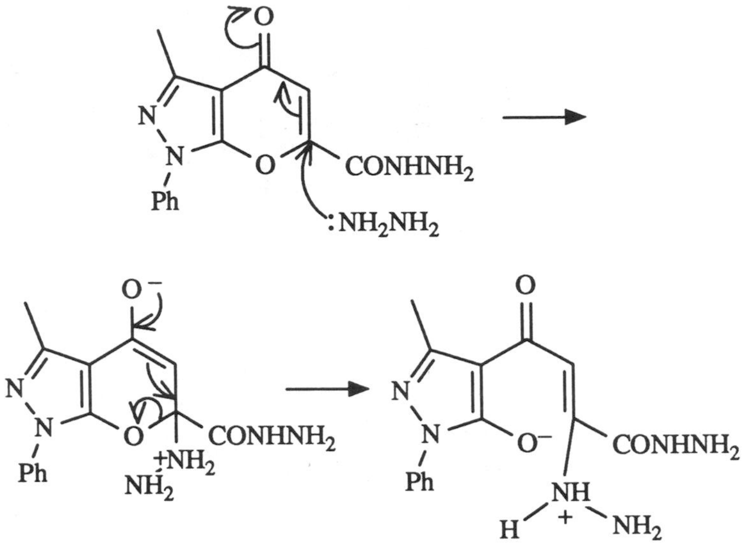

3-61

The enolate ion, **3-61**, will pick up a proton from solvent, and the hydrazinium ion will lose a proton to a base, such as a molecule of hydrazine. The distance between the groups in the molecule makes an intramolecular transfer of a proton quite unlikely. The nucleophilic hydrazine group in the neutral intermediate, **3-62**, can react intramolecularly with the electrophilic carbon of the carbonyl group.

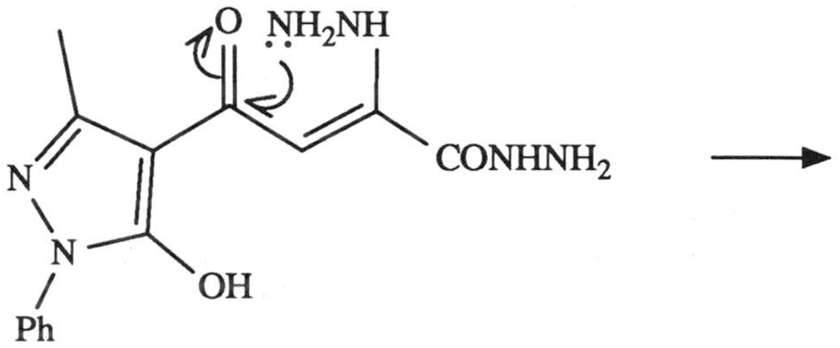

3-62

Again proton loss from the positive nitrogen and pickup of a proton at the negative oxygen can take place. In the final step, base promoted elimination of water occurs. The driving force for loss of water is formation of an aromatic ring.

Do not write intramolecular loss of water in the last step. The mechanism shown is better for two reasons. First, the external base, hydrazine, is a much better base than the hydroxyl group, and second, most eliminations go best when the proton being removed and the leaving group are *anti* to one another.

The following would not be a good step:

All of the atoms involved in the step lie in a plane (the C=N nitrogen is sp^2 hybridized), but the amino group must interact with the p orbitals of the double bond, which are perpendicular to the plane defined by the atoms.

3-19.e. Bromine is not present in the product, so one reaction is probably nucleophilic substitution at the carbon bearing the bromine. A simple analysis, by numbering or inspection, also reveals that the methylene group in the imine-ester is substituted twice in the reaction. Thus, removal of a proton from the methylene group is a good first step, followed by a nucleophilic substitution reaction.

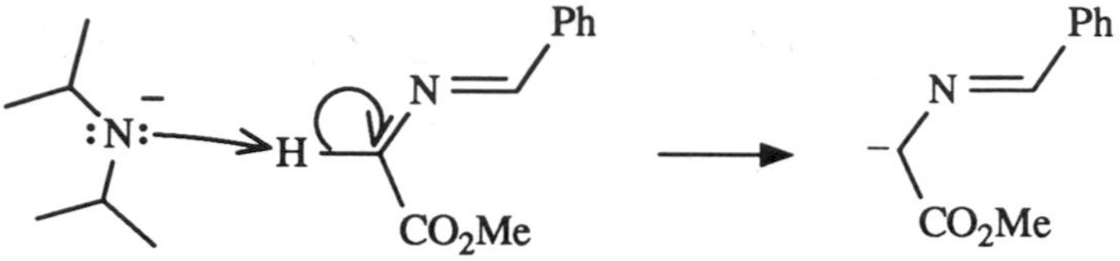

3-63

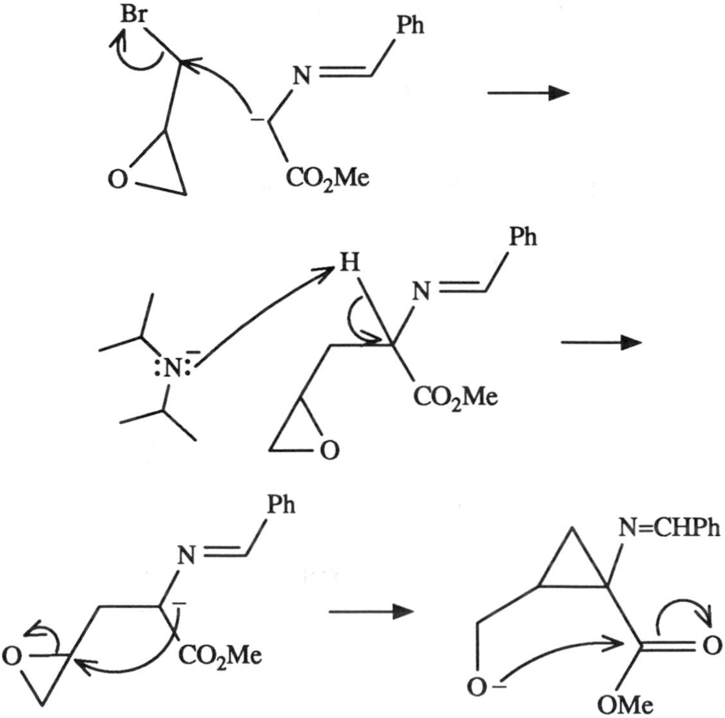

3-64

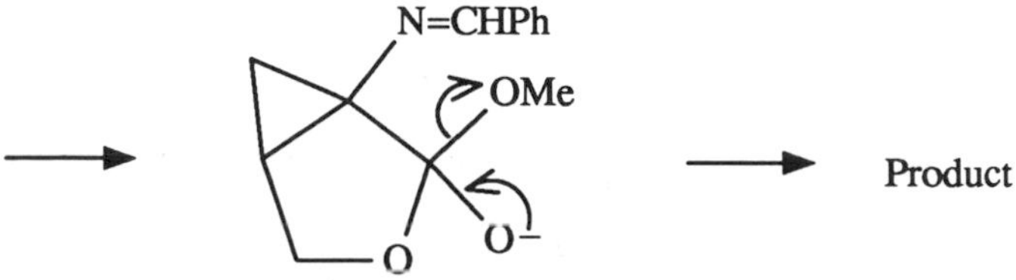

Another possible nucleophilic reaction of anion **3-64** would be at the other electrophilic carbon of the epoxide:

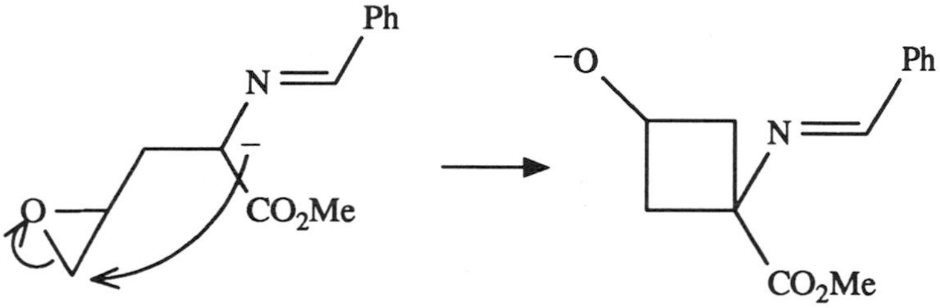

3-64

This reaction might have been favored for two reasons: reaction with the epoxide, an S_N2 reaction, should occur best at the least hindered carbon; formation of a four-membered ring would be enthalpy favored, because it would have less strain energy than the three-membered ring. The fact that the three-membered ring is actually formed must mean that the reaction is directed by entropy.

Another possible mechanism starts with nucleophilic reaction by **3-63** at the epoxide. The alkoxide ion than displaces bromide to produce a new epoxide.

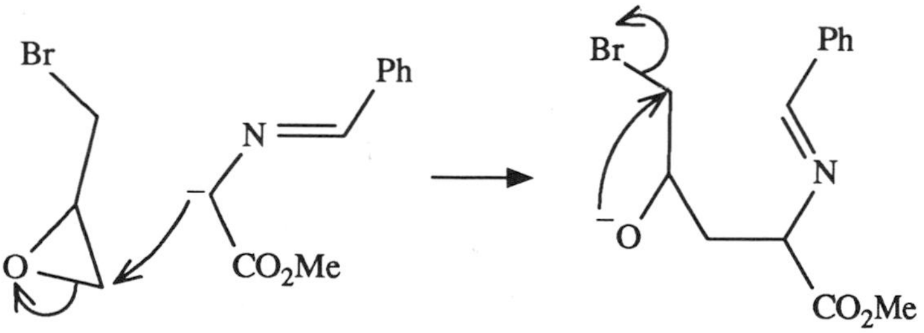

3-63

Proton removal followed by nucleophilic reaction gives a cyclopropane.

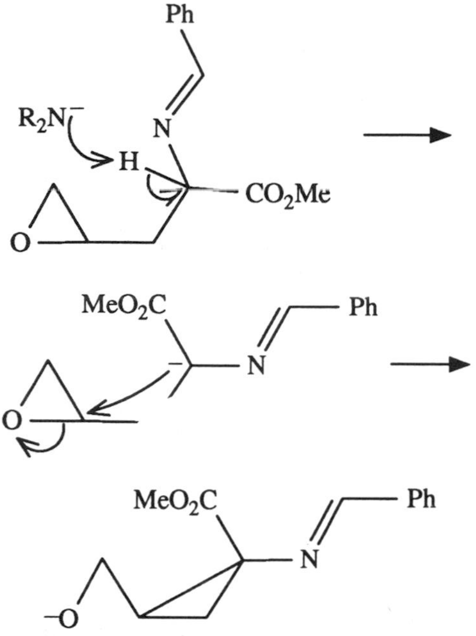

3-65

The nucleophilic alkoxide, **3-65**, reacts intramolecularly with the carbon of the ester functional group. The resulting intermediate loses methoxide giving the product.

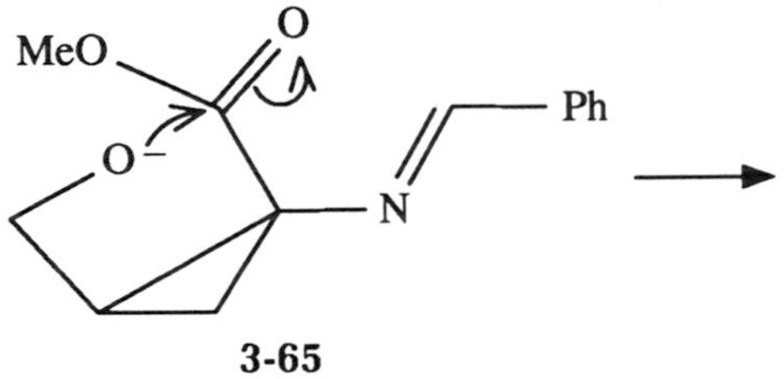

3-65

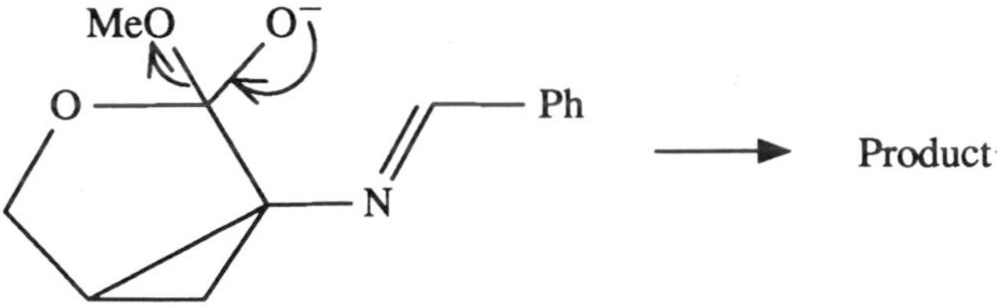

Product

Initial reaction of **3-63** at the CH of the epoxide (rather than the CH_2) is at the more hindered carbon which is not preferred for an S_N2 reaction.

3-20. The reaction conditions, sodamide in ammonia, as well as the lack of electron-withdrawing groups directly attached to the aromatic ring, suggest the aryne mechanism (Mechanism 2) rather than the nucleophilic aromatic substitution (Mechanism 1). There are also several other problems with Mechanism 1.

Step (1), the nucleophilic addition of the anion to the aromatic ring, is less likely than the aryne formation [step (7)] because the intermediate anion is not very stable. The methoxy and bromine substituents can remove electron character from the ring only by inductive effects. Nucleophilic aromatic substitution (except at elevated temperatures) ordinarily requires substituents which withdraw electrons by resonance. The usual position for nucleophilic reaction in this mechanism is at the carbon bearing the leaving group, in this case, bromide. However, in some other mechanism, involving several steps, addition of the nucleophile at other positions might be acceptable.

For step (2) the arrow between the two structures should be replaced with a double-headed arrow, indicating that these are resonance structures.

In step (3) cyanide can act as a leaving group, but it is not a very good one. Furthermore, the product contains a cyano group. Thus, another problem with this step is that if cyanide leaves the molecule, it will be diluted by the solvent to such a low concentration that subsequent addition will occur very slowly. (However, that doesn't mean it can't happen.)

In step (4) the elimination of HBr is reasonable: it gives an aromatic system, and *syn* E-2 elimination is a common reaction.

In step (5) direct substitution by cyanide ion at an sp^2 center, as shown, is an unlikely process.

In step (6) removal of a proton from ammonia, by an anion which is much more stable than the amide ion, is very unlikely; this would be a very unfavorable equilibrium. In other words, the product forming step, like step (11), would occur on workup.

Other comments:

(1) Writing the aryne formation as a two step process would be acceptable.

(2) Direct substitution of cyanide for bromine on the ring is not a good mechanistic step. Initially, nucleophilic substitution at an sp^2-hybridized carbon is unlikely.

(3) Writing an aryne intermediate and then using it as a nucleophile is unlikely. Because of the high s character in the orbitals forming the third bond, arynes tend to be electrophilic, not nucleophilic.

(4) Other mechanisms, which involve formation of a carbanion in the ring, are also subject to the criticism that this carbanion is not stabilized by strongly electron-withdrawing substituents on the ring.

3-21.   There are two apparent ways that triphenylphosphine can react with the diethyl azocarboxylate. One is nucleophilic reaction at the electrophilic carbonyl carbon, and the other is 1,4 addition. Since the ester groups are intact in the hydrazine product, the 1,4 addition is more likely. The intermediate anion can be protonated by the carboxylic acid, the carboxylic acid, giving a salt. The electrophilic phosphorus atom then undergoes nucleophilic reaction with the alcohol. An intriguing aspect of this reaction is the fate of the proton on the alcohol. There is no strong base present in the reaction mixture. A good possibility is that the carboxylate anion removes the proton from the alcohol as it's reacting with the phosphorus.

How do we rationalize what appears to be a trimolecular reaction? Because the solvent is a nonpolar aprotic solvent, the phosphonium carboxylate must be present as an ion pair and can be considered as a single entity. The carboxylate ion may also be properly situated to remove the proton.

Finally, the carboxylate anion acts as a nucleophile to displace triphenylphosphine oxide and give the inverted product.

$$PhCO_2 \overset{CH_3}{\underset{C_6H_{13}}{\rule[0.5ex]{0pt}{0pt}\Big|}}\!\!-H \quad + \quad {}^-O \longrightarrow \overset{+}{P}Ph_3$$

An alternate mechanism, in which the alcohol reacts with the phosphonium salt without assistance from the carboxylate anion, is less likely, because the intermediate produced has two positive centers adjacent to each other.

Chapter 4

Reactions Involving Acids and Other Electrophiles

4-1 Carbocations

Reactions in acid often involve the formation of carbocations, trivalent positive carbons, as intermediates. The order of stability of carbocations containing only alkyl substituents is $3°>2°>1°>CH_3$. This order of stability is due to two important features:

(1) The sp^3-hybridized carbons of the alkyl substituents are electropositive relative to the sp^2-hybridized carbocation; thus the flow of electron character is from the alkyl group toward the carbocation. This is called the inductive effect.

(2) The alkyl substituents are polarizable; polarizability is the measure of the ability of an electron cloud to distort itself. Such distortion leads to distribution of the charge and thus its stabilization. Large groups are more polarizable than small ones.

A third feature which can stabilize a carbocation is delocalization by the resonance effect. Thus carbocations which are benzylic or allylic are stabilized by the overlap of the cationic p orbital with the adjacent π system (For example, see Problem 1-4 c.). Carbocations with α-amino, α-alkoxy, and similar substituents are stabilized by overlap of a filled p orbital on the heteroatom with the empty p orbital on the carbocation.

Cations at sp^2- or sp-hybridized carbons are especially unstable: in general, the more s character in the orbital, the less stable the cation.

4-1-1 FORMATION OF CARBOCATIONS

Example 4-1. Ionization.

When a compound undergoes unimolecular ionization to a carbocation and a leaving group, the process is called S_N1 or E-1, depending on the final product formed; that is, either a substitution or an elimination product. In both cases, the rate-determining step is the ionization, not the product-forming step. An example of a unimolecular ionization is the loss of water from a protonated alcohol.

Other ionization reactions don't require acid catalysis when a good ionizing solvent and good leaving group are present.

Example 4-2. Addition of electrophiles to a π bond, the proton as an example.

Intermediate carbocations are often produced by addition of a proton to a π bond.

Example 4-3. Protonation of a carbonyl group.

In acid, carbonyl compounds are in equilibrium with their protonated counterpart. Such protonation is often the first step in nucleophilic addition or substitution. For aldehydes and ketones, the protonated carbonyl group is a resonance hybrid of two forms: one with positive charge on the carbonyl oxygen and one with positive charge on the carbonyl carbon.

When esters are protonated at the carbonyl group, there are three resonance forms: two corresponding to those above, and a third with positive charge on the alkyl oxygen.

Example 4-4. Reaction of a carbonyl compound with a Lewis acid.

Carbonyl groups form complexes or intermediates with Lewis acids like $AlCl_3$, BF_3, and $SnCl_4$. For example, in the Friedel-Crafts acylation reaction in nonpolar solvents, an aluminum chloride complex of an acid chloride is often the acylating agent. Because of the basicity of ketones, the products of the acylation reaction are also complexes. For more detail on electrophilic aromatic substitution, see Section 4-4.

4-1-2 REARRANGEMENT OF CARBOCATIONS

A major attribute of carbocations is their propensity to rearrange. This is in sharp contrast to carbanions which rearrange much less readily. Under usual reaction conditions, the rearrangement of carbocations is from those of lesser stability to those of the same or greater stability. However, under very strongly acidic conditions or at very high temperatures, rearrangements from more stable to less stable carbocations may occur. Rearrangement usually involves an alkyl, phenyl or hydride shift to the carbocation from an adjacent carbon. In problems involving these rearrangements the method of numbering both the starting material and product, introduced in Chapter 2, can be very helpful.

Example 4-5. A hydride shift in the rearrangement of a carbocation.

Treatment of isobutyl alcohol with HBr and H_2SO_4 at elevated temperature leads to t-butyl bromide. In the first step the hydroxyl group is protonated by the sulfuric acid to convert it into the better leaving group, the water molecule. Water then leaves giving the 1° isobutyl carbocation, **4-1**.

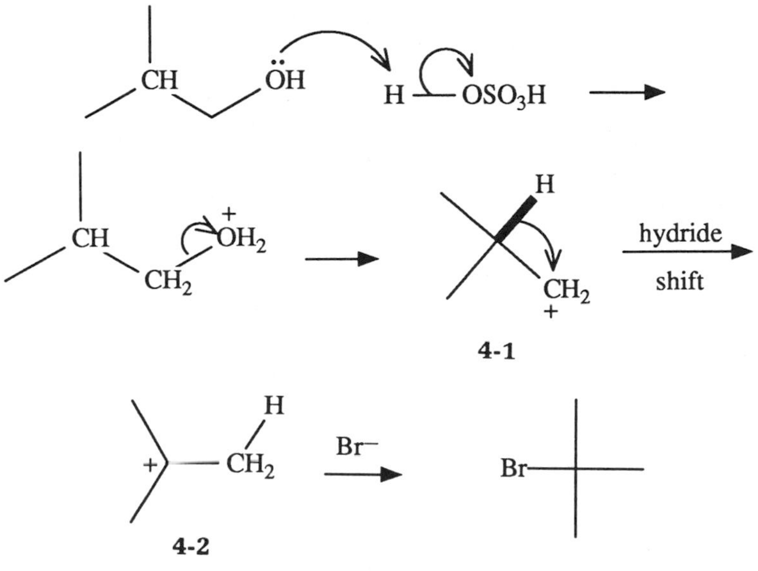

4-1

4-2

The hydrogen and the electrons in its bond to carbon, highlighted in **4-1**, move to the adjacent carbon. Now the carbon from which the hydride left is deficient by one electron and is thus a carbocation. Since the new carbocation, **4-2**, is 3°, the molecule has gone from a relatively unstable 1° carbocation to the much more stable 3° carbocation. In the final step, the nucleophilic bromide ion reacts with the positive electrophilic 3° carbocation to give the alkyl halide product.

Example 4-6. An alkyl shift in the rearrangement of a carbocation.

Consider the following reaction.

(This example is derived from the following reference: Corona, T.; Crotti, P.; Ferretti, M.; Macchia, F. *J. Chem. Soc. Perkin I* **1985**, 1607–1616.)

Since the product has neither a t-butyl group nor a six-membered ring, a rearrangement must have taken place. Numbering both the starting material and the product helps to visualize what takes place during the course of the reaction.

4-3   4-4

The three methyls in **4-3** are given the same number, because they are chemically equivalent. Numbering **4-4** to give the least rearrangement is very straight-forward. Comparison of the numbering in **4-3** to that in **4-4** shows that one of the methyl groups shifts from carbon 8 to carbon 7 and that the bond between carbons 2 and 7 is broken. Since a rearrangement to carbon 7 takes place, that carbon must have a positive charge during the course of the reaction.

Starting the mechanism is also simple. Because the oxygen is the only basic atom in the molecule, its protonation must be the first step.

The protonated epoxide, unstable because of the high strain energy of the three-membered ring, will open readily. There are two possible modes of ring opening:

4-5

or

4-6

Because **4-6**, a tertiary carbocation, is more stable than **4-5**, a secondary cation, **4-6** would be expected to be formed preferentially. (However, if the tertiary carbocation didn't lead to the product, we would go back to consider the secondary cation). The formation of **4-6** also appears to lead toward the product, because the carbon which bears the positive charge is number 7 in **4-3**. The 3° carbocation can undergo rearrangement by a methyl shift to give another 3° carbocation.

methyl shift

Now, one of the lone pairs of electrons on oxygen facilitates breaking the bond between C-2 and C-7; the formation of two new π bonds compensates for the energy required.

4-7

The only remaining step is deprotonation of the protonated aldehyde, **4-7**, to give the neutral product. This would occur during the workup, probably in a mild base like sodium bicarbonate.

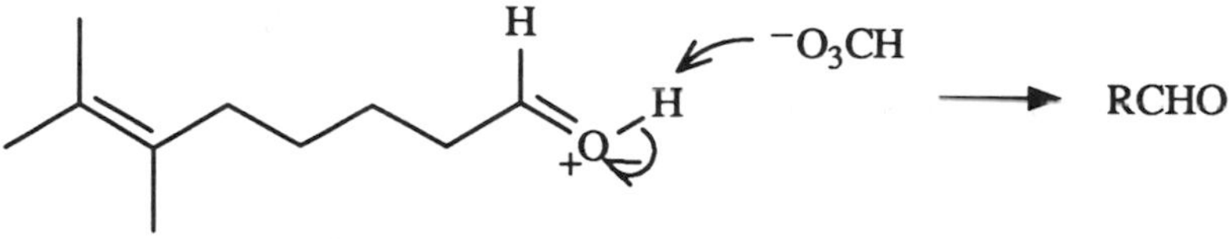

$$\longrightarrow \quad RCHO$$

Example 4-7. The dienone-phenol rearrangement.

The name for this rearrangement is derived from the fact that the starting material is a dienone and the product is a phenol. Since alkyl shifts of ring carbons take place, the cyclic structure itself is modified. Consider the following reaction.

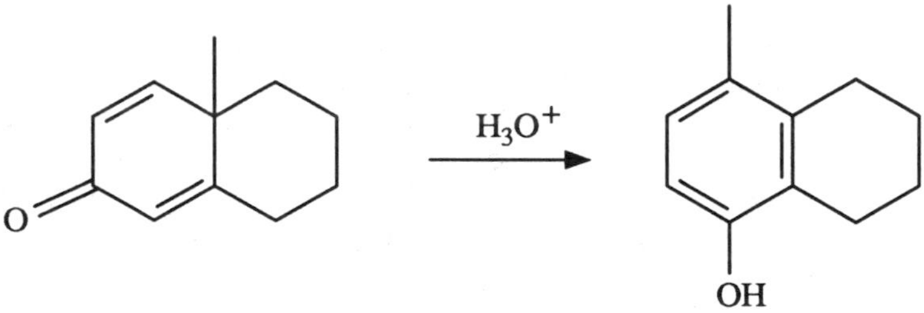

The first step in the mechanism for this reaction is protonation of the most basic atom in the molecule, the oxygen of the carbonyl.

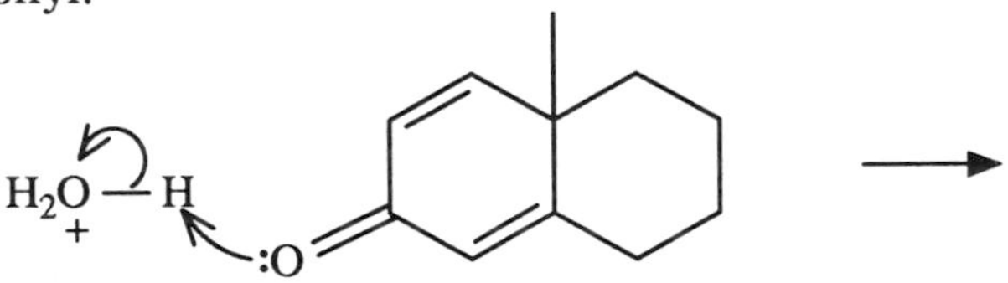

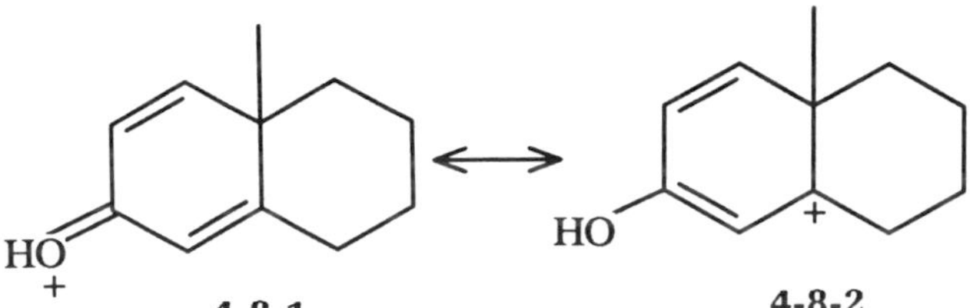

4-8-1   4-8-2

The intermediate carbocation, **4-8,** undergoes an alkyl shift to give another resonance-stabilized carbocation, **4-9.**

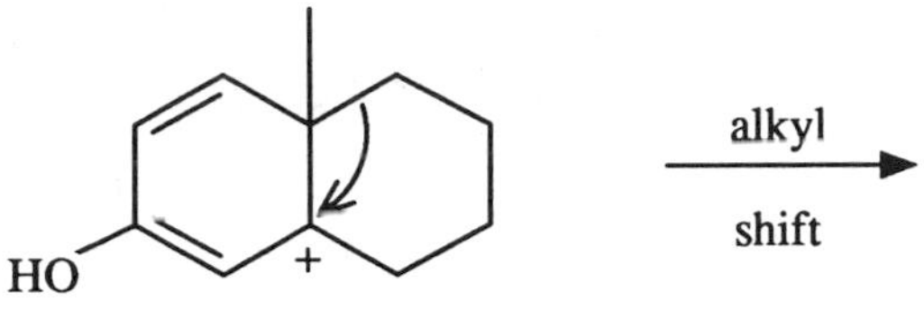

4-8-2

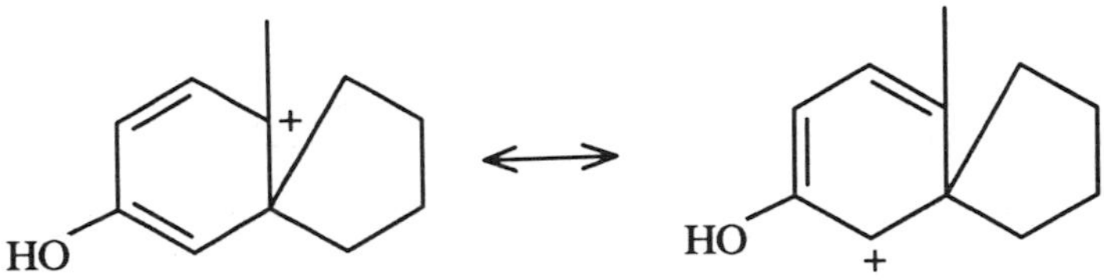

alkyl

shift

4-9-1   **4-9-2**

Finally **4-9** undergoes another alkyl shift, followed by loss of a proton, to give the product.

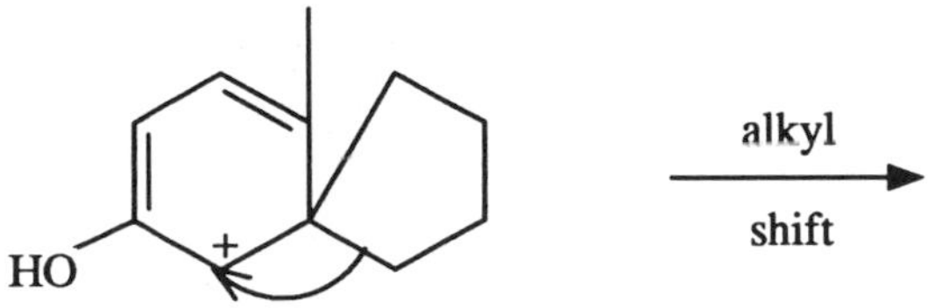

4-9

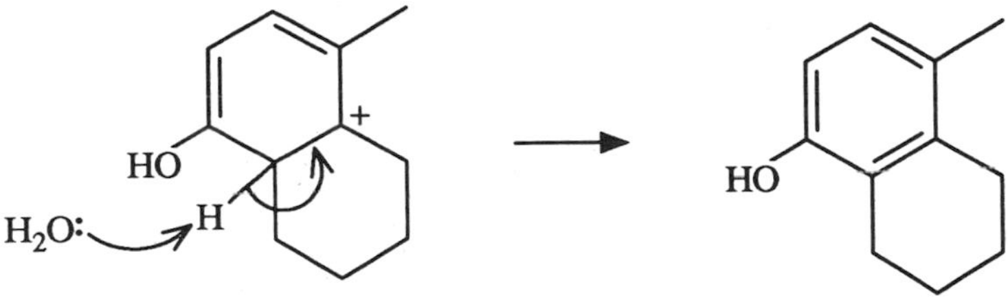

alkyl

shift

What initially appears to be a complicated reaction is the result of a series of simple steps. For other examples of this reaction, see Miller, B. *Acct. Chem. Res.* **1975,** *8,* 245-256.

Problem 4-1. Write step-by-step mechanisms for the following transformations.

a.

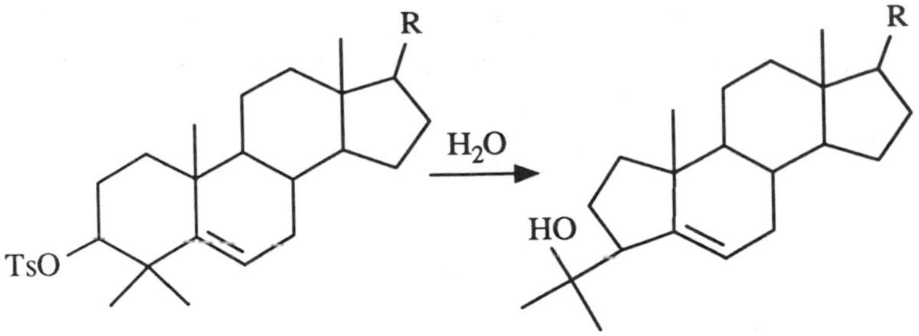

b.

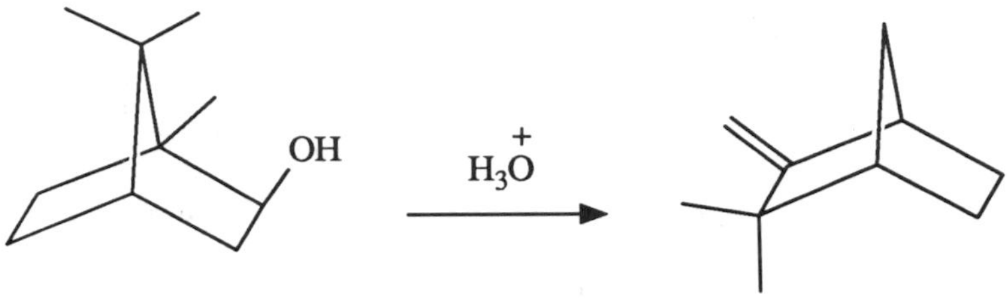

Many common rearrangement reactions are related to the rearrangement of 1,2-dihydroxy compounds to carbonyl compounds. These reactions are often called pinacol rearrangements, because one of the first examples was the transformation of pinacol to pinacolone:

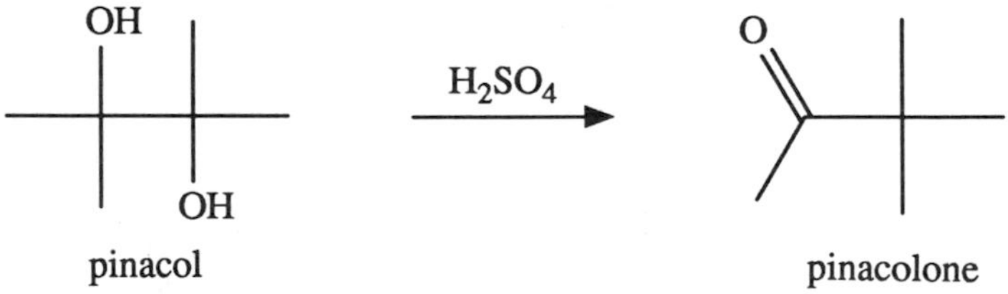

pinacol   pinacolone

Example 4-8. The rearrangement of 1,2-diphenyl-1,2-ethanediol to 2,2-diphenylethanal.

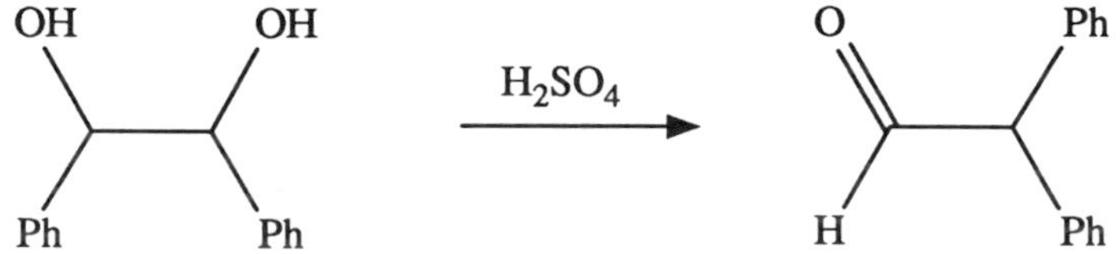

The mechanism of this reaction involves formation of an intermediate carbocation, a 1,2-phenyl shift, and loss of a proton to form the product:

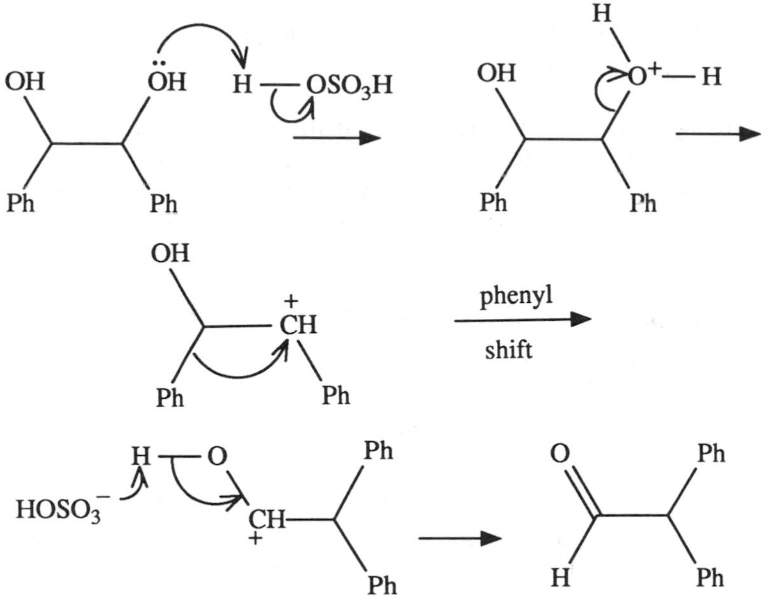

In reactions of this type, if the starting material is not symmetrical, there is the possibility of forming more than one initial carbocation. The more stable carbocation is usually formed. However, the propensity of one group to migrate in preference to another, referred to as migratory aptitude, often depends on reaction conditions.

Problem 4-2. Write a step-by-step mechanism for the transformation of pinacol to pinacolone in the presence of sulfuric acid.

Problem 4-3. Write a step-by-step mechanism for the following reaction.

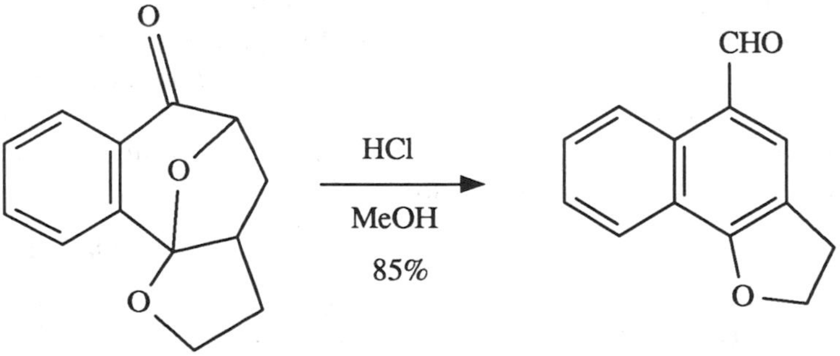

Padwa, A.; Carter, S. P.; Nimmesgern, H.; Stull, P. D. *J. Am. Chem. Soc.* **1988,** *110,* 2894-2900.

4-1-3 CATIONIC REARRANGEMENTS INVOLVING ELECTRON-DEFICIENT NITROGEN

In strong acids such as thionyl chloride or phosphorus pentachloride, an oxime will react to give a rearranged amide. This is known as the Beckmann rearrangement. When the reaction gives products other than amides, these products are referred to as abnormal products. One such abnormal pathway is illustrated in Problem 4-4.

Example 4-9. Beckmann rearrangement of benzophenone oxime.

The overall reaction is as follows:

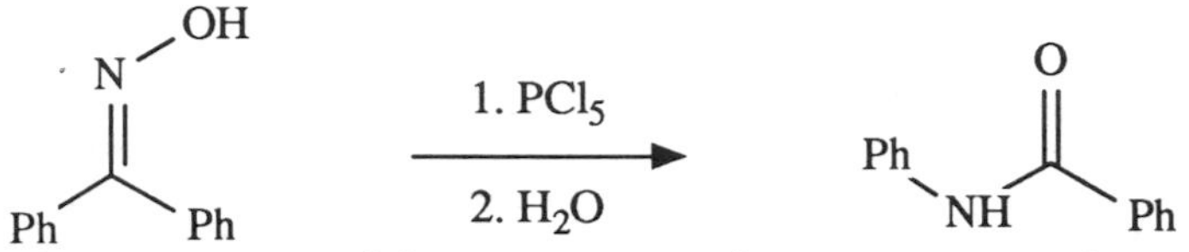

The first two steps of the reaction mechanism convert the original oxime to derivative, **4-10**. This process converts the hydroxyl group of the oxime into a better leaving group.

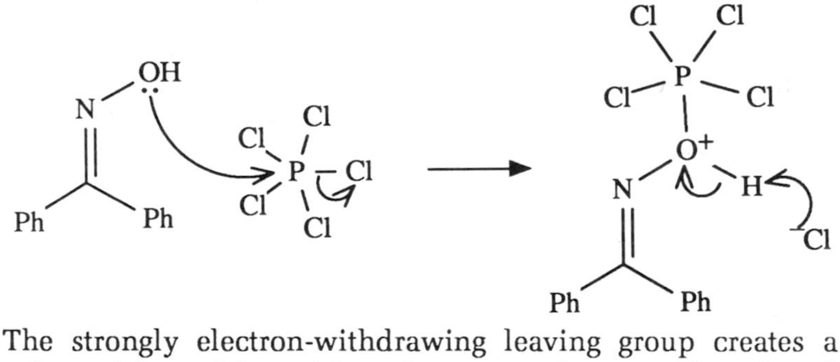

The strongly electron-withdrawing leaving group creates a substantial partial positive charge on nitrogen. The phenyl group, on the side opposite the leaving group, moves with its pair of electrons to the electron deficient nitrogen as the leaving group leaves. Then the resulting ions collapse to form **4-11**. During work-up, **4-11** will undergo hydrolysis and tautomerization to the final product, the amide.

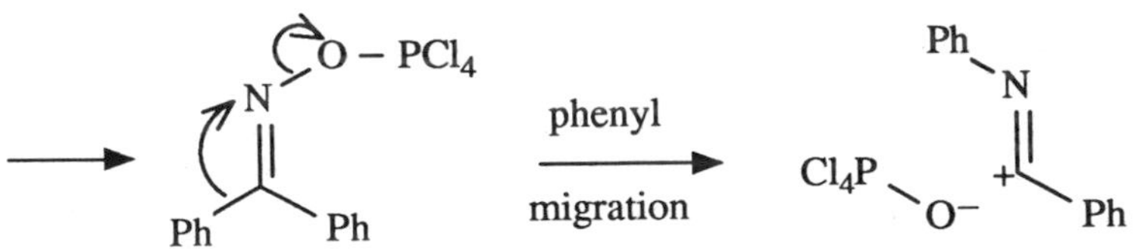

4-10

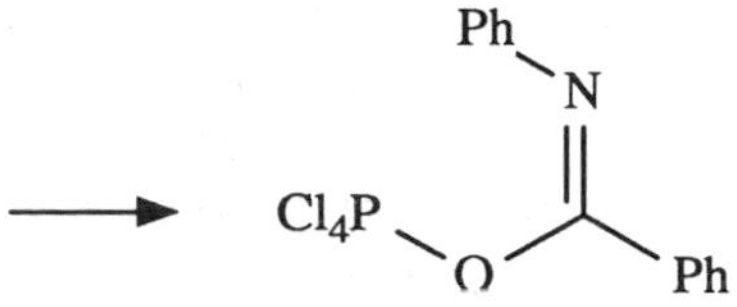

$$\text{Ph}-\text{N}=\underset{\text{Ph}}{\overset{}{\text{C}}}-\text{O}-\text{PCl}_4$$

4-11

Problem 4-4. Write a step-by-step mechanism for the following reaction.

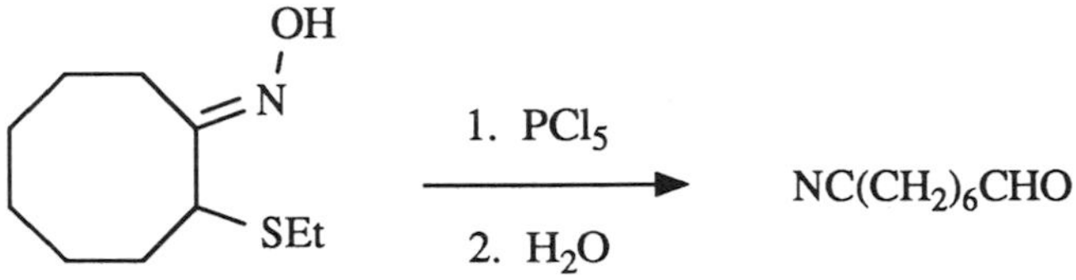

$$\text{(cyclooctane ring with } =N\text{—OH and —SEt substituents)} \quad \xrightarrow[\text{2. H}_2\text{O}]{\text{1. PCl}_5} \quad NC(CH_2)_6CHO$$

Ohno, M.; Naruse, N.; Torimitsu, S.; Teresawa, I. *J. Am. Chem. Soc.* **1966**, *88*, 3168–69.

4-2 Electrophilic Addition

Addition of electrophiles is a typical reaction of aliphatic π bonds. (See Example 4-2.) Such additions involve two major steps: (1) addition of the electrophile to the nucleophilic π bond to give an intermediate carbocation, and (2) reaction of the carbocation with a nucleophile. Typical electrophiles are bromine; chlorine; a proton supplied by HCl, HBr, HI, H_2SO_4, or H_3PO_4; Lewis acids; and carbocations. The nucleophile in step (2) is often the anion associated with the electrophile; e.g., bromide, chloride, iodide, etc., or a nucleophilic solvent like water or acetic acid. Because the more stable of the two possible carbocations is predominantly formed as the intermediate in step (1), electrophilic additions are often regiospecific.

Example 4-10. Regiospecificity in electrophilic additions.

When HI adds to a double bond, the proton acts as an electrophile, giving an intermediate carbocation which then reacts with the nucleophilic iodide ion to give the product. In the reaction of HI with 1-methylcyclohexene, there is only one product, 1-iodo-1-methylcyclohexane; no 1-iodo-2-methyl-cyclohexane is formed.

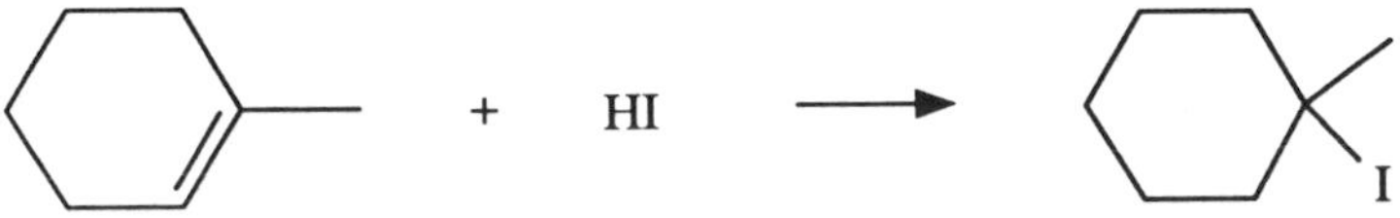

$$\text{(1-methylcyclohexene)} \quad + \quad HI \quad \longrightarrow \quad \text{(1-iodo-1-methylcyclohexane)}$$

This reaction is said to be regiospecific, because although the iodide might occupy the ring position at either end of the original double bond, only one of these products is actually formed. The rationale for this regiospecificity is that the proton adds to form the more stable 3° carbocation, **4-12**, and not the 2° carbocation, **4-13**.

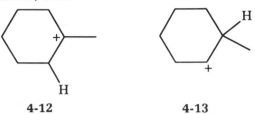

           4-12   **4-13**

In some cases an unusual 3-membered ring intermediate is formed. When this intermediate is stable under the reaction conditions, an *anti* addition of electrophile and nucleophile takes place. For example, the bromination of *cis-* and *trans-*2-butene occurs stereospecifically with each isomer. From *cis-*2-butene, the products are **4-16** and **4-17**, enantiomers which are formed in equal amounts. This results from electrophilic addition of bromine to the top and bottom faces of the alkene to give intermediate bromonium ions, **4-14** and **4-15** in equal amounts. These then react with the nucleophile, at either carbon of the bromonium ion, on the side opposite the bromine, to give the product dibromides.

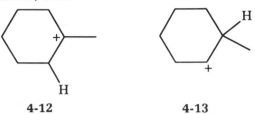

                        4-14   **4-15**

          4-14   **4-16**

4-14   **4-17**

Note that **4-16** and **4-17** are mirror images and nonsuperimposable. In a like manner the reaction of **4-15** with bromide ion also gives **4-16** and **4-17**.

From *trans*-2-butene, *anti* reaction of bromide at either carbon of the bromonium ion **4-18**, gives only **4-19**, which is a *meso* compound; it has a mirror plane (in the totally eclipsed conformation).

4-18   **4-19**

In the addition of bromine to *cis*- and *trans*-2-butene, the stereochemistry of each product reflects the fact that the two bromines were introduced into the molecule on opposite faces of the original double bond.

Under some experimental conditions, electrophilic additions of either Cl^+ or a proton may form stable three-membered ring intermediates. Thus, often when a double bond undergoes stereospecific *anti* addition, formation of a three-membered ring intermediate, analogous to the bromonium ion, is part of the mechanism.

Sometimes *syn* addition to a double bond may occur. Such reactions usually occur in very nonpolar media. For example, the chlorination of indene in carbon tetrachloride gives *cis*-1,2-dichloroindane.

4-20

The mechanism for this reaction probably involves reaction of the nucleophilic alkene with the electrophilic chlorine to form an ion pair intermediate, **4-20,** composed of a positive organic ion and a negative chloride ion. In the nonpolar solvent, the attraction between the positive center and the chloride ion is so strong that the chloride ion cannot move to the other side of the double bond. The ion pair forms because, in nonpolar solvents, the carbocation and the chloride ion are not as strongly stabilized by solvation as they are in more polar solvents.

There are also substrates and reaction conditions which lead to products from both *syn* and *anti* additions. Such additions may be the result of equilibration of relatively stable intermediate ions or may involve a mechanism in which two molecules of the species supplying the electrophile are involved in the reaction.

Electrophilic additions in nucleophilic solvents often give a mixture of products, because the nucleophile derived from the electrophilic reagent and the solvent compete for the intermediate carbocation. For example, the bromination of styrene in acetic acid gives both dibromo and bromoacetoxy products:

Syn and *anti* isomers of both products are formed, so the reaction is not stereospecific. (The squiggly lines in the products mean that any stereochemistry is possible.) On the other hand, the formation of only one of the possible regioisomers of the bromoacetoxy compound, **4-21** but no **4-22,** means that the reaction is 100% regiospecific. This is a reflection of the fact that the carbocation at the phenyl end of the molecule is considerably more stable than the 1° carbocation that would be formed at the other end.

Problem 4-5. Write step-by-step mechanisms for the following transformations.

a.

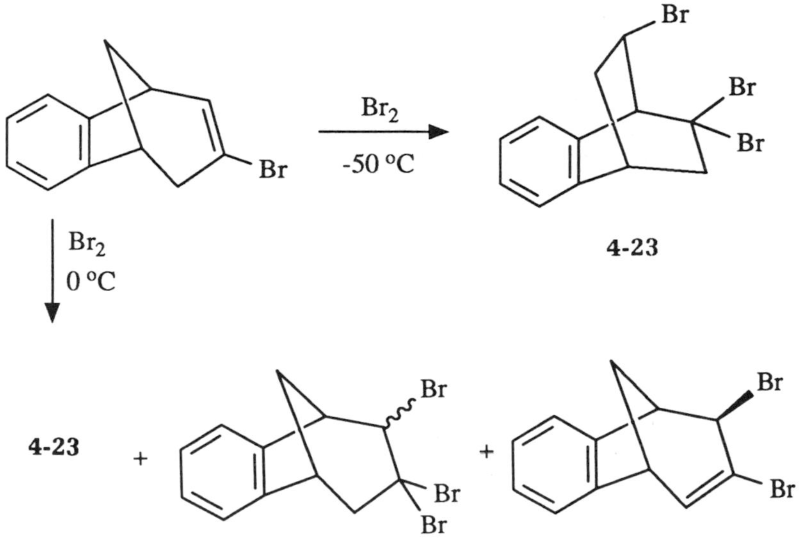

4-23

4-23 +

4-24   **4-25**

At −50°C, **4-23** is the only product; at 0°C, **4-23** is still the major product, but several other products are also produced. The squiggly line to the bromine at the 2-position means that both the *exo* and *endo* isomers are produced.

Harmandar, M.; Balci, M. *Tetrahedron Lett.*, **1985**, *26*, 5465-5468.

b.

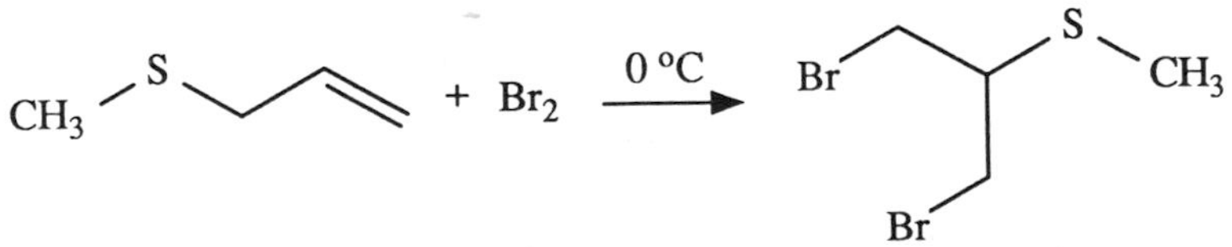

Bland, J. M.; Stammer, C. H. *J. Org. Chem.* **1983**, *48*, 4393-4394.

4-3 Acid-Catalyzed Reactions of Carbonyl Compounds

Several examples of the importance of acid catalysis have already been given. Example 2-5 gives one of the steps in the acid-catalyzed formation of an ester, and Example 3-13 shows the acid-catalyzed mechanism for the formation of a hydrazone.

4-3-1 HYDROLYSIS OF DERIVATIVES OF CARBOXYLIC ACIDS

Acidic hydrolysis of all derivatives of carboxylic acids gives the corresponding carboxylic acid as the product. These hydrolyses can be broken down into the following steps.

(1) Protonation of the oxygen of the carbonyl group. This enhances the electrophilicity of the carbonyl carbon, increasing its reactivity with nucleophiles.

(2) The oxygen of water acts as a nucleophile and adds to the carbonyl carbon.

(3) The oxygen of the water, which has added and which is positively charged, loses a proton.

(4) A leaving group leaves. In the case of acid halides, the leaving group leaves directly; in the case of esters or amides, the leaving group leaves after prior protonation.

(5) A proton is lost from the protonated carboxylic acid.

Example 4-11. Hydrolysis of an amide.

The overall reaction is as follows:

$$CH_3CONH_2 \ + \ H_3O^+ \ \longrightarrow \ CH_3CO_2H \ + \ ^+NH_4$$

The mechanism of this reaction follows:

(1) The initial protonation of the carbonyl oxygen gives a cation, a resonance hybrid with positive character on carbon
and nitrogen as well as oxygen.

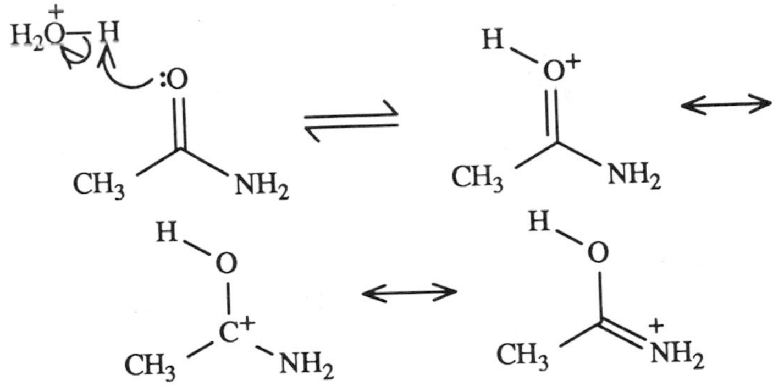

(2) The electrophilic cation reacts with the nucleophilic oxygen
of water; (3) loss of a proton gives **4-26**:

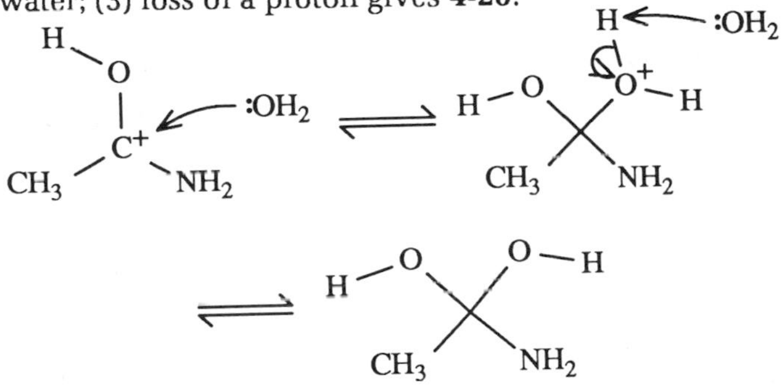

4-26

(4) The neutral intermediate, **4-26**, can be protonated on either
oxygen or nitrogen, but only protonation on nitrogen leads
to product formation. Notice that the NH$_2$ group is now an
amine and is much more basic than the NH$_2$ group of the
starting amide.

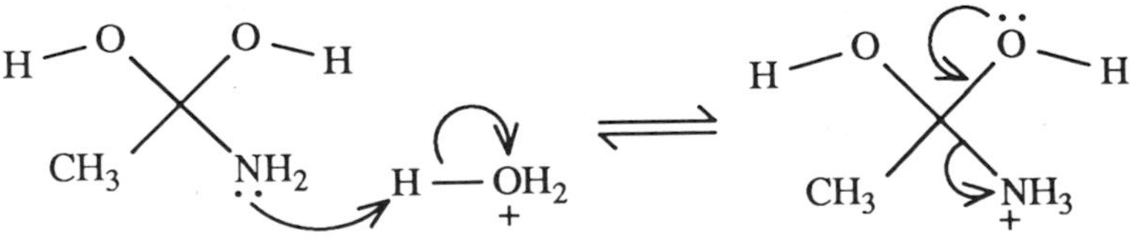

Ammonia is the leaving group; $^-NH_2$ would be a very poor leaving group. While the loss of ammonia is potentially a reversible process, the protonation of ammonia to give the ammonium ion occurs much more rapidly in the acidic medium. Thus, the loss of ammonia is irreversible, not because the addition of ammonia in the reverse process is not energetically favorable, but because there is no ammonia present. (5) Finally, a proton is removed from the protonated carboxylic acid to give the carboxylic acid product.

Problem 4-6. Write step-by-step mechanisms for the following transformations:

Hauser, F. M.; Hewawasam, P.; Baghdanov, V. M. *J. Org. Chem.* **1988**, *53*, 223-224.

Serafin, B.; Konopski, L. *Pol. J. Chem.* **1978**, *52*, 51-62.

4-3-2 OTHER ACID-CATALYZED REACTIONS

The formation and hydrolysis of acetals and orthoesters are acid-catalyzed processes.

Example 4-12. Hydrolysis of ethyl orthoformate to ethyl formate.

An example of an orthoester is the starting material in the following reaction.

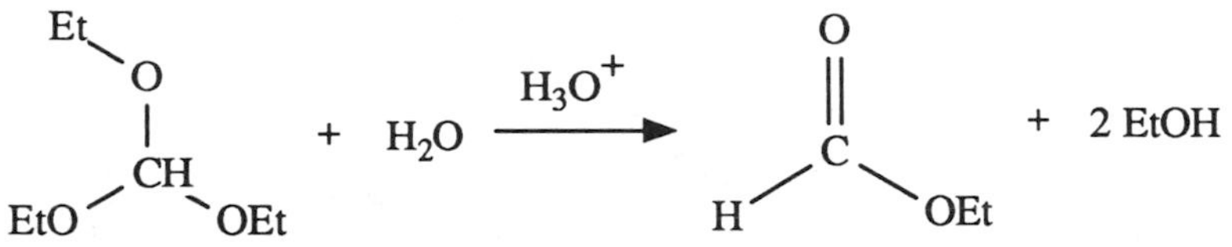

$$
\begin{array}{ccc}
\underset{\text{EtO}}{\overset{\text{Et}}{\diagdown}}\!\!\overset{\text{O}}{\underset{\text{CH}}{|}}\!\!\overset{}{\diagup}\!\text{OEt} & + \ H_2O \ \xrightarrow{\ H_3O^+\ } & \underset{H}{\overset{O}{\diagdown}}\!\!\overset{\|}{C}\!\!\overset{}{\diagup}\!\text{OEt} \ + \ 2\,\text{EtOH}
\end{array}
$$

As in the case of other acid-catalyzed hydrolyses, the first step involves protonation of the most basic atom in the molecule.

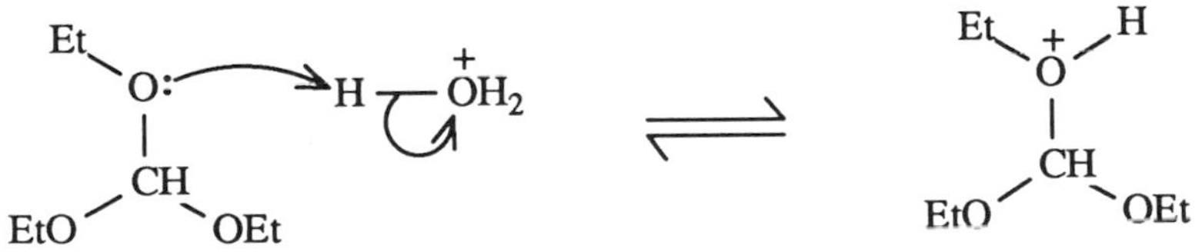

$$
\underset{\text{EtO}}{\overset{\text{Et}}{\diagdown}}\!\!\overset{\text{O:}}{\underset{\text{CH}}{|}}\!\!\overset{}{\diagup}\!\text{OEt} \longrightarrow H \!-\! \overset{+}{O}H_2 \ \ \rightleftharpoons \ \ \underset{\text{EtO}}{\overset{\text{Et}}{\diagdown}}\!\!\overset{\overset{+}{O}\diagdown H}{\underset{\text{CH}}{|}}\!\!\overset{}{\diagup}\!\text{OEt}
$$

This has the function of creating a better leaving group; that is, ethanol will be the leaving group rather than ethoxide ion.

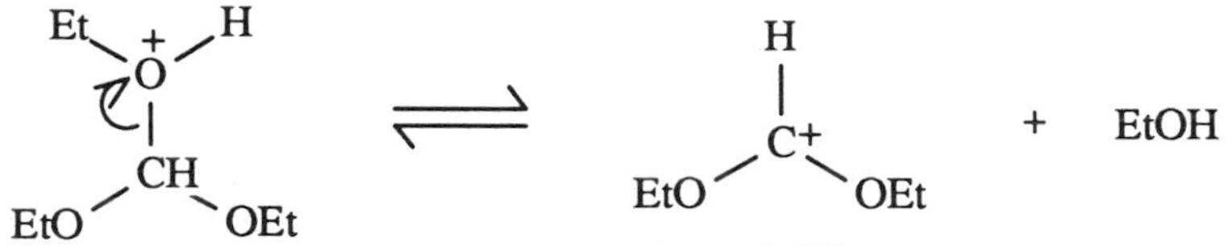

$$
\underset{\text{EtO}}{\overset{\text{Et}}{\diagdown}}\!\!\overset{\overset{+}{O}\diagdown H}{\underset{\text{CH}}{|}}\!\!\overset{}{\diagup}\!\text{OEt} \ \ \rightleftharpoons \ \ \underset{\text{EtO}}{\overset{H}{\diagdown}}\!\!\overset{|}{\underset{}{\overset{+}{C}}}\!\!\overset{}{\diagup}\!\text{OEt} \ + \ \text{EtOH}
$$

4-27

The electrophilic carbocation, **4-27**, reacts with nucleophilic water. This is like the reverse of the above reaction, except that water is the nucleophile rather than ethanol. Since water is present in large excess over ethanol, this reaction occurs preferentially and shifts the equilibrium toward the hydrolysis product. The protonated intermediate loses a proton to give **4-28**.

4-28

The neutral intermediate, **4-28**, can be protonated on either a hydroxyl or ethoxy oxygen. Protonation on the hydroxyl oxygen is simply the reverse of the deprotonation step. That is, of course, a reasonable step although it goes toward starting material. However, protonation on the oxygen of the ethoxy group will lead toward product.

Loss of ethanol, followed by removal of a proton by water, gives the product ester.

Consideration of the overall reaction, given in the first equation of this example, aids in understanding why the hydrolysis conditions yield the ester and do not favor starting material. Water is present on the left hand side of the equation and ethanol on the right hand side. Thus, an excess of water would shift the equilibrium to the right and an excess of ethanol would shift the equilibrium to the left. In fact, in order to get the reaction to go to the left, water must be removed as it is produced. Depending on the reaction conditions, the hydrolysis may proceed further to the corresponding carboxylic acid.

The reverse of the reactions depicted in the answer to Problem 3-14a is an example of the acidic hydrolysis of an enamine, a reaction which is very similar to the hydrolysis depicted in Example 4-12.
Electrophilic addition to α-β-unsaturated carbonyl compounds is analogous to electrophilic addition to unconjugated double bonds, except that the electrophile adds to the carbonyl oxygen, the most basic atom in the molecule. After that, the nucleophile adds to the β-carbon, and the resulting intermediate enol tautomerizes to the more stable carbonyl compound. These reactions may also be considered as the electrophilic counterpart of the nucleophilic Michael and 1,4-addition reactions discussed in Section 3-3-4, pp. 137–148.

Example 4-13. Electrophilic addition of HCl to acrolein.

The overall reaction is as follows:

The first step in a mechanism is reaction of the nucleophilic oxygen of the carbonyl group with the positive end of the HCl molecule.

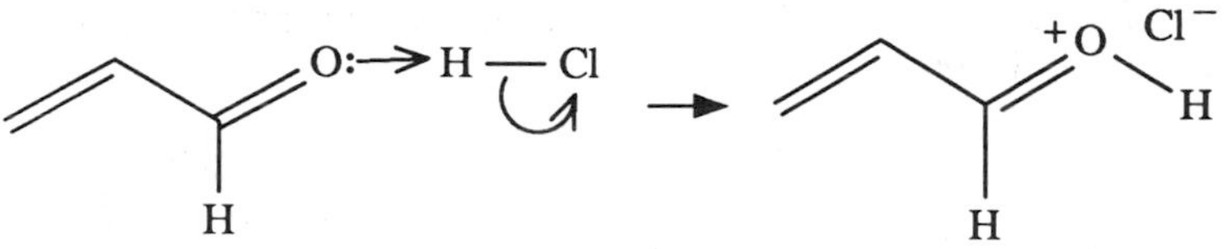

The resulting cation is a resonance hybrid with a partial positive charge on carbon as well as on oxygen:

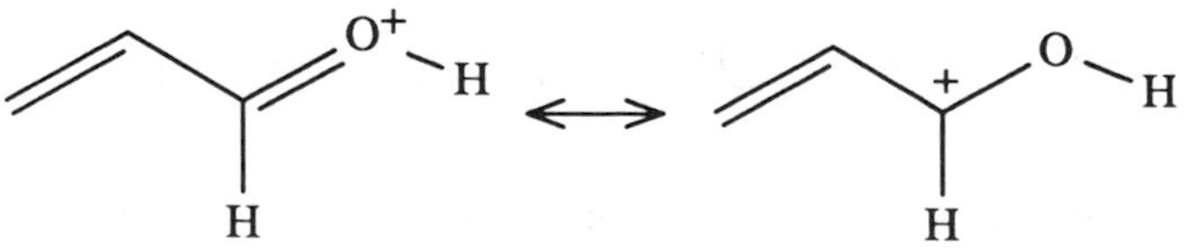

The electrophilic β-position now reacts with the nucleophilic chloride ion, giving an enol which then tautomerizes to the keto form.

This reaction is the acid-catalyzed counterpart of a 1,4-addition reaction to an α,β-unsaturated carbonyl compound. Chloride ion, without an acid present, will not add to acrolein. That is, chloride ion is not a strong enough nucleophile to drive the reaction to the right. However, if the carbonyl is protonated, the intermediate cation is a stronger electrophile and will react with chloride ion.

Problem 4-7. Rationalize the particular regiochemistry of the protonation shown in Example 4-13 in comparison to protonation at other sites in the molecule.

Problem 4-8. Write a mechanism for the following tautomerization in the presence of anhydrous HCl.

Problem 4-9. Write a step-by-step mechanism for the following transformation.

4-4 Electrophilic Aromatic Substitution

The interaction of certain electrophiles with an aromatic ring leads to substitution. Such electrophilic reactions involve a carbocation intermediate which gives up a stable positively-charged species, usually a proton, to a base, to give back an aromatic compound. Typical electrophiles include chlorine and bromine, activated by interaction with a Lewis acid for all but highly reactive aromatic compounds; nitronium ion; SO$_3$; the complexes of acid halides and anhydrides with Lewis acids (See Example 4-4) or the cations formed when such complexes decompose: RC=O or ArC=O; and carbocations.

Example 4-14. Electrophilic substitution of toluene by sulfur trioxide.

In this reaction the aromatic ring is a nucleophile and the sulfur of sulfur trioxide is an electrophile.

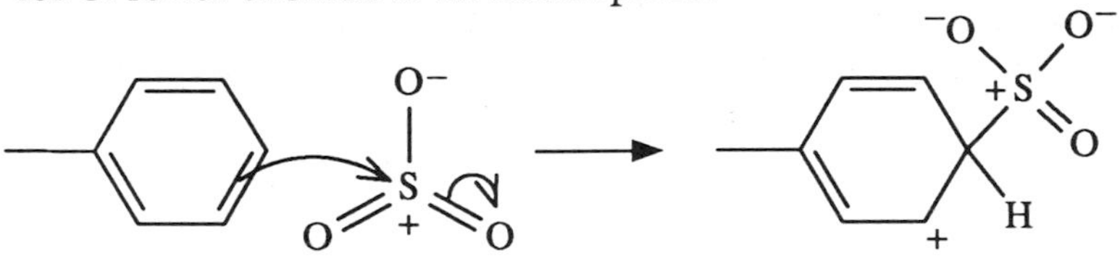

The positive charge in the ring is stabilized by resonance.

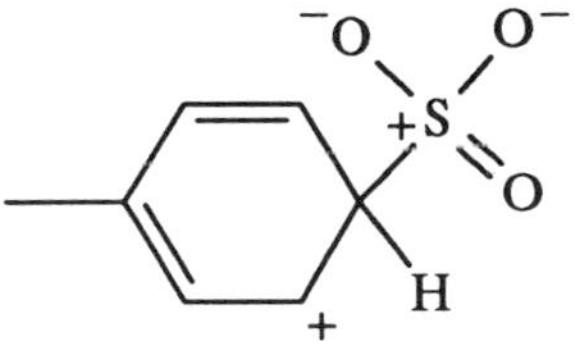

4-29-1 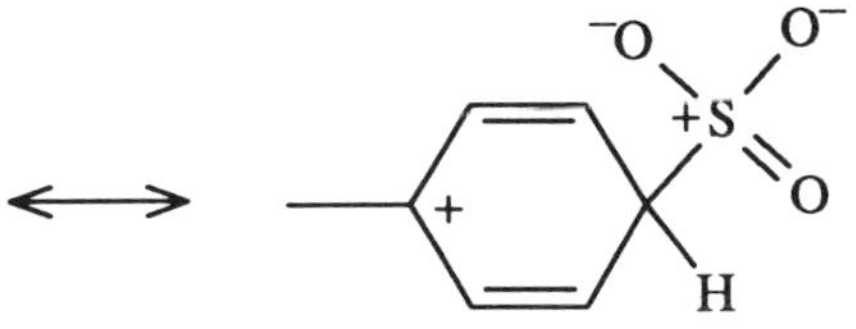 **4-29-2**

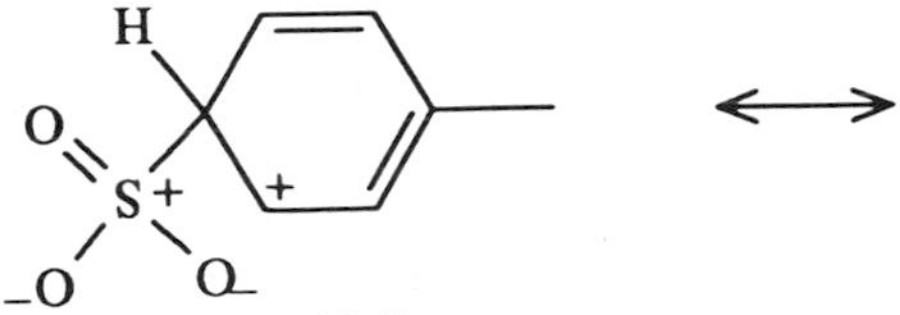

4-29-3

A sulfonate anion, acting as a base, can remove a proton from the intermediate to give the product:

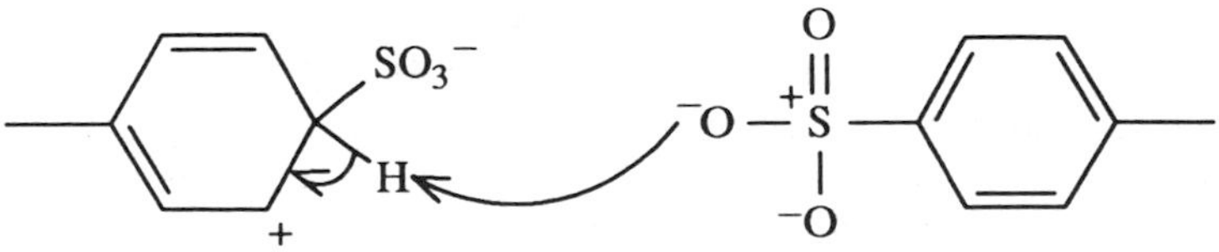

$$\longrightarrow \quad -\!\!\!\bigcirc\!\!\!- \; \overset{\overset{O^-}{|}}{\underset{\underset{O}{\|}}{S^+\!-\!O^-}} \quad + \quad HO_3S - Ar$$

Because the aromatic ring acts as a nucleophile, the reaction rate will be enhanced by electron-donating, and slowed by electron-withdrawing substituents. Furthermore, the intermediate cation is especially stabilized by an adjacent electron-donating group as in resonance structure, **4-29-2**. Because of this, electrophiles react at positions *ortho-* or *para-* to electron-donating groups. Such groups are said to be *ortho* and *para* directing substituents. Conversely, because electron-withdrawing groups destabilize a directly adjacent positive charge, the electrophile will react at the *meta* position in order to avoid this destabilization. A review of directing and activating/deactivating effects of various substituents is given in Table 4-1.

TABLE 4-1

Influence of Substituents in Electrophilic Aromatic Substitution

---

Strongly Activating and *ortho, para*-Directing

  —NR$_2$, —NRH , —NH$_2$, —O$^-$, —OH

Moderately Activating and *ortho, para*-Directing

  —OR , —NHCOR

Weakly Activating and *ortho, para*-Directing

  —R , —Ph ,

Weakly Deactivating and *ortho, para*-Directing

  —F   —Cl , —Br ,

Strongly Deactivating and *meta*-Directing

  —$\overset{+}{S}R_2$, —$\overset{+}{N}R_3$, —NO$_2$, —SO$_3$H , —CO$_2$H , —CO$_2$R ,

  —CHO , —COR , —CONH$_2$, —CONHR , —CONR$_2$, —CN

The effect of fluorine, chlorine or bromine as a substituent is unique: the ring is deactivated but the entering electrophile is directed to the *ortho* and *para* positions. This can be explained by an unusual competition between resonance and inductive effects. In the starting material, halogen-substituted benzenes are deactivated more strongly by the inductive effect than they are activated by the resonance effect. However, in the intermediate carbocation, halogens stabilize the positive charge by resonance more than they destabilize it by the inductive effect.

> *Problem 4-10. For the following reactions, explain the orientations in the product by drawing resonance forms for possible intermediate carbocations and rationalize their relative stabilities.*
>
> *a. The reaction in Example 4-14*

b.

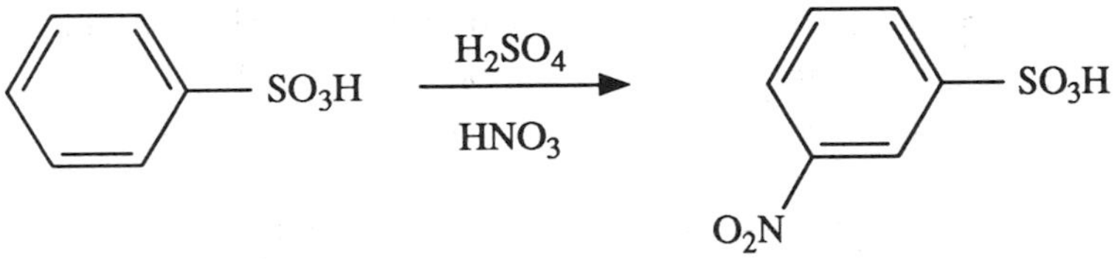

c.

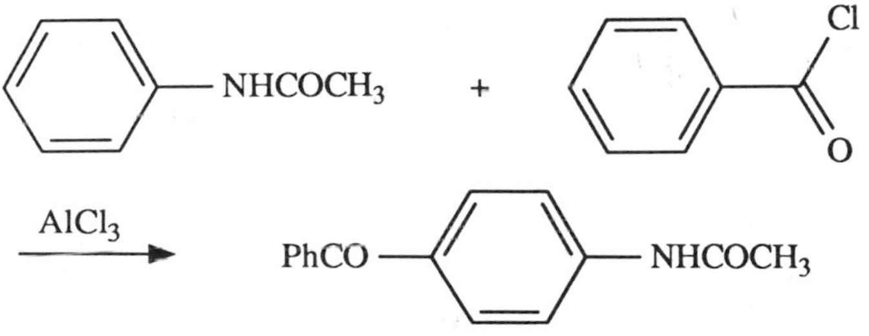

Example 4-15. A metal-catalyzed, intramolecular, electrophilic aromatic substitution.

Write a mechanism for the following transformation.

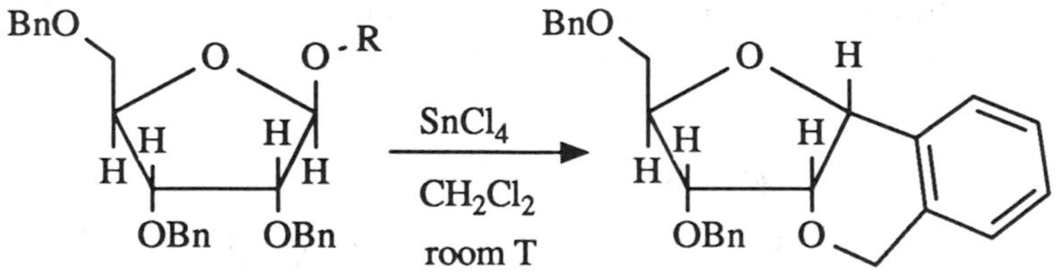

Bn = benzyl

Tin (IV) chloride can undergo nucleophilic substitution which

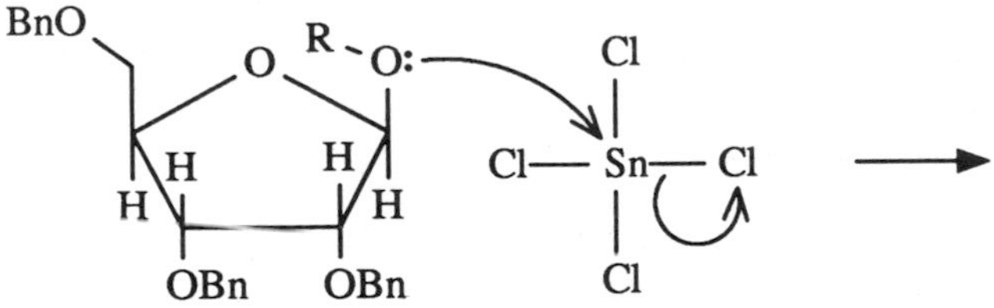

converts the acetal OR group into a better leaving group. (The positively charged oxygen of the OR group of an acetal is a much better leaving group than the oxygen of an ether, because the resulting cation is stabilized by resonance interaction with the remaining oxygen of the original acetal.) After the leaving group leaves, the aromatic ring of a benzyl group at position 2 in **4-30** (suitably situated geometrically for this interaction) acts as a nucleophile toward the positive center. The resulting carbocation then loses a proton to give the product:

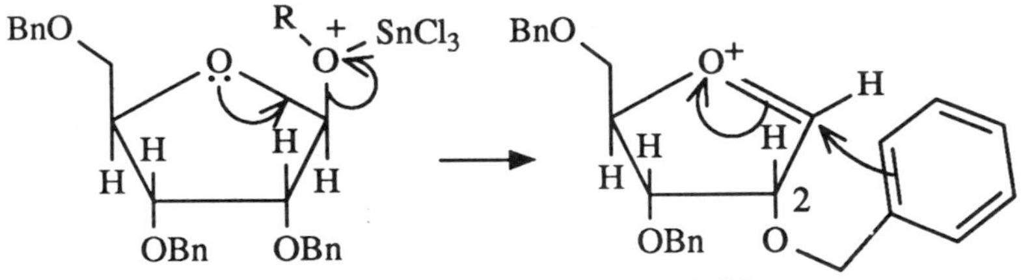

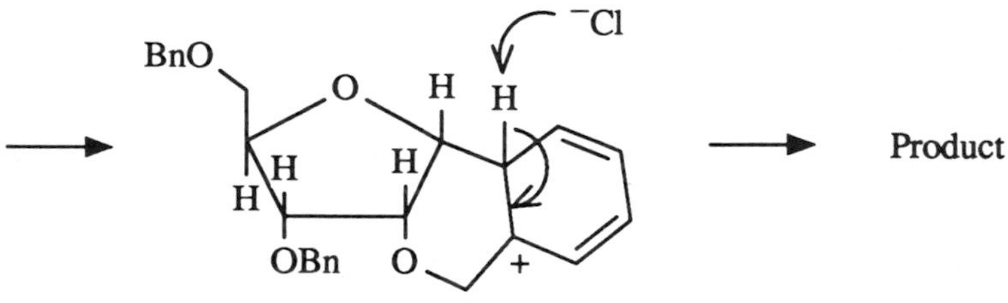

Such intramolecular electrophilic aromatic substitution reactions are common, especially when 5- or 6-membered rings are formed.

Martin, O. R. *Tetrahedron Lett.* **1985**, *26*, 2055–2058.

Problem 4-11. Write step-by-step mechanisms for the following transformations:

a.

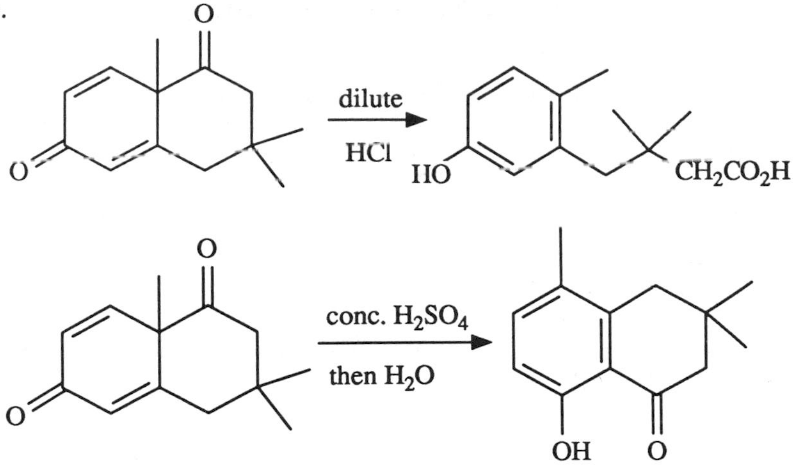

The reaction in dilute HCl was discussed in another context in Example 2-9, p. 86. Modify the mechanism given there, to account for the result shown here for concentrated sulfuric acid.

Waring, A. J.; Zaidi, J. H. *J. Chem. Soc. Perkin Trans I* **1985**, 631–639.

b.

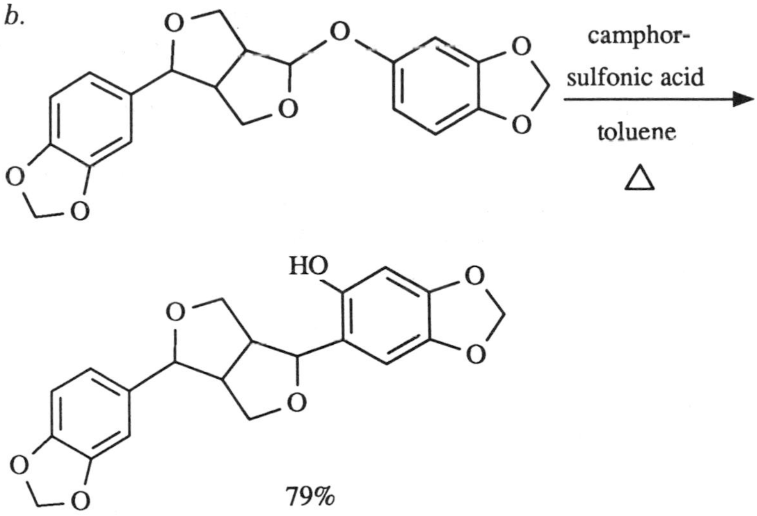

79%

Fukuda, Y.; Isobe, M.; Nagata, M.; Osawa, T.; Namiki, M. *Heterocycles* **1986**, *24*, 923-926.

4-5 Carbenes, Nitrenes and Nitrenium Ions

Carbenes, nitrenes and nitrenium ions are very reactive intermediates. Some have been isolated in matrices at low temperature, but with few exceptions (see *Chem. Eng. News*, January 28, 1991, p. 19–20), they are very short-lived at ambient temperatures. Although carbenes and nitrenes are often generated in basic media, they usually act as electrophiles.

A carbene is a neutral divalent carbon, containing two electrons which are not shared with another atom. When these electrons are in the same orbital, the carbene is called a singlet; when they are in different orbitals, the carbene is said to be a triplet.

singlet carbene   triplet carbene

Note that calculation of the formal charge on either of these carbons gives zero; the singlet carbene **is not** a carbanion. The chemical reactivity of some carbenes indicates that they are coordinated with a metal salt. Since they are somewhat different in structure, yet exhibit similar reactivity to carbenes, they are called carbenoids. See **4-31** below.

A nitrene is a neutral univalent nitrogen, which contains two lone pairs of electrons not shared with other atoms. A nitrenium ion is a positively-charged divalent nitrogen with one lone pair of electrons; thus it is isoelectronic with a carbene.

phenylnitrene   nitrenium ion

Example 4-16. Generation of carbenes from alkyl halides and base.

(1) Dichlorocarbene can be formed from aqueous KOH and chloroform:

$$HO^- \quad H - CCl_3 \longrightarrow \underset{Cl}{\overset{Cl}{\underset{}{\overset{\_}{C}}}}Cl \longrightarrow Cl \overset{..}{\frown} Cl + Cl^-$$

The base removes the acidic proton from chloroform. The resulting anion then loses chloride ion, giving a divalent carbon with two unshared electrons. In this case, the unshared electrons are paired (occupy the same orbital); i.e., dichlorocarbene is a singlet. The carbon of the carbene has no charge.

A carbenoid is formed when potassium t-butoxide reacts with benzal bromide in benzene. The reactivity of the carbenoid, **4-31**, is similar to that of phenylbromocarbene. The exact nature of the interaction between the carbenoid carbon and the metal halide, in this case potassium bromide, is not known.

4-31

(2) Some alkyl halides react with alkyl lithiums, under aprotic conditions, to give carbenoids by a process called halogen-metal exchange.

4-32

In some cases a proton may be removed from carbon by the alkyl lithium instead of the exchange reaction. For details see Kobrich, G. *Angew. Chem. Int. Ed. Eng.* **1972**, *11*, 473–485.

Example 4-17. Generation of ICH₂ZnI from methylene iodide and zinc-copper couple.

When diiodomethane is treated with zinc-copper couple, a carbenoid is formed.

$$CH_2I_2 \quad + \quad Zn\text{-}Cu \quad \longrightarrow \quad ICH_2ZnI$$

4-33

This reaction is called the Simmons-Smith reaction; **4-33**, other carbenoids, and carbenes are very useful in the synthesis of cyclopropanes. (See Example 4-19.)

Example 4-18. Generation of carbenes from diazo compounds.

Loss of nitrogen from a diazo compound can be effected by heat, light or a copper catalyst. This gives either a carbene (heat or light) or carbenoid (copper).

Electrophilic addition of carbenes to carbon-carbon double and triple bonds has been extremely useful synthetically. In many cases, the reaction goes with 100% stereospecificity, so that the stereochemistry about a double bond in the starting material is maintained in the product.

Example 4-19. Addition of singlet dichlorocarbene to *cis*-2-butene and *trans*-2-butene.

The addition of dichlorocarbene to the double bond of *cis*-2-butene goes with 100% stereospecificity; that is, the only product is *cis*-1,2-dimethyl-3,3-dichlorocyclopropane. The addition of dichlorocarbene to *trans*-2-butene gives only the corresponding *trans* isomer. The stereospecificity of the reaction has been interpreted to mean that dichlorocarbene is a singlet and that both ends of the double bond react simultaneously, or nearly so, with the carbene.

In cases where addition is not 100% stereoselective, it is rationalized that the reaction proceeds through a triplet carbene, or diradical. Under these circumstances, it is not anticipated that the formation of the two new carbon-carbon bonds would be in concert.

Example 4-20. A nonstereospecific addition of a carbene.

The mechanism for this reaction would be written as stepwise addition to the double bond by the diradical carbene.

Note that the arrows used to show flow of radicals have only a half head. Also the intermediate in the reaction is a diradical. Rotation about the highlighted single bond takes place at such a rate that the stereochemistry of the starting olefin is lost.

Other nucleophiles, with which carbenes react, include hydroxide, thiolate and phenoxide. The reaction with phenoxide is the classical Reimer-Tiemann reaction.

Example 4-21. The Reimer-Tiemann reaction.

Phenols react with chloroform in the presence of hydroxide ion in water to give o- and p-hydroxybenzaldehydes. The first steps of the reaction are: (1) the formation of dichlorocarbene, as shown in Example 4-16; (2) nucleophilic reaction of the phenoxide with the electrophilic carbene; and (3) hydrolysis.

Problem 4-12. Complete the mechanism of Example 4-21, showing the steps leading from the intermediate carbanion to the product o-salicylaldehyde.

Problem 4-13. Write a step-by-step mechanism for the following transformation:

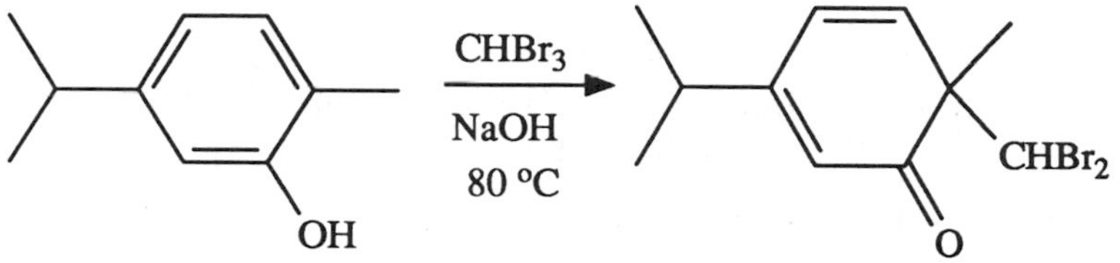

Wenkert, E.; Arrhenius, T. S.; Bookser, B.; Guo, M.; Mancini, P. *J. Org. Chem.* **1990**, *55*, 1185–1193.

Another reaction of carbenes is insertion in which the carbene inserts itself between two atoms. Insertions have been observed into C-H, C-C, C-X, N-H, O-H, S-S, S-H, and M-C bonds, among others. The mechanism of the process is often concerted. A three-center transition state is usually written for the concerted mechanism:

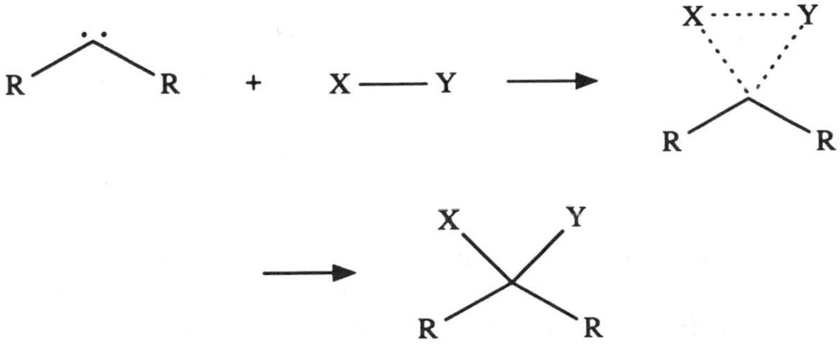

Example 4-22. A synthetically useful insertion reaction.

The insertion reaction of **4-32**, whose generation was shown above, into the C1-C2 bond, led to a 70% yield of cyclo-heptanone. Other homologues gave even higher yields of ring-expanded products.

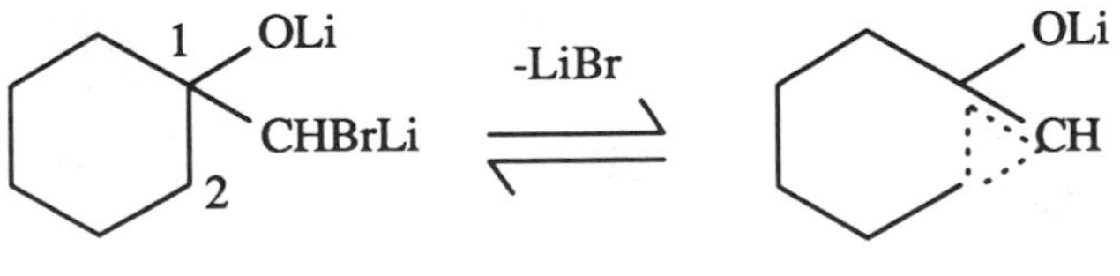

4-32

4-34

The intermediate enolate, **4-34**, forms the corresponding ketone in acid. From Taguchi, H.; Yamamoto, H.; Nozaki, H. *J. Am. Chem. Soc.* **1974**, *96*, 6510–6511.

Example 4-23. Generation of nitrenes.

Common methods for generating nitrene intermediates are the photolysis or thermolysis of azides. Generation of nitrenes from acyl azides can only be effected photochemically; thermolysis of an acyl azide gives the corresponding isocyanate.

The reactivity of nitrenes is similar to that of carbenes. They readily add to double bonds to give the corresponding three-membered heterocycles, aziridines. Insertion reactions are also common.

Example 4-24. Generation and reaction of a nitrenium ion.

Silver ion-assisted loss of chloride ion from the starting material gives a nitrenium ion which acts as an electrophile toward the aromatic ring to give **4-35**.

4-35

Kawase, M.; Kitamura, T.; Kikugawa, Y. *J. Org. Chem.* **1989**, *54*, 3394–3403.

Problem 4-14. Write step-by-step mechanisms for the following transformations:

Capozzi, G.; Chimirri, A.; Grasso, S.; Romeo, G. *Heterocycles* **1984**, *22*, 1759-1762.

b.

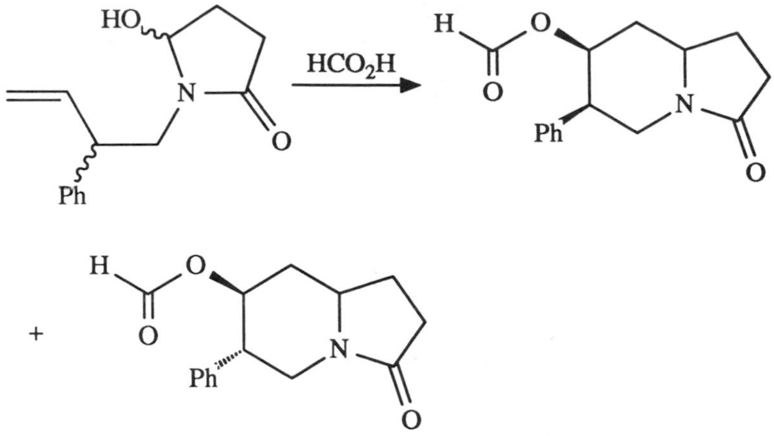

Ent, H.; de Koning, H.; Speckamp, W. N. *J. Org. Chem.* **1986**, *51*, 1687-1691.

c.

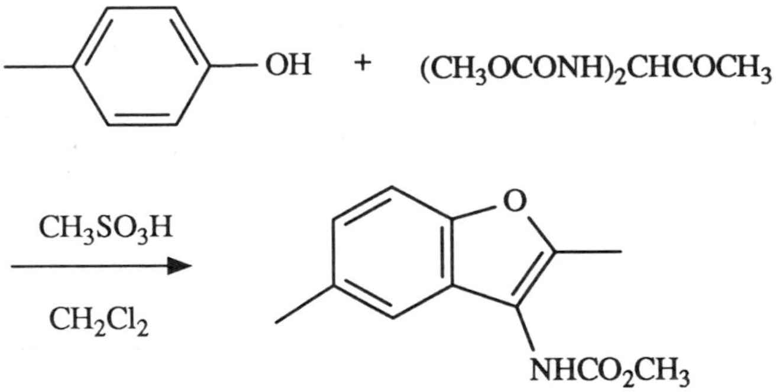

$$\text{—} \overset{}{\underset{}{\bigcirc}} \text{—OH} \quad + \quad (CH_3OCONH)_2CHCOCH_3$$

$$\xrightarrow[\text{CH}_2\text{Cl}_2]{\text{CH}_3\text{SO}_3\text{H}}$$

Ben-Ishai, D.; Denenmark, D.; *Heterocycles* **1985**, *23*, 1353–1356.

d.

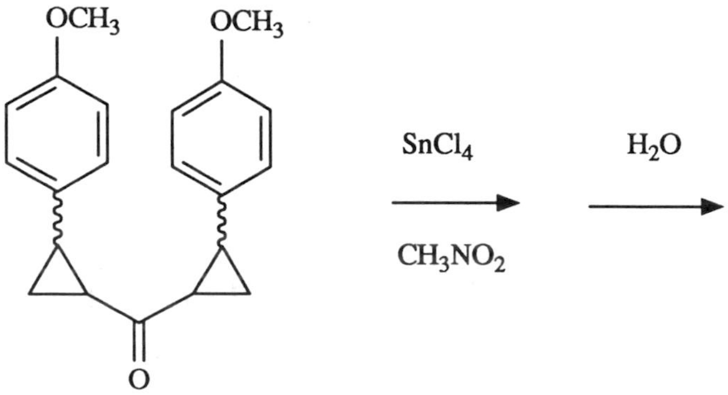

$$\xrightarrow[\text{CH}_3\text{NO}_2]{\text{SnCl}_4} \qquad \xrightarrow{\text{H}_2\text{O}}$$

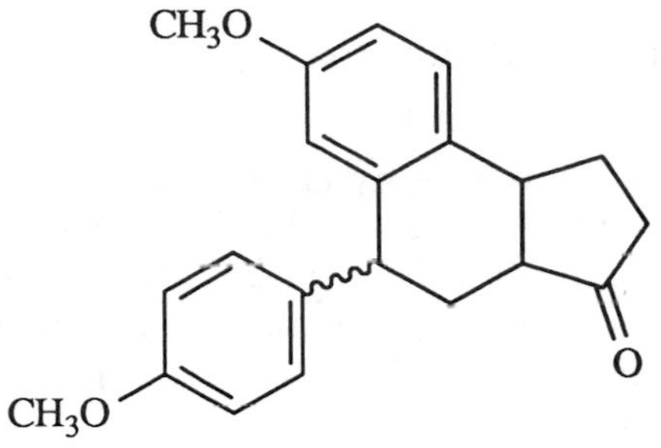

Hantawong, K.; Murphy, W. S.; Boyd, D. R.; Ferguson, G.; Parvex, M. *J. Chem. Soc. Perkin Trans. II* **1985**, 1577-1582.

e.

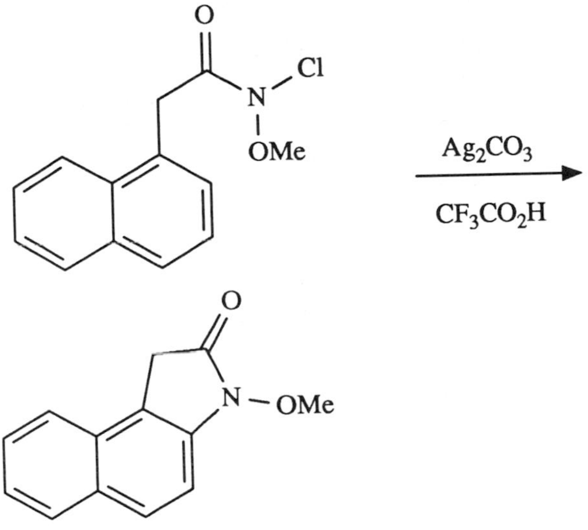

Kikugawa, Y.; Kawase, M. *J. Am. Chem. Soc.* **1984**, *106*, 5728-5729.

f.

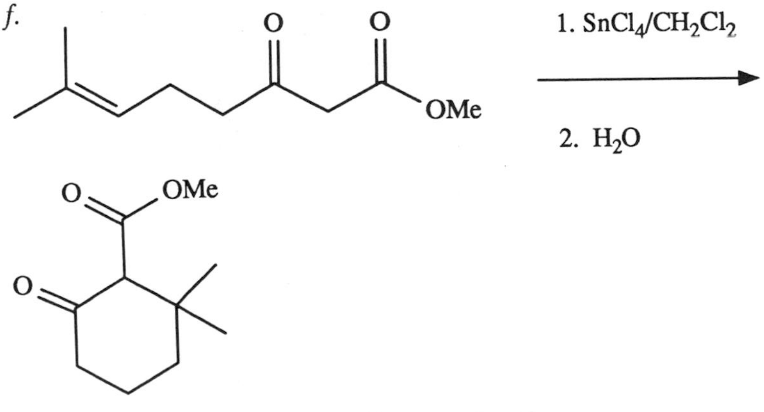

White, J. D.; Skeean, R. W.; Trammell, G. L. *J. Org. Chem.* **1985**, *50*, 1939-1948.

Answers to Problems

4-1.a. This is an ionization, followed by an alkyl shift and nucleophilic reaction of solvent with the electrophilic
carbocation intermediate. In other words, it is an example
of an S_N1 reaction with rearrangement. As noted in Table
3-1, the tosylate anion is an excellent leaving group.

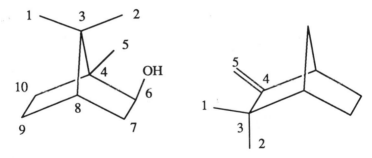

4-1.b. Since the hydroxyl group is lost and an alkene is formed, the mechanism appears to be a carbocation rearrangement, followed by loss of a proton. Numbering the carbons in starting material and product is helpful. Numbering the first five carbons in the product is straightforward because of the methyl carbons.

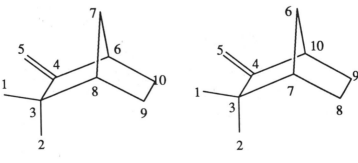

Minimizing bond reorganization allows only two ways to number the remaining carbons of the product:

                    4-36   **4-37**

A major difference between **4-36** and the starting material
is that, at carbon 10 (C-10), a bond to C-4 has been replaced
by a bond to C-6. This is equivalent to an alkyl shift of C-10
from C-4 to C-6. In **4-37** at C-6, the bond to C-4 has been replaced by a bond to C-10. This is equivalent to a alkyl shift
of C-6 from C-4 to C-10. However, the carbocation will be
formed at C-6, and thus this carbon cannot be the one
which shifts. Thus **4-36** must be the product.

4-2.   The mechanism of this reaction involves protonation of
oxygen and loss of water to form a carbocation. Because the
molecule is symmetrical, the hydroxyls are equivalent;
thus protonation of either oxygen leads to the product.

The subsequent alkyl shift gives a carbocation, **4-38**, which
is resonance stabilized by interaction with the adjacent
oxygen. Loss of a proton then leads to the product.

4-38-1

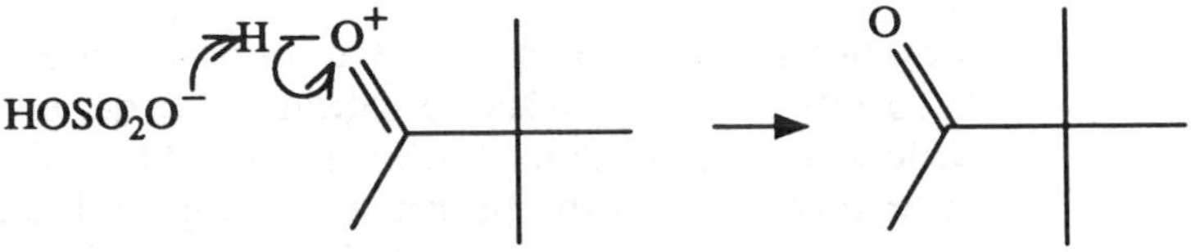

4-38-2

4-3.   There are several possible mechanistic variations for this reaction. Because the starting material contains several basic oxygens and acid is one of the reagents, protonation is a reasonable first step. Dimethyl ether is only 0.5 pK units more basic than acetophenone. (See Table 1-2.) An acetal would be expected to be less basic than an ether because of the electron-withdrawing effect of the second oxygen. Thus, protonation at both the acetal oxygen of the center ring and the carbonyl oxygen will be examined. (Protonation of the other acetal oxygen is unlikely to lead to product, because the ring containing this oxygen is unrearranged in the final product.)

SEQUENCE 1

In this sequence initial protonation of the acetal oxygen in the center ring is considered.

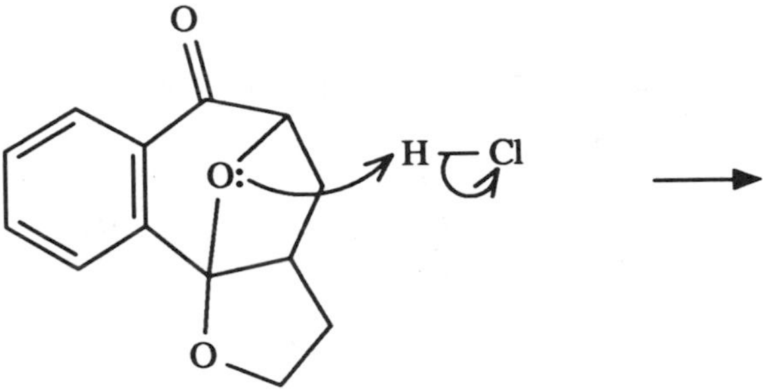

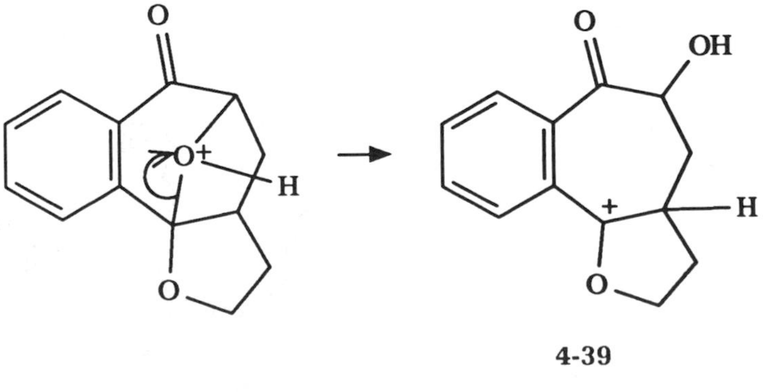

4-39

Ring-opening of the protonated intermediate could produce either cation **4-39**, or cation **4-40**, depending upon
which bond is broken. Carbocation **4-39** is stabilized by
resonance with both the aromatic ring and the adjacent
ether oxygen. Carbocation **4-40** is a hybrid of only two resonance forms, and **4-40-2** is quite unstable because the positive oxygen does not have an octet of electrons. Thus, the
first ring opening reaction is preferred.

4-40-1   **4-40-2**

Reaction of **4-39** continues with elimination of a proton to
form a double bond, followed by protonation of the carbonyl group.

4-39

Then an alkyl shift occurs, followed by a deprotonation.

Finally, a dehydration takes place, forming the product.

4-41

In this mechanism methanol was consistently used as the base, because data in Table 1-2 show that it is a stronger base than chloride ion.

SEQUENCE 2

This mechanism starts with protonation of the carbonyl oxygen to produce cation **4-42**.

4-42   **4-43-1**

An alkyl shift gives cation **4-43**, which is resonance-stabilized. This cation then undergoes bond cleavage to form **4-44**, stabilized in the same manner as **4-39** in Sequence 1. Loss of a proton gives **4-41**, which is also an intermediate in Sequence 1.

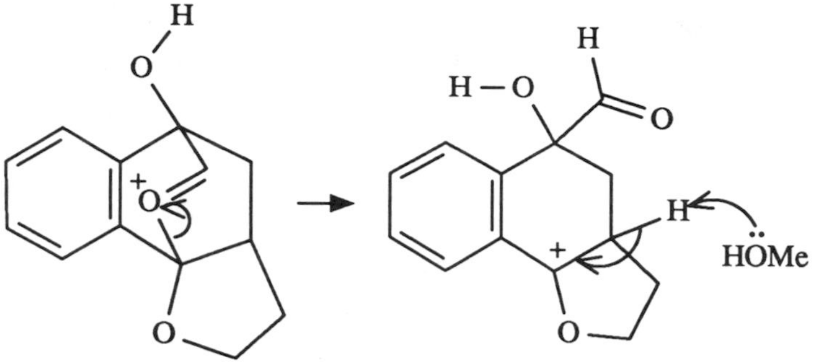

         4-43-2   **4-44**

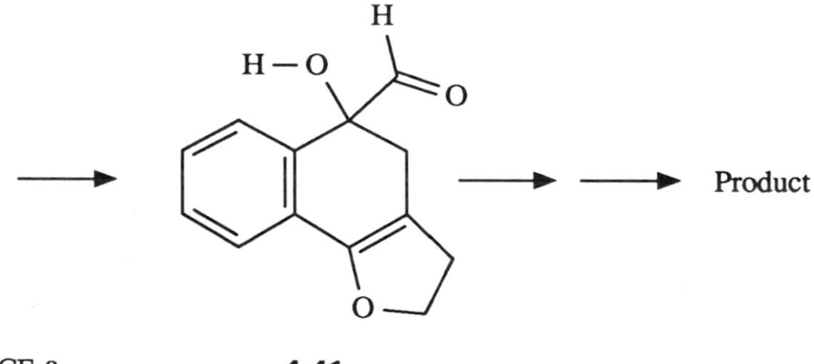

                                    Product

SEQUENCE 3   **4-41**

Intermediate **4-42** of Sequence 2 undergoes a hydride shift, instead of an alkyl shift, to give the following:

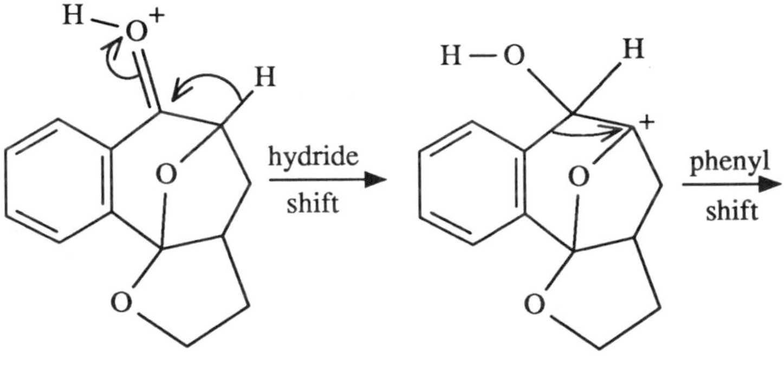

         4-42   **4-45**

Because an unshared pair of electrons on the oxygen bonded to the carbocation in intermediate **4-45** may not overlap well with the positive carbon, this intermediate may not be stabilized enough to make this mechanism viable. A phenyl shift gives the cation, **4-46**, which can lose a proton to give the aldehyde functional group. The acetal oxygen in the center ring can be protonated and open to give cation, **4-44**, which behaves as shown above in Sequence 2.

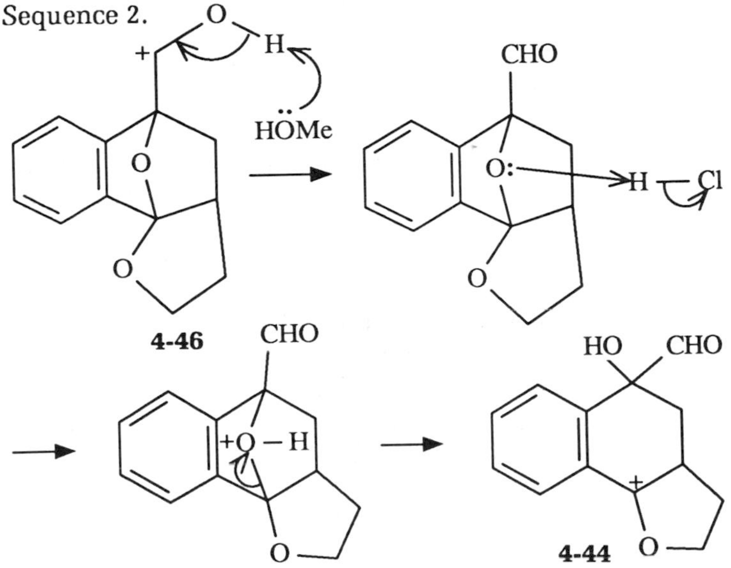

An alternative way to show the phenyl shift in **4-45** is by a stepwise process. The intermediate, **4-47**, is called a phenonium ion; such ions have often been observed in reactions when phenyl groups interact with carbocations. It can rearrange to give **4-46**, which continues to product as discussed above.

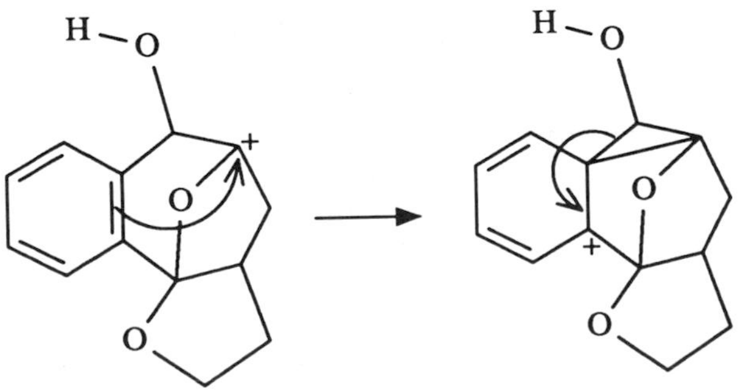

4-46

In summary, the phenyl shift involved in the transformation of **4-45** to **4-46** may involve either one or two steps.

4-4.   This reaction is a fragmentation, which often accompanies a Beckmann rearrangement. Reaction of the hydroxyl group with phosphorus pentachloride converts it into a better leaving group. In **4-48**, instead of an alkyl group migration, the ring bond on the side opposite the leaving group cleaves, producing a ring-opened cation, **4-49**. This mimics the stereochemical requirement of the Beckmann rearrangement.

4-48   **4-49**

The electrophilic cation, **4-49**, reacts with a nucleophile at carbon to give the neutral derivative, **4-50**.

4-49   **4-50**

The product, **4-50**, will readily undergo hydrolysis when the reaction is worked up in water. This hydrolysis is acid-catalyzed because the phosphorus compounds in the reaction mixture are acidic.

One student wrote an alternative mechanism which involved ring expansion of **4-48** rather than fragmentation.

4-48

This is a good step: it is the normal Beckmann rearrangement. However, to get from this intermediate to the open chain product took a fairly large number of "creative" steps. Applying Ockham's Razor suggests that the initial fragmentation occurs early.

4-5.a.   The formation of **4-23** could involve formation of a bromonium ion, **4-51**, similar to **4-14**, **4-15**, and **4-18**. This ion undergoes an aryl shift and then reacts with bromide to give the product. The bromonium ion must form on the *exo* side for the rearrangement to take place; that is, such rearrangements must have the migrating aryl group enter from the side opposite the leaving bromo group.

4-51

4-23

In the last step above the bromide ion enters on the side opposite the phenonium ion in analogy to the stereochemistry of reaction of bromide ion with a bromonium ion.

There are several possible explanations for the different result at higher temperature. One is that the bromonium ion forms on both the *exo* and *endo* sides. Reaction of the *exo* bromonium ion gives **4-23**, as before, and **4-25** is formed in an elimination reaction.

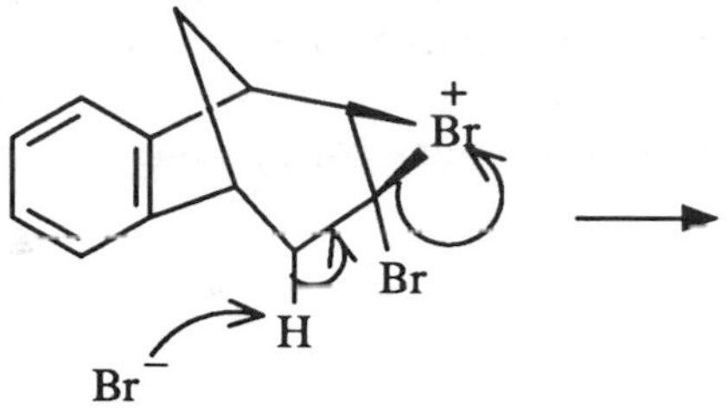

Reaction of the *endo* bromonium ion gives the isomers, **4-24.**

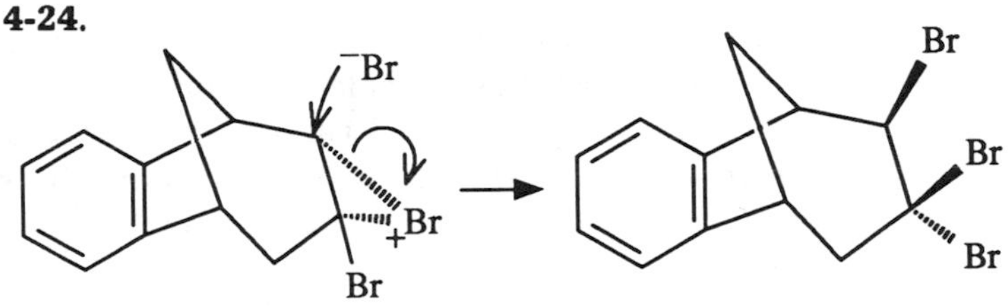

4-24

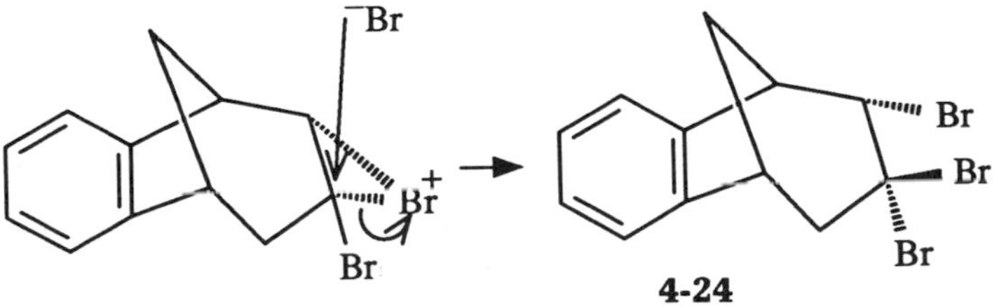

4-24

Another explanation of the result is that only the *exo* bromonium ion forms and, at the higher temperature, in addition to the rearrangement, the elimination occurs as well as reaction of bromide ion from the *endo* side giving the two isomers, **4-24**.

4-5.b.   The displacement of Br by sulfur in the intermediate bromonium ion, **4-52**, is similar to the ring closure in Problem 3-4a.

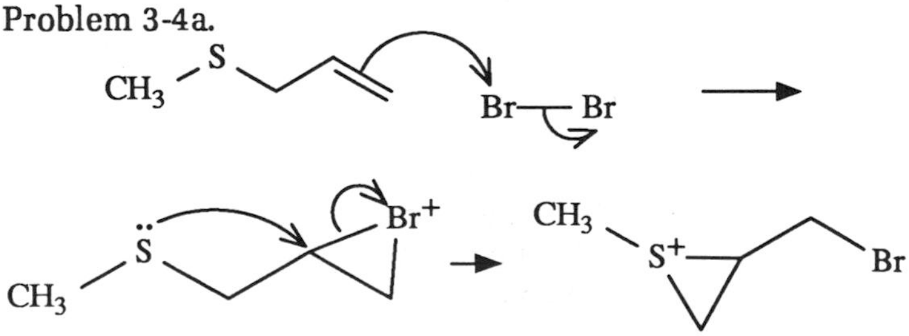

4-52   **4-53**

Though, by convention, mechanisms are usually not written from doubly charged resonance forms of neutral starting material, the initial reaction to give **4-52** might also be written from resonance form, **4-54**:

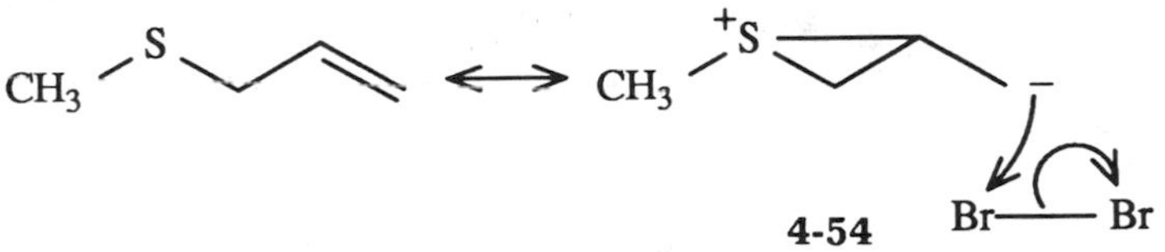

4-54

At low temperatures, the three-membered ring episulfonium ion, **4-53**, which resembles a bromonium ion, reacts with bromide only at the 1° carbon. The product of this S_N2 reaction is determined by the reaction rate, which is faster at the 1° carbon than at the 2° carbon. This is a reaction in which product formation is rate-controlled.

4-53   **4-55**

At elevated temperatures, **4-55** is in equilibrium with **4-53**. Under these conditions, **4-53** also reacts at the 2° carbon to give **4-56**. Since **4-56** is more stable than **4-55**, its reverse reaction back to **4-53** is slower, and **4-56** accumulates. Thus, at higher temperatures, product formation is equilibrium-controlled.

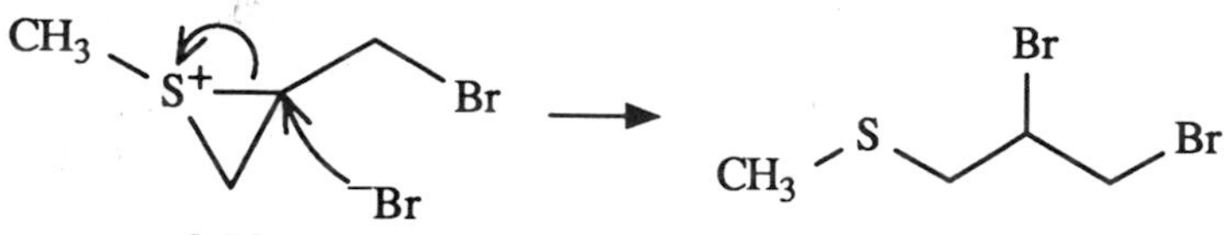

4-53   **4-56**

4-6.a.   This reaction involves the loss of the diethylamino group of the amide, and ring closure with the aldehyde, to form the hydroxy-lactone product. Because the timing of the ring closure is open to question, two possible mechanistic sequences are given.

SEQUENCE 1

In this mechanism, the second ring is formed early. The carbonyl oxygen of the amide is protonated; then the oxygen of the aldehyde adds to this protonated functional group to form a new ring.

The four steps, after participation of the aldehyde oxygen, are: nucleophilic reaction of water with the most electrophilic carbon; loss of a proton; protonation of the nitrogen; and loss of diethylamine. This order of steps avoids formation of more than one positive charge in any intermediate.

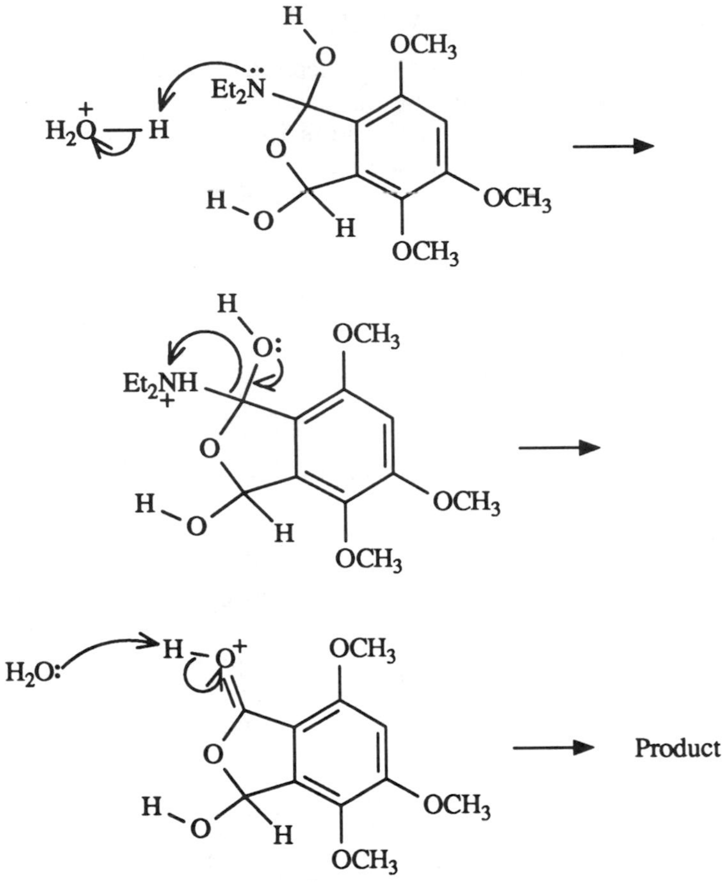

4-57

This mechanism seems better than that of Sequence 2, because no tetrahedral intermediate is formed at the amide carbon prior to cyclization. The amide position is sterically congested, because it is flanked by *ortho* substituents. A tetrahedral intermediate (sp^3-hybridized carbon) adds to the congestion, and might be of such high energy that its formation would be unlikely.

SEQUENCE 2

Another possibility is hydrolysis of the amide to a carboxylic acid followed by closure with the aldehyde. The hydrolysis of the amide mimics the steps in Example 4-11.

Since it is easier to protonate an aldehyde than a carboxylic acid (compare benzoic acid and benzaldehyde in Table 1-2), the ring closure would best be written as protonation of the aldehyde oxygen followed by nucleophilic reaction of the carboxylic acid carbonyl oxygen. The carbonyl oxygen acts as the nucleophile because a resonance-stabilized cation is produced. If the hydroxyl oxygen acts as a nucleophile, the cation is not resonance stabilized.

4-57

The last step is the same as that for Sequence 1.

4-6.b.   Both the ester and nitrile undergo hydrolysis. (There are many acids present; H_3O^+ and H_2SO_4 can be used interchangeably.) Normally, esters are hydrolyzed more rapidly than nitriles, so the ester will be hydrolyzed first.

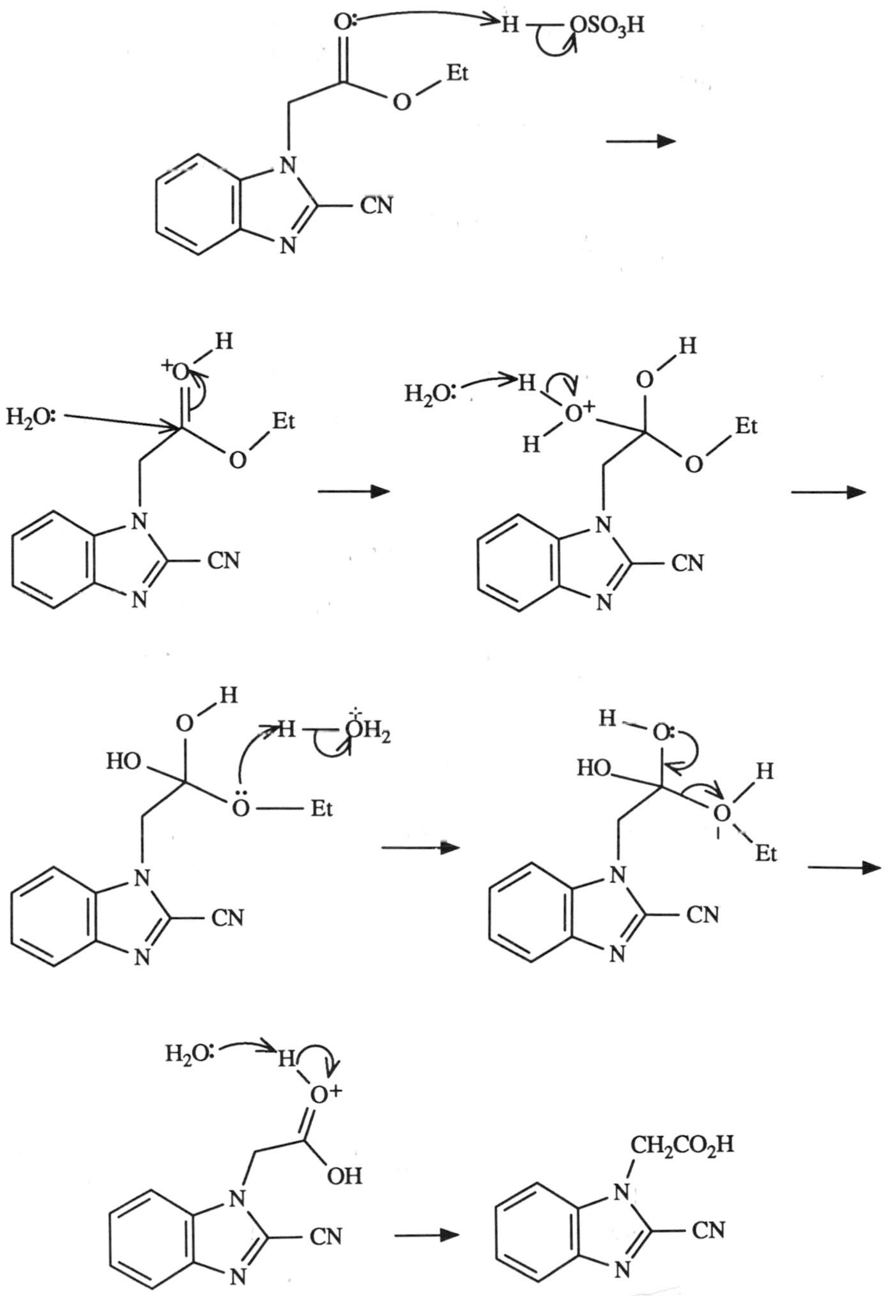

O:
H
OSO$_3$H
Et
O
N
CN
N
H
+O
H$_2$O:
Et
O
N
CN
N
H$_2$O:
H
O+
H
O
H
Et
O
N
CN
N
O
H
HO
H
ÖH$_2$
Ö
Et
N
CN
N
H
O:
HO
H
O
Et
N
CN
N
H$_2$O:
H
O+
OH
N
CN
N
CH$_2$CO$_2$H
N
CN
N

Then hydrolysis of the nitrile will produce the imino form of the amide, which readily tautomerizes by the usual mechanism to form the amide.

$$CH_2CO_2H$$

If reaction conditions are controlled, the hydrolysis of nitriles can be stopped at the amide stage. In this case once the amide is produced, the nitrogen will react as a nucleophile with the protonated carbonyl of the acid. Because a six-membered ring transition state is produced, this intramolecular reaction is very favorable.

Reaction of the nitrile, acting as a nucleophile, with the electrophilic protonated ester, is unlikely as a ring-forming reaction. The linearity of the nitrile group means that there can be little overlap between its lone pair of electrons and the p orbital on the carbon of the carbonyl group.

4-7. None of the cations produced by protonation at the carbons of acrolein are as stable as the cation produced by protonation at oxygen. Let's look at the possibilities.

Protonation of the aldehyde carbonyl carbon gives cation **4-58**. This is a very unstable intermediate: it is not stabilized by resonance, and the positive oxygen lacks an octet. Notice that the double bond and the positive oxygen are not conjugated, because they are separated by an sp^3-hybridized carbon.

4-58   **4-59**

The intermediate produced by protonation at the α carbon also gives a very unstable 1° cation, **4-59**, which is not stabilized by resonance. Protonation at the β carbon gives a delocalized cation, **4-60**, but the resonance form, **4-60-2**, with a positive charge on oxygen is especially unstable, because the oxygen does not have an octet of electrons.

4-60-1   4-60-2

4-8.   HCl is a catalyst for the transformation.

4-9.   By analogy to Example 4-13 and Problem 4-7, protonation
at the carbonyl group, rather than at the double bond, of the
starting material will give a more stable cation. (That is not
to say that all reactions occur *via* a mechanism involving
formation of the most stable carbocation. But if the most
stable cation leads to product, that's the one to use.)

4-61-1   4-61-2

The initially formed carbocation, **4-61**, can undergo a hydride shift to give a 3° carbocation. The oxygen of the OH or OD group reacts as a nucleophile with the electrophilic carbocation. The oxygen which reacts must be the one on the same side of the double bond as the carbocation, so that the double bond remains *cis* within the ring. Therefore, it is the OD oxygen, not the OH oxygen, which reacts in **4-62**.

4-61   hydride
          ————→
          shift

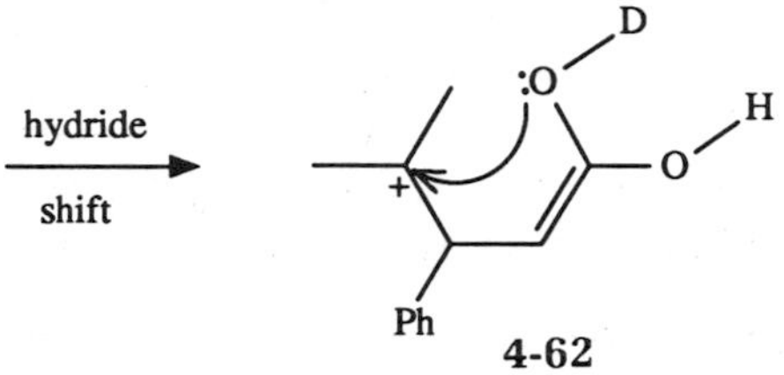

4-62

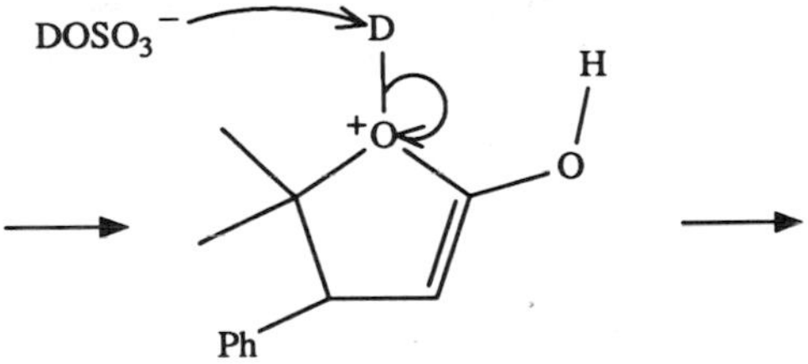

Intermediate **4-63** undergoes acid-catalyzed tautomerization to give the product.

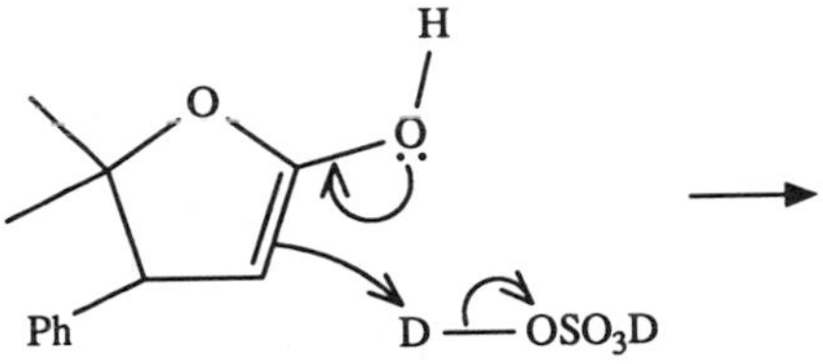

4-63

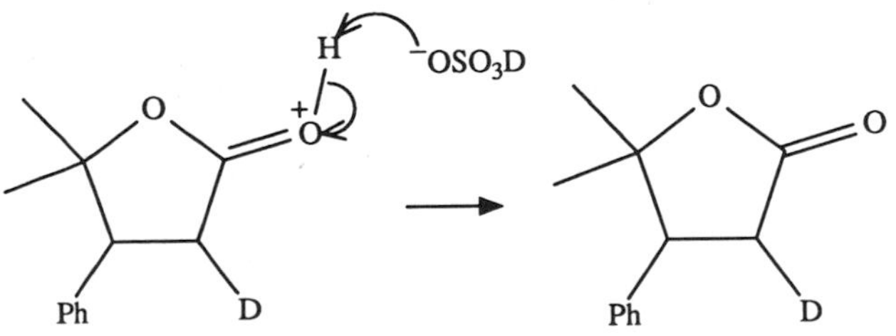

There is another cation that could undergo the same rearrangement as above. This cation could be formed by direct protonation of the double bond with Markovnikoff regio-specificity:

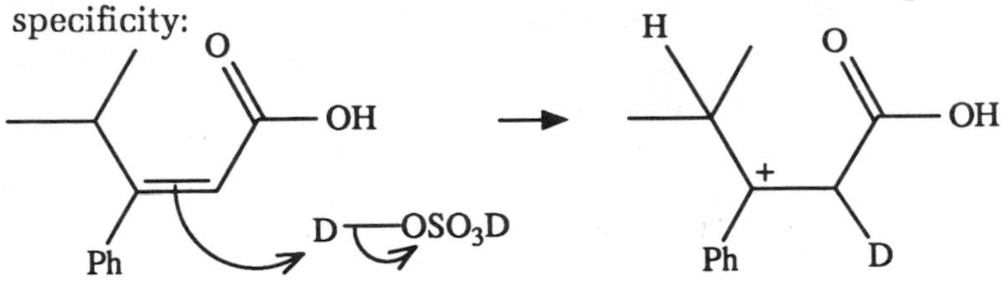

The remaining steps in this mechanism would be very similar to those of the first. However, the first mechanism is better, because the initially formed carbocation is more highly stabilized by resonance and thus should form faster.

4-10.a.   The intermediate, formed by reaction of the electrophile at the *para* position, has three resonance forms. Form **4-64** is especially stable, because the positive charge is next to an alkyl group which stabilizes positive charge by the inductive effect and by polarizability.

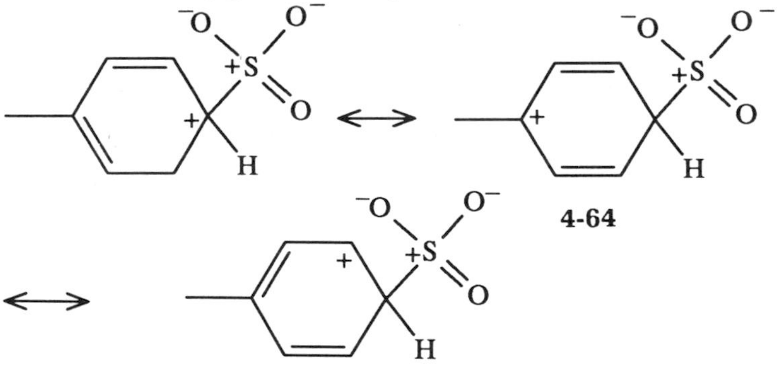

4-64

Reaction of the electrophile at the *ortho* position would give an intermediate of essentially the same stability. However, reaction of the electrophile at the *meta* position gives an intermediate, **4-65**, in which the positive character cannot be located at the alkyl group position. Thus, this intermediate is not as stable and is not formed as rapidly as the intermediates from either *ortho* or *para* attack.

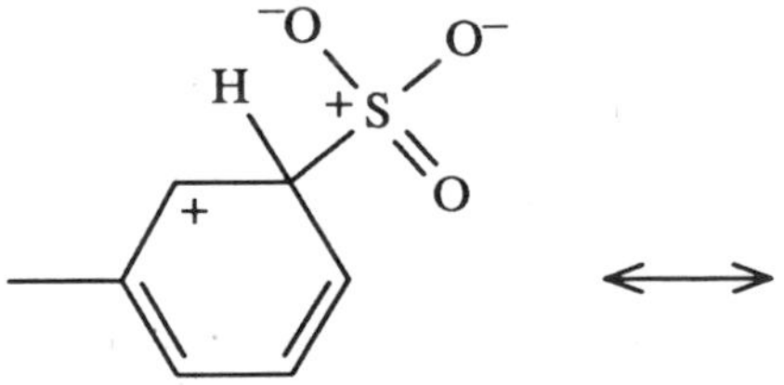

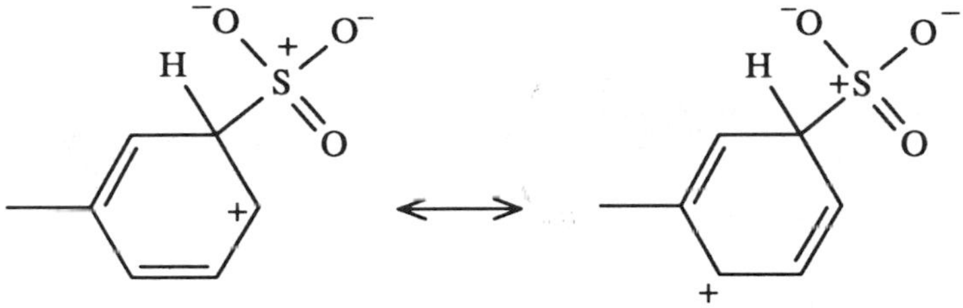

4-65

4-10.b. The intermediate, formed by reaction of a nitronium ion at the *meta* position, does not have positive charge at the sulfonic acid group position. This is favorable, because sulfur has a positive charge, and location of two positive charges close to one another would be destabilizing.

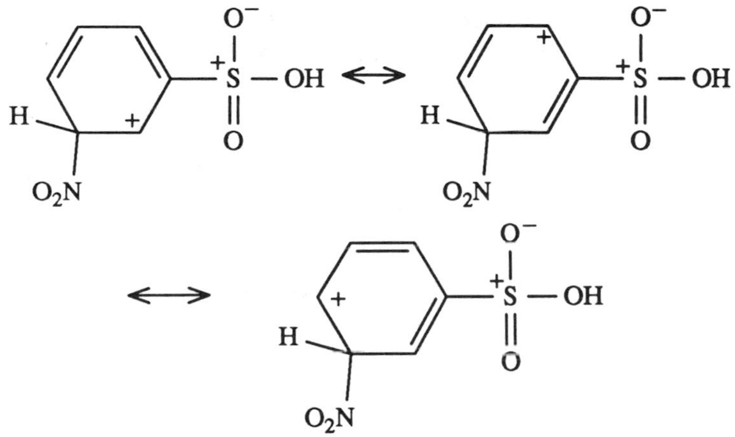

The resonance hybrids for electrophilic attack at the *ortho* or *para* position are not as stable because, in each case, one of the three resonance forms does have the destabilizing resonance form. For example, for *para* attack of the electrophile, **4-66-2**, is especially unstable.

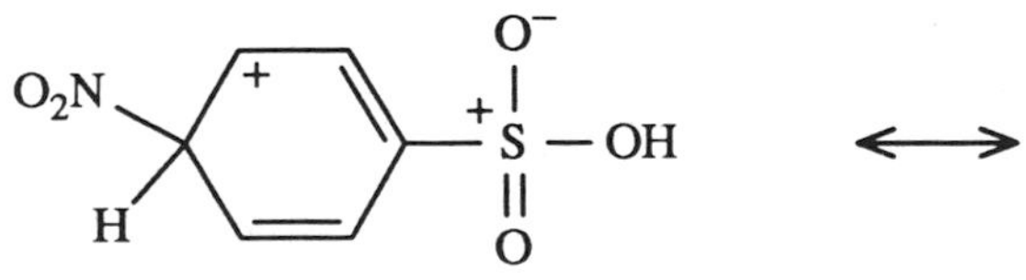

4-66-1

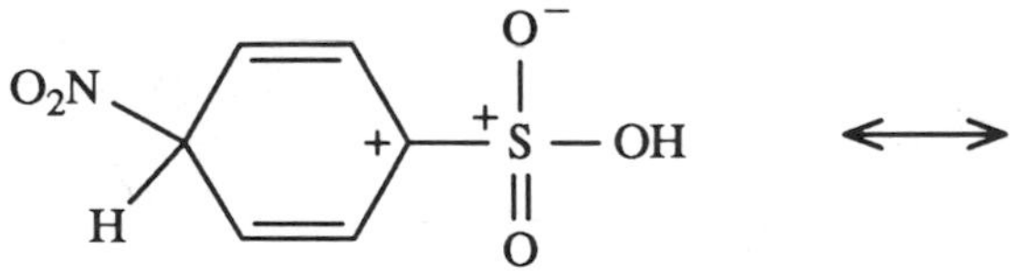

4-66-2

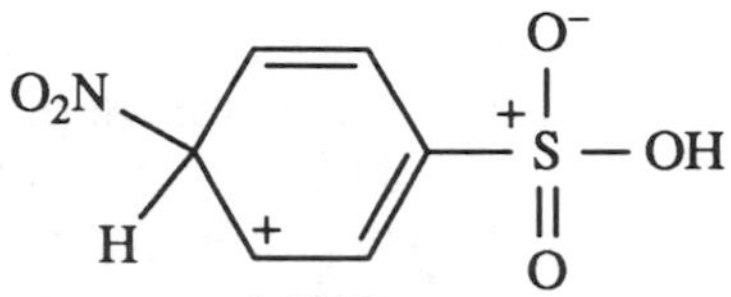

4-66-3

4-10.c. The intermediate, formed by reaction of the electrophile at the *para* position has a resonance form, **4-67-2**, with positive character at the substituent position. This means that the unshared pair of electrons on the nitrogen of the acetamido group can overlap with the adjacent positive charge, as shown by the fourth resonance form, **4-67-4**.

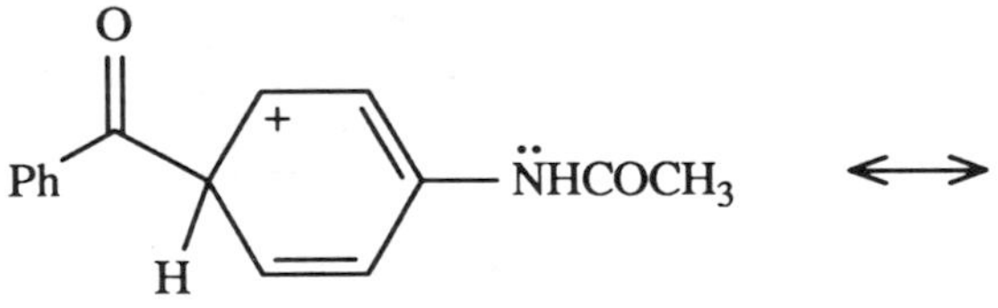

4-67-1

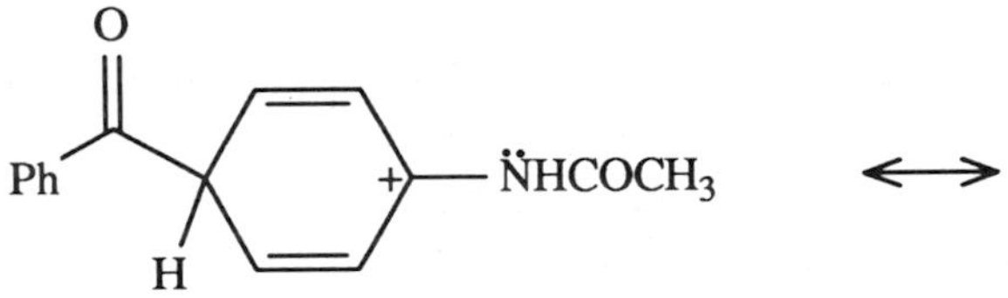

4-67-2

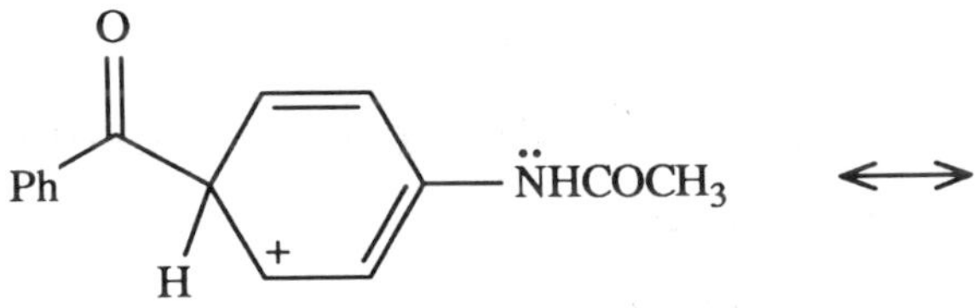

4-67-3

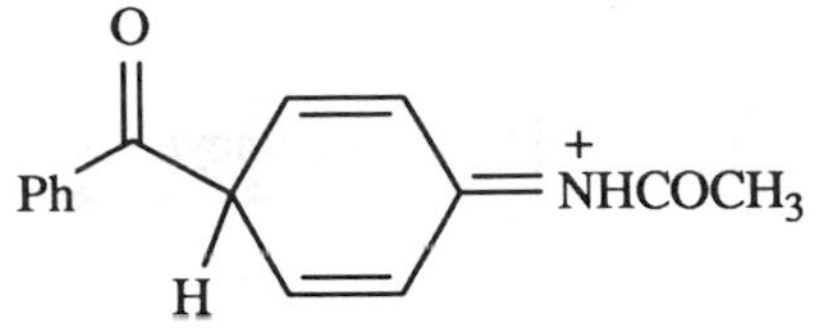

4-67-4

This extra delocalization of the positive charge adds to the stability of the intermediate. On the other hand, if the electrophile reacts at the *meta* position, the positive charge cannot be placed on the nitrogen.

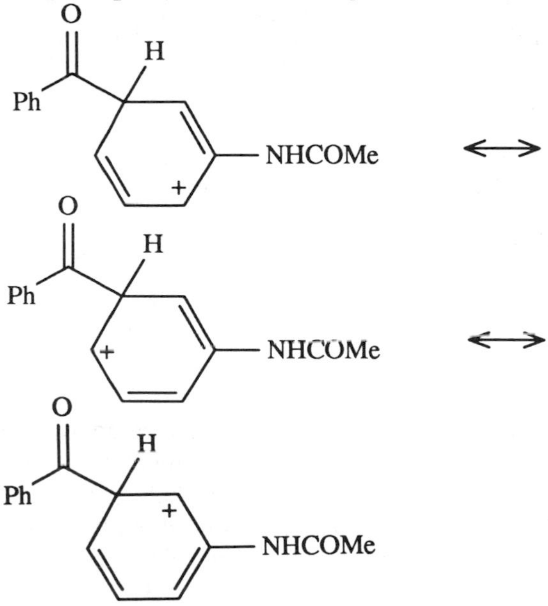

4-11.a. The mechanism in sulfuric acid might occur by the steps typical for a dienone-phenol rearrangement. Protonation of the A-ring carbonyl gives a more highly resonance-stabilized intermediate than protonation of the B-ring carbonyl.

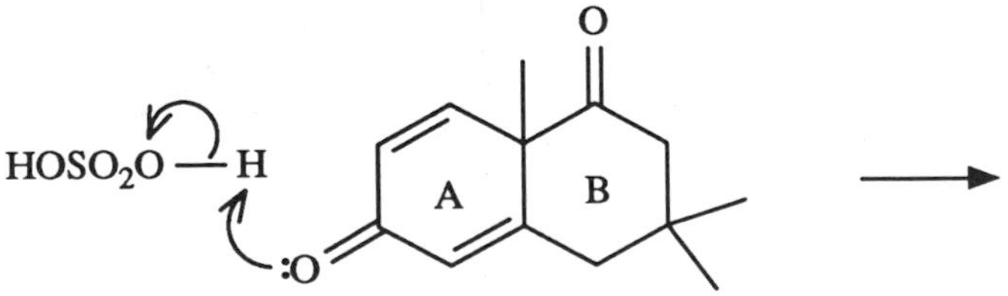

4-68

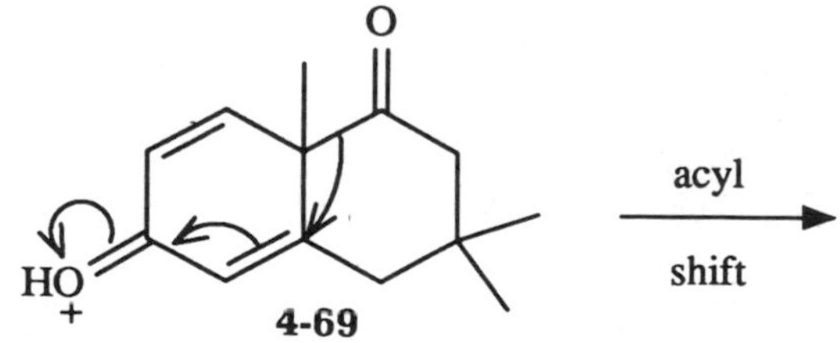

4-69

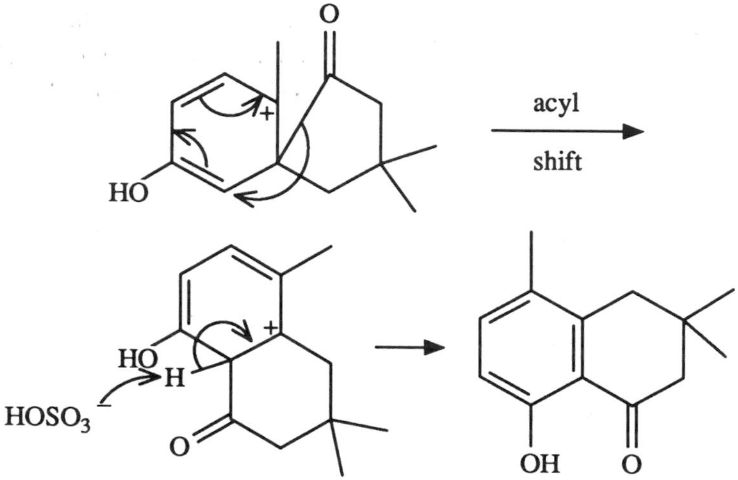

4-70

Formation of the product, **4-71**, formed in dilute HCl requires cleavage of the B ring. A mechanism involving the
formation of the hydrate of the keto group in the B ring,
followed by ring opening, is shown below. (Example 2-9, p.
86 shows a slightly different possibility.)

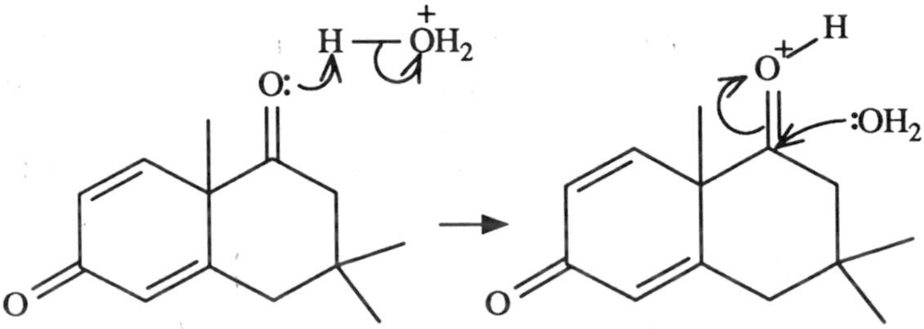

4-71

The mechanism in dilute acid suggests an alternative route in the stronger acid, sulfuric acid, in which initial cleavage of the B ring occurs. In this pathway the intermediate cation **4-68** can ring open to give the acylium ion **4-72**.

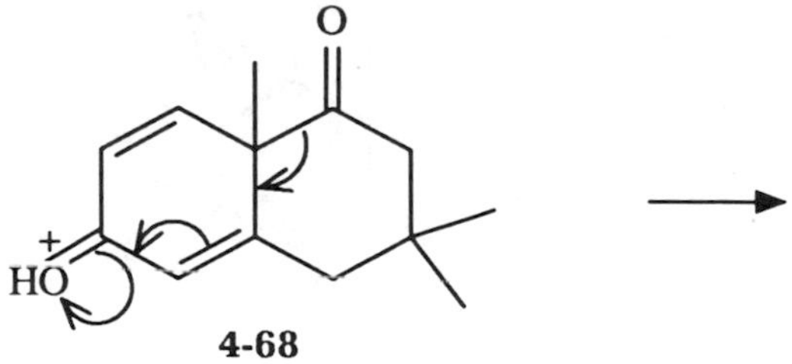

4-68

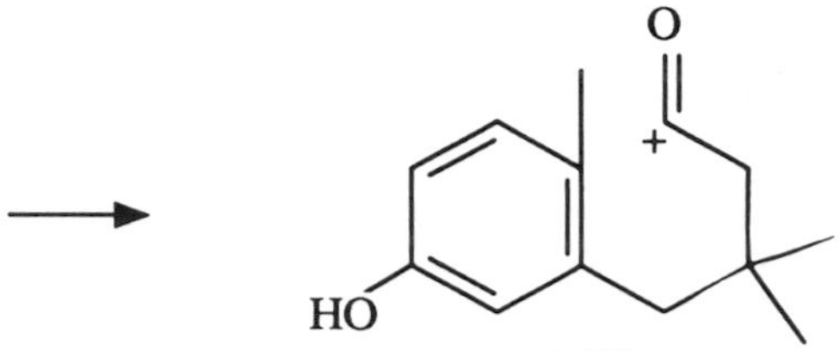

4-72

Rewriting **4-72**, with the acylium ion in proximity to the position *ortho* to the hydroxyl group, clarifies the intramolecular acylation reaction which forms a 6-membered ring.

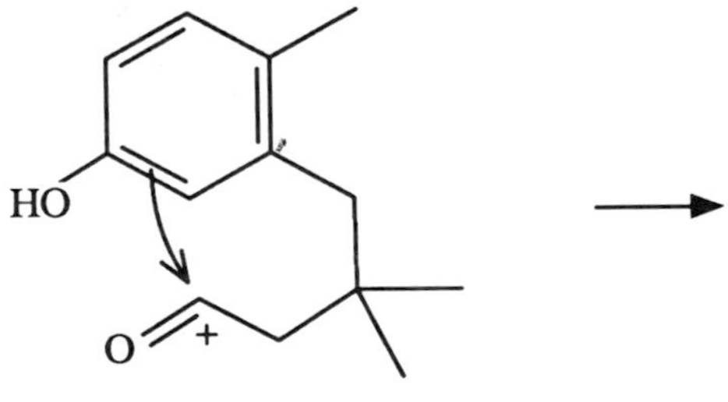

4-72

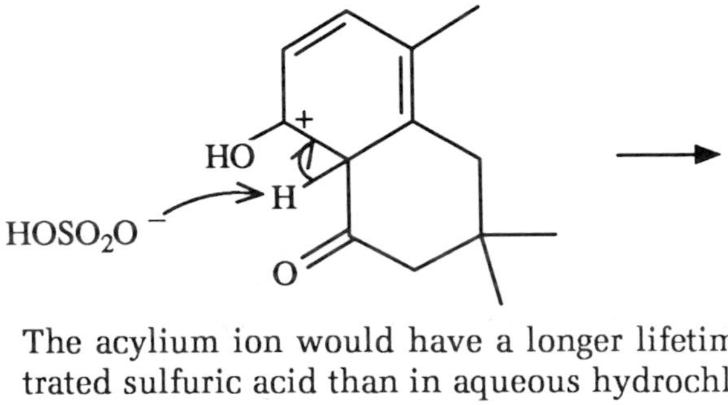

4-70

The acylium ion would have a longer lifetime in concentrated sulfuric acid than in aqueous hydrochloric acid, because the concentration of water in the former acid is extremely low. In aqueous hydrochloric acid, the starting ketones are more likely to be in equilibrium with their hydrates, and the intermediate acylium ion might never form. If it does form, it will react so rapidly with the nucleophilic oxygen of water that the electrophilic aromatic substitution cannot occur.

The experimental data, given in the paper cited, support the ring opening of the B ring under both sets of conditions. On treatment with sulfuric acid, **4-71**, reacts to give **4-70** at the same rate that **4-68** reacts to give **4-70**. However, the mechanism involving successive acyl shifts appears to occur in trifluoroacetic acid. In this medium **4-68** rearranges to **4-70**; but **4-71** cannot be an intermediate, because it does not give **4-70**.

4-11.b. The reaction proceeds by a cleavage-recombination reaction.

4-73   **4-74**

Support for the formation of **4-73** and **4-74** comes from a crossover experiment reported in the paper. In the presence of *m*-cresol, **4-75** was obtained.

4-75

This product is formed by capture of the intermediate carbocation **4-73** by the added *m*-cresol. Loss of a proton gives **4-75**. Furthermore, as the concentration of *m*-cresol increases, the amount of **4-75** increases. This supports the idea that the *m*-cresol is competing with **4-74** for **4-73**. The term, crossover experiment, refers to the fact that **4-73** reacts with an external reagent rather than **4-74**, its co-cleavage product.

The fact that crossover occurs rules out the following mechanism to the product:

4-12. The intermediate anion picks up a proton; removal of the proton on the sp^3-hybridized carbon α to the carbonyl group gives the phenolate ion. The phenolate loses chloride ion to give **4-76** which undergoes addition of hydroxide at the carbon at the *ortho* position. Loss of the remaining chloride and removal of a proton gives the product phenolate.

4-76

Acidic workup will give the phenol itself.

4-13. Dibromocarbene can be produced from bromoform and base:

Phenolate, formed in the basic medium, reacts as a nucleophile with the electrophilic carbene. The resulting anion is
protonated by water to give the product.

Problem 3-4b, p 118, gives a subsequent reaction of this
product and its literature reference.

An alternative mechanism, with the ring acting as an
electrophile and the carbene as a nucleophile would be incorrect. The dihalocarbenes are quite electrophilic. Also,
since the phenol readily forms a salt with sodium hydroxide, the aromatic ring would not be electrophilic.

4-14.a. Trifluoromethanesulfonic acid is a very strong acid, and
the most basic atom in the amide is the carbonyl oxygen.
Protonation of the carbonyl group is a likely process but it
is not one which leads to product. That is, protonation of
the hydroxyl oxygen is on the pathway to product while
protonation of the carbonyl oxygen is not. The nitrogen
bearing the protonated hydroxyl group can act as an
electrophile, and the phenyl ring, situated to form a favorable six-membered ring, can act as a nucleophile.

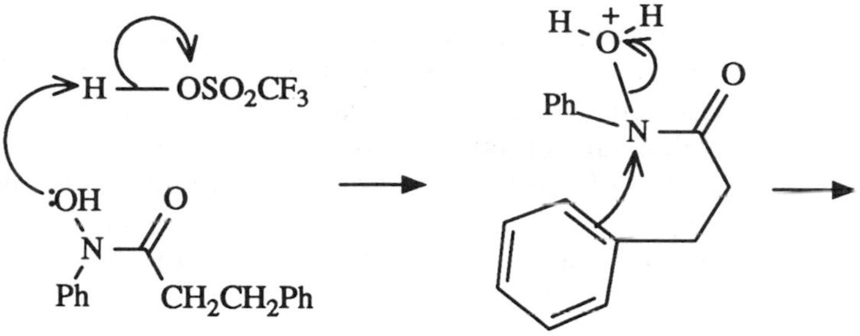

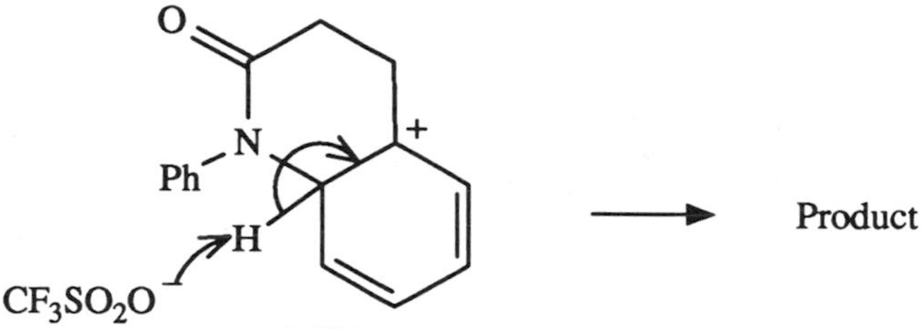

4-77

In an alternate mechanism, water leaves the protonated starting material prior to involvement of the nucleophile (the aromatic ring). This gives a nitrenium ion, which acts as the electrophile. Cyclization leads to the same intermediate carbocation, **4-77**, as above.

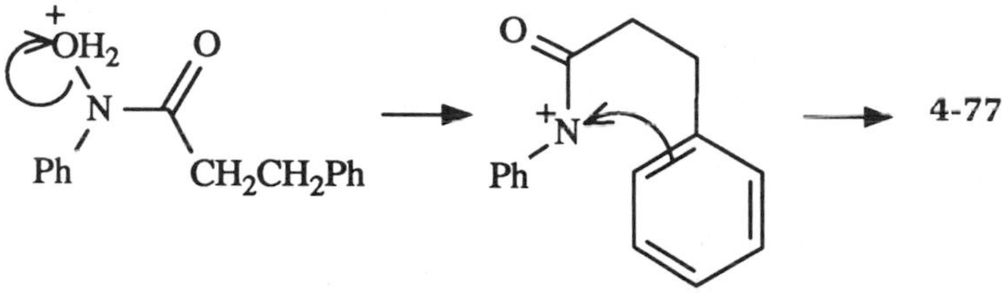

4-77

The following reaction, which goes to 76% yield in trifluoromethanesulfonic acid, provides support for initial formation of positive nitrogen. (Endo, Y.; Ohta, T.; Shudo, K.; Okamoto, T. *Heterocycles* **1977**, *8*, 367–370.)

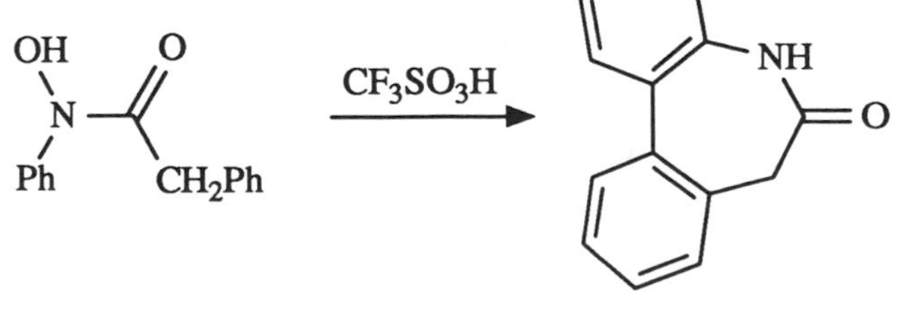

This reaction is an electrophilic aromatic substitution in which one benzene ring acts as the electrophile and the other benzene ring acts as the nucleophile. In order to develop substantial positive charge in the electrophilic ring (a driving force for the reaction), resonance must occur. This requires prior loss of water. The intermediate nitrenium ion and the direction of the substitution reaction are shown below.

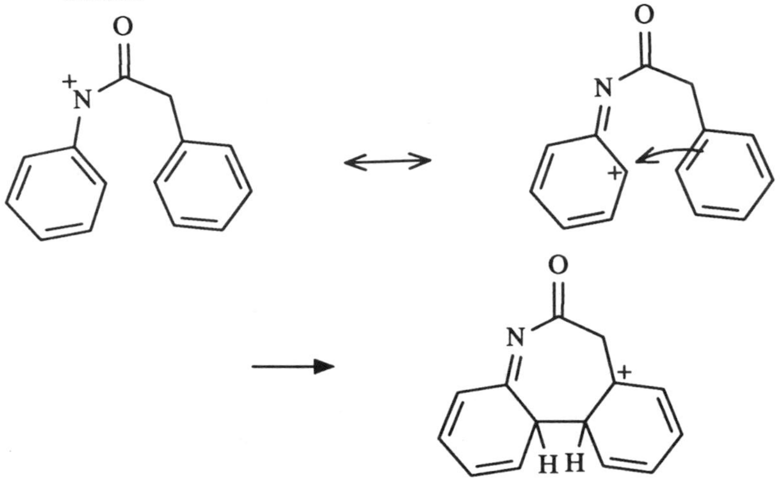

Formation of the product requires loss of a proton to regenerate a neutral compound and tautomerization to regenerate the other benzene ring.

4-14.b.

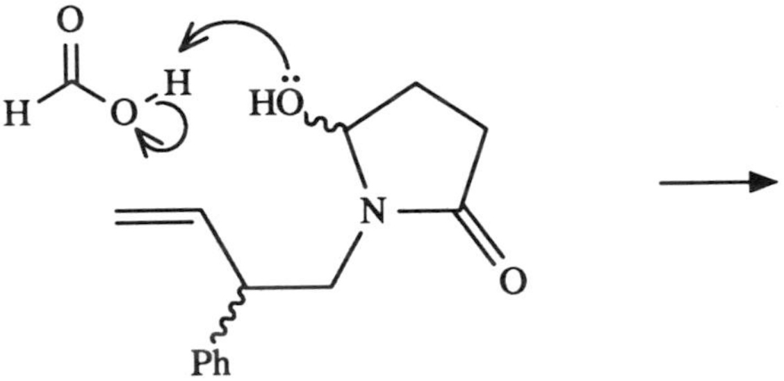

Nucleophilic reaction of formic acid on either side of the electrophilic carbocation in the bicyclic intermediate leads to the two products.

4-14.c. There are several different mechanistic sequences that might lead to the product.

SEQUENCE 1

The two equivalent amide carbonyl oxygens are the most basic atoms in the starting materials. Protonation of one of these oxygens gives a carbocation, **4-78**, which is stabilized by de-localization onto oxygen and nitrogen as well as carbon.

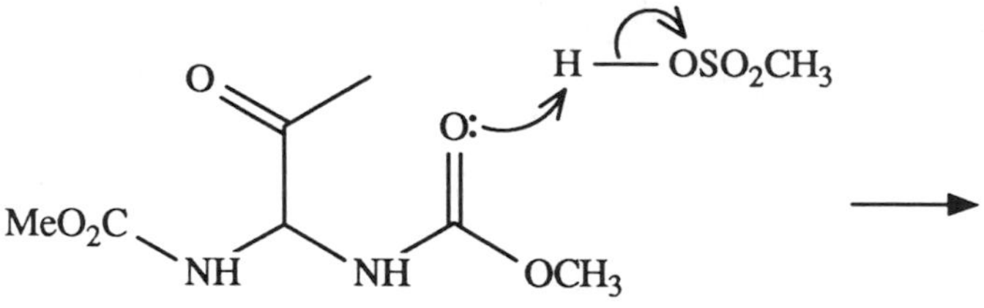

4-78-1

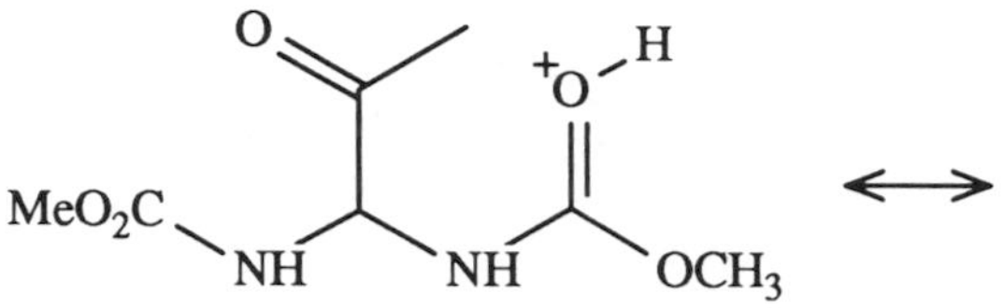

4-78-2

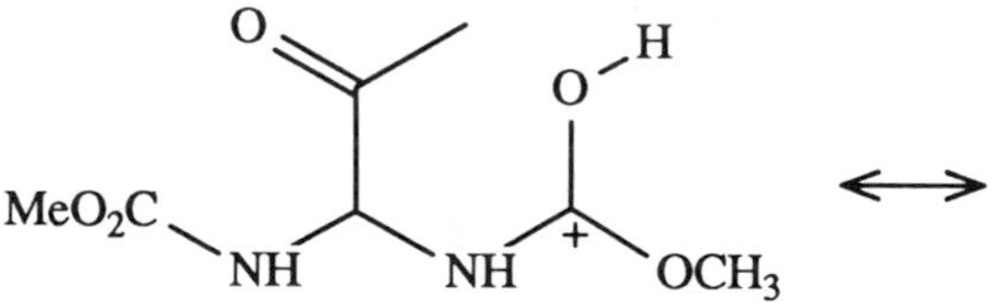

4-78-3

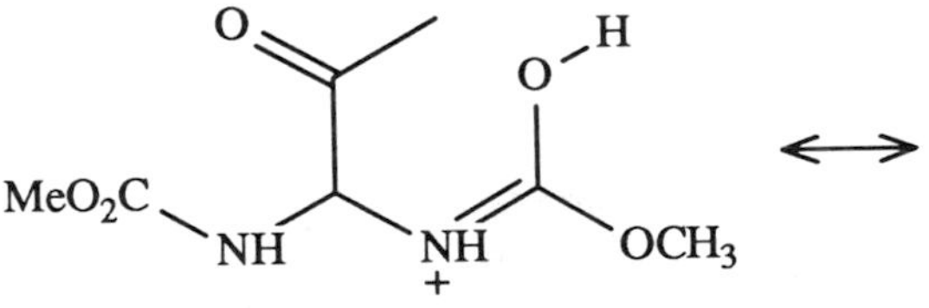

4-78-4

The intermediate, **4-78**, can decompose, giving a new electrophile with which the nucleophilic phenol then reacts.

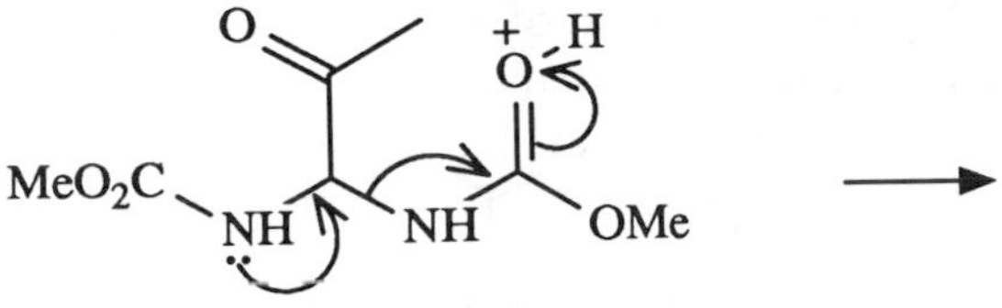

4-78

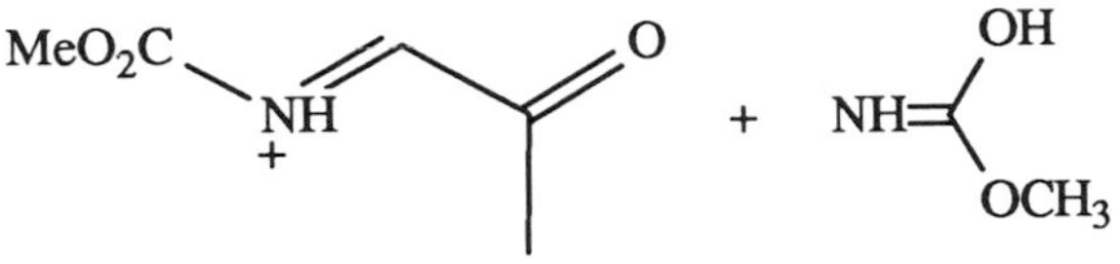

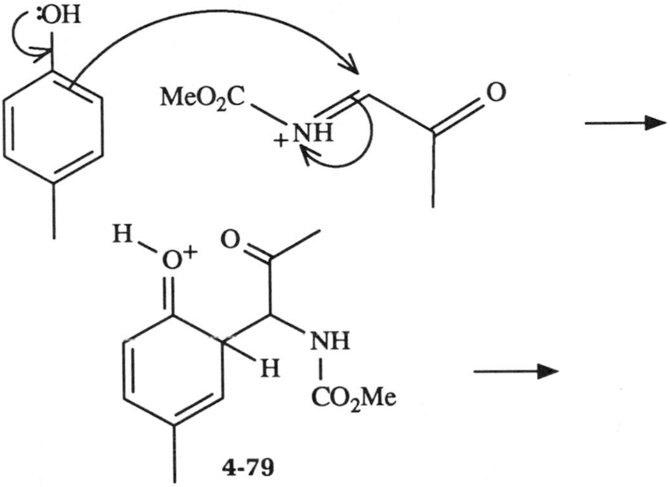

4-79

An alternative mechanism to **4-79** might be protonation of one of the nitrogens in the starting material, followed by a nucleophilic reaction of the phenol with the carbon bearing the protonated leaving group (S_N2 reaction). However, because this carbon is quite sterically congested, this is not an attractive option.

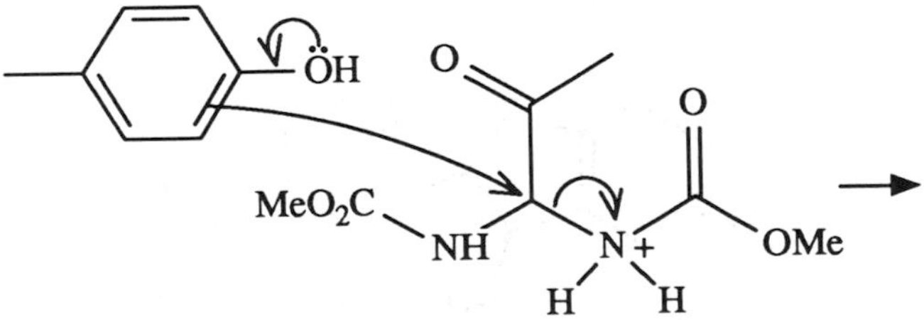

After aromatization of **4-79** and protonation of the keto carbonyl, cyclization can occur:

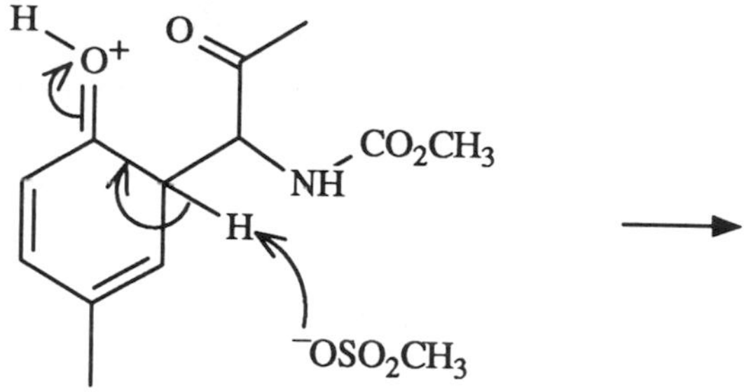

4-79

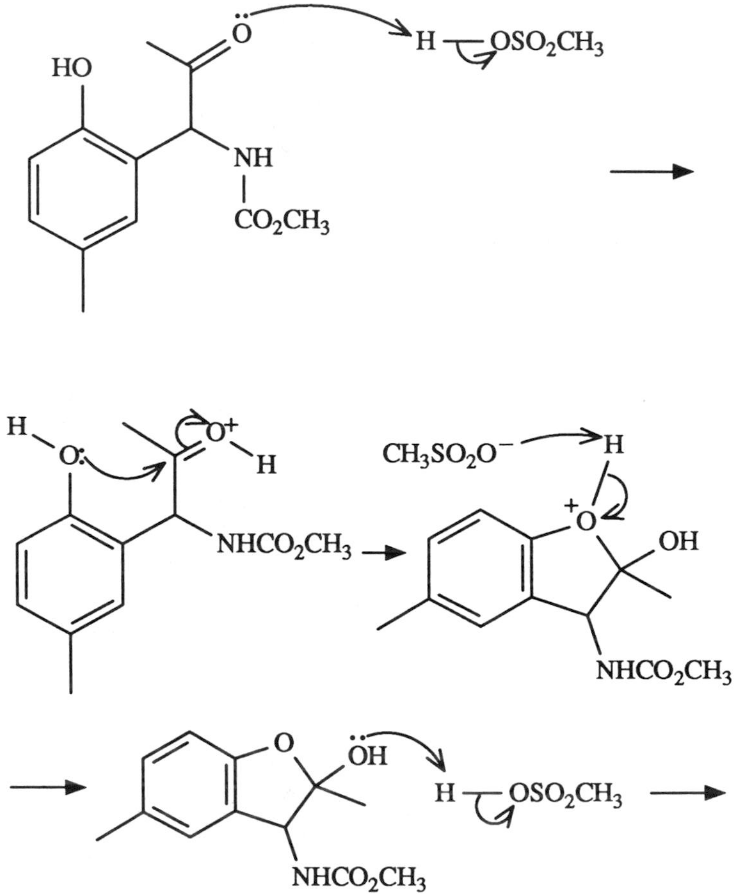

4-80

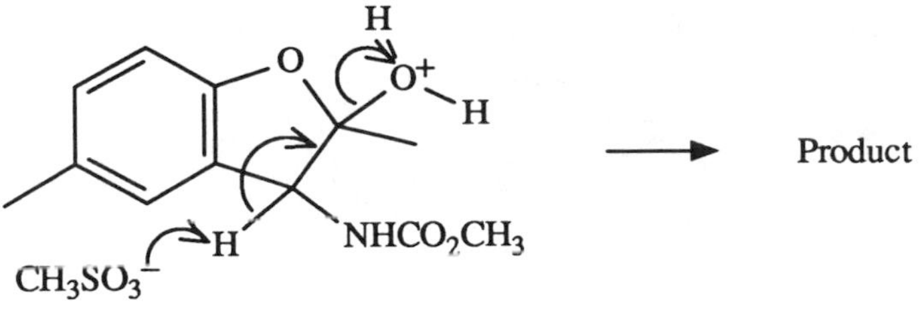

→ Product

SEQUENCE 2

If the keto group in the starting material were protonated, the phenolic oxygen could react as a nucleophile at the carbonyl carbon.

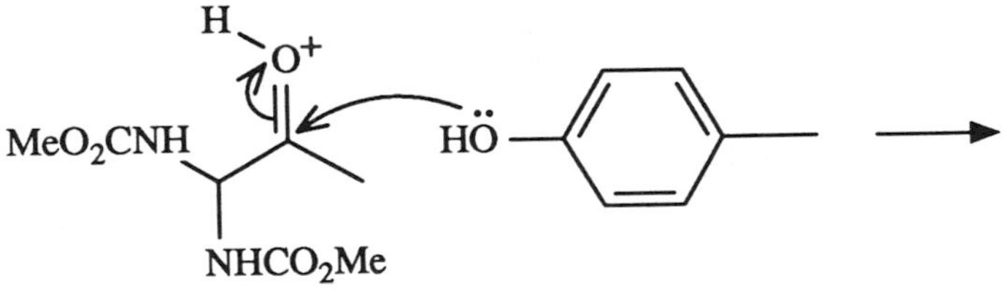

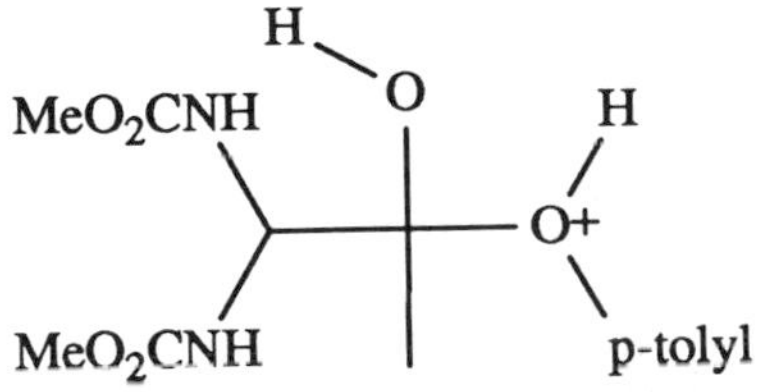

Following proton loss from the positive oxygen and protonation of one of the amide nitrogens, intermediate **4-81** is formed. This intermediate can lose methyl carbamate to form **4-82** in which the ring acts as a nucleophile.

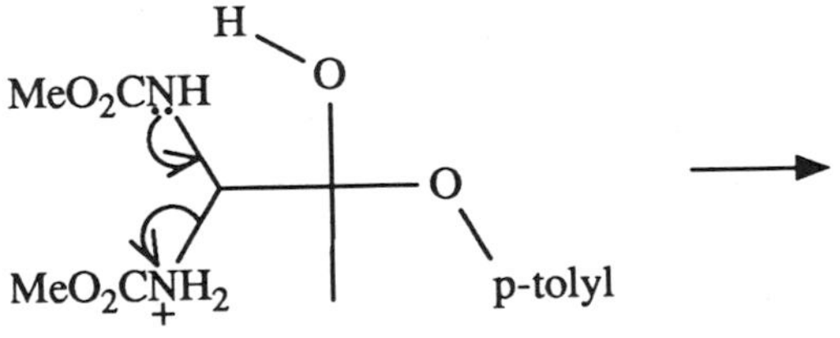

4-81

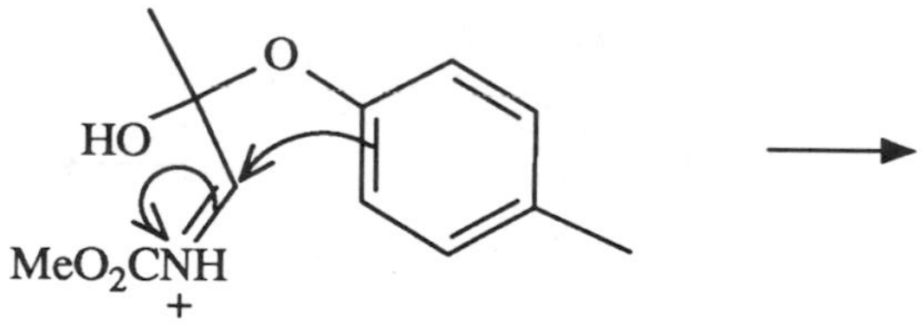

4-82

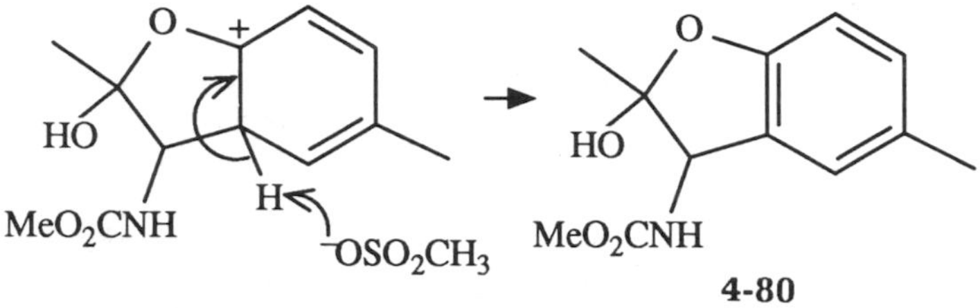

4-80

Loss of a proton gives **4-80** which undergoes dehydration as in Sequence 1.

4-14.d. Either the product of a nucleophilic displacement at tin or an expanded orbital on tin can be written as the tin intermediate.

SEQUENCE 1

Ring opening occurs readily in the cyclopropyl-substituted cation, **4-83**, to give **4-84**. The driving forces are release of the strain energy of the 3-membered ring and the stabilization of the resulting cation by the *p*-methoxyphenyl group.

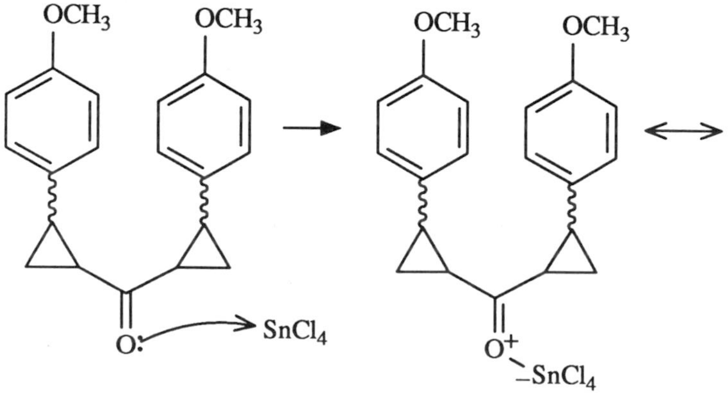

4-83-1

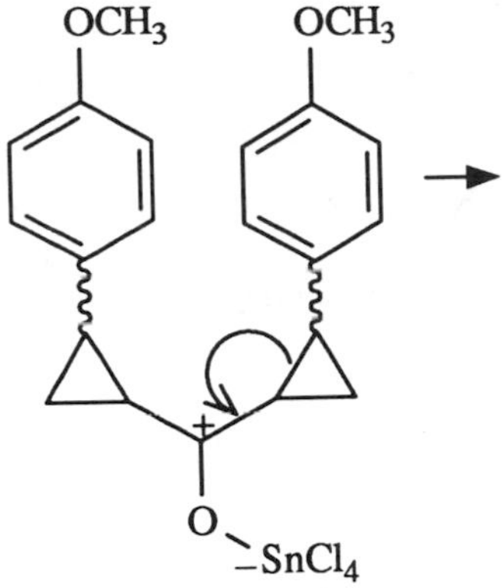

4-83-2

Intermediate **4-84** must equilibrate with the *cis* isomer, **4-85**, which undergoes a further ring opening and cyclization to **4-86**, with the same driving forces as the reaction of **4-83** to give **4-84**.

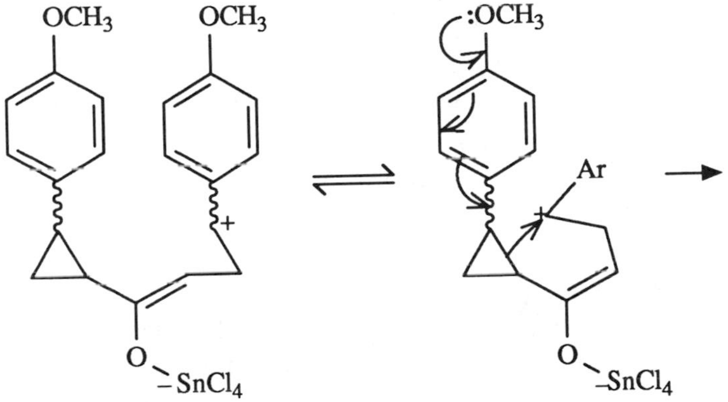

4-84   **4-85**

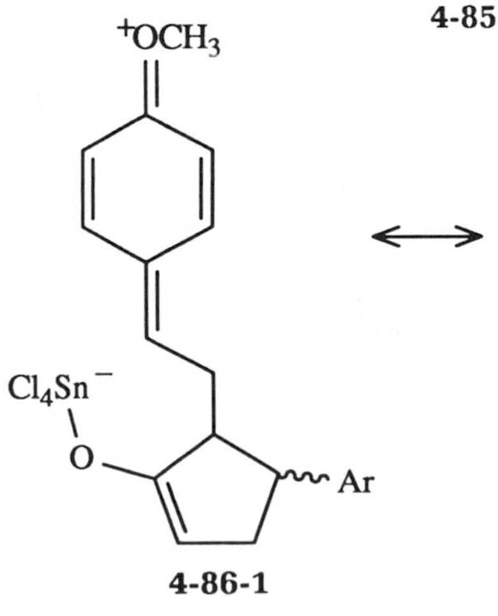

4-86-1

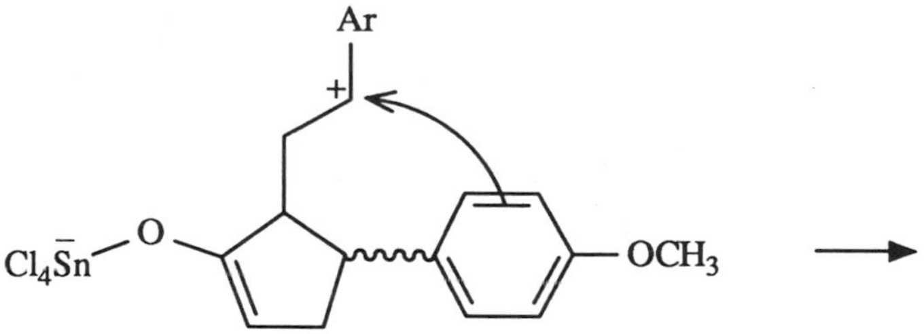

4-86-2

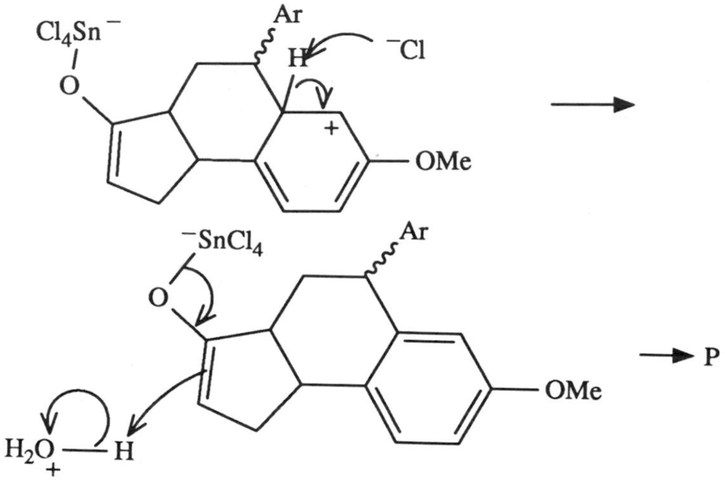

SEQUENCE 2

An alternative mechanism involves formation of an eight-membered ring intermediate from **4-85**:

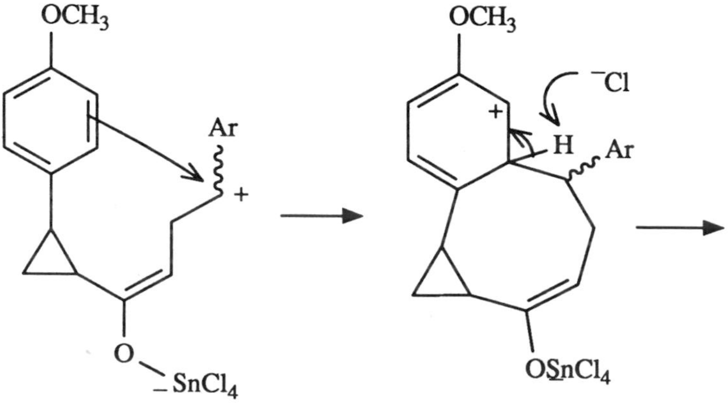

4-85

Formation of the eight-membered ring is entropically less favorable than the first mechanism. Another advantage of the first mechanism is that the methoxy group participates in resonance stabilization of the positive charge produced when the five-membered ring is formed. Direct participation of the methoxy group is not possible in the intermediates formed in the second mechanism.

The source of chloride ion, used as a base in these mechanisms, would be the following equilibrium:

$$\text{RO}\overline{\text{S}}\text{nCl}_4 \quad \rightleftharpoons \quad \text{ROSnCl}_3 \quad + \quad \text{Cl}^-$$

Nitromethane is solely the solvent for the reaction.

4-14.e. Silver ion coordinates with the chlorine and increases its ability to act as a leaving group. After the positive center, a nitrenium ion, has been created, an intramolecular electrophilic aromatic substitution takes place.

4-14.f.   The nucleophilic ester carbonyl oxygen reacts with the electrophilic tin (IV) chloride. Loss of a proton from the initially formed intermediate gives **4-87**.

The HCl, which is produced, can add in a regiospecific manner to the isolated double bond giving a 3° carbocation. An intramolecular nucleophilic reaction with this cation generates the product.

4-87

The following would not be a mechanistic step, because the anion which is generated is so unstable. Also, we would not expect such a strong base to be formed in a strongly acidic medium.

4-87

Chapter 5

Radicals and Radical Anions

5-1 Introduction

Radicals are species which contain one or more unpaired electrons. This chapter will cover only those species which contain a single unpaired electron. The unpaired electron is represented by a dot.

Radical mechanisms are portrayed in one of two ways. The most usual method is to write each individual step without using arrows. This leads to a series of equations which depict the order of events and assume one-electron transfers are taking place. A second method is to use arrows with half heads: ⟶ ; the half head denotes a single electron while, as you know, a whole head on an arrow denotes two electrons.

5-2 Formation of Radicals

Many radicals are produced by homolytic cleavage of bonds. The energy for such bond breaking comes from thermal or photochemical energy or by electron-transfer reactions effected by either inorganic compounds or electrochemistry. These are the kinds of reactions which begin (or initiate) reactions which proceed by a radical mechanism. Compounds which readily produce radicals are called initiators or free radical initiators.

Radicals produced from chlorine and bromine can be generated photochemically and/or thermally. Because of the lower reactivity of bromine relative to chlorine atoms, brominations are often done in the presence of both heat and light.

$$Cl_2 \xrightarrow{\text{light}} 2\ Cl\cdot$$

$$Br_2 \xrightarrow[\text{heat}]{\text{light}} 2\ Br\cdot$$

Many peroxides and azo compounds can be heated to generate radicals. Peroxides decompose readily because of the weak O-O bond, while azo compounds cleave readily because of the driving force provided by the formation of the stable nitrogen molecule. Common examples of these decompositions are shown below. (The $t_{1/2}$ is the time it takes half of the material to decompose.)

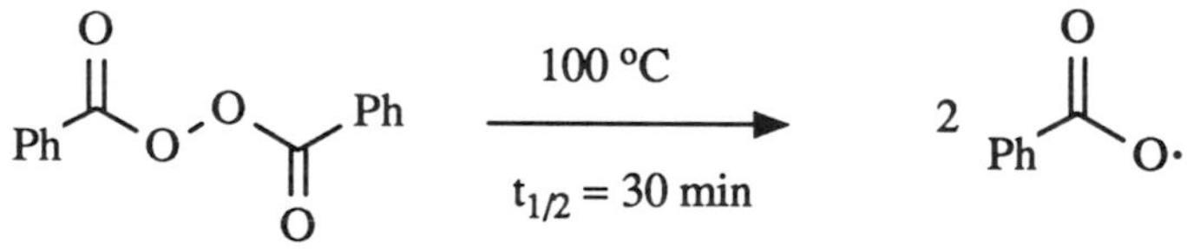

benzoyl peroxide

$$\text{t-BuOOt-Bu} \xrightarrow[t_{1/2} = 1\ h]{150\ ^\circ C} 2\ \text{t-BuO}\cdot$$

di-t-butylperoxide

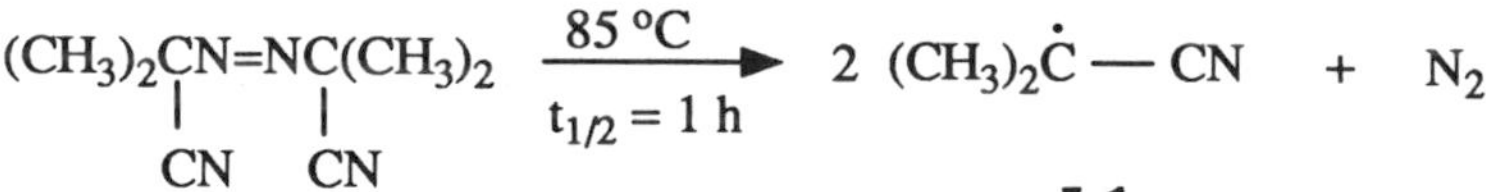

5-1

AIBN = azobisisobutyronitrile

Many radicals remove hydrogen atoms from organic molecules. This process is called hydrogen abstraction. Ordinarily, the energy necessary to break a C-H bond is so large that reaction cannot occur unassisted, and some bond making occurs in the transition state. In other words, the rate of elimination of atomic hydrogen, H•, is very slow, and the formation of atomic hydrogen is not written in ordinary radical reactions. Some radicals are so stable that they cannot abstract hydrogen from most organic compounds. An example of such a radical is **5-1**, formed from AIBN. Thus, when side reactions of hydrogen abstraction are to be avoided, AIBN can be a good choice of initiator.

Because most radicals are electrophilic, the effects of structure on their rate of formation are very similar to those for the formation of carbocations. Thus, the rates of hydrogen abstraction are 1°<2°<3°, corresponding to the order of stability of the resulting radicals: 1°<2°<3°. Also, it is relatively easy to abstract a hydrogen from an allylic or benzylic position, because the resulting radicals are stabilized by delocalization. In contrast, it is quite difficult to abstract vinyl and aromatic hydrogens because the electrophilicity of the resulting radicals is increased because of the higher s character of an sp^2-hybridized orbital relative to an sp^3-hybridized orbital. In these cases, the sp^2-hybridized orbital is perpendicular (orthogonal) to the π system, and thus the radical **cannot** be stabilized by resonance. Finally, it is difficult to abstract a hydrogen from the alcohol functional group: the resulting alkoxy radical is quite unstable because of the high electronegativity of oxygen.

Reactive radicals are often produced by abstraction of a halogen atom from a substrate. A commonly used halogen-abstracting reagent is tri-n-butyltin radical, formed from tri-n-butyltin hydride using AIBN as an initiator. AIBN generates **5-1,** which abstracts a hydrogen atom from the tri-n-butyltin hydride, generating the tri-n-butyltin radical. This tin radical can abstract a halogen from a variety of substrates: alkyl, olefinic, or aryl chlorides, bromides, or iodides, generating the corresponding radical and tri-n-butyltin halide.

$$(CH_3)_2\overset{\bullet}{C} - CN + (n\text{-Bu})_3SnH \longrightarrow (CH_3)_2CH - CN + (n\text{-Bu})_3Sn\cdot$$

5-1

$$(n\text{-Bu})_3Sn\cdot + RBr \longrightarrow (n\text{-Bu})_3SnBr + R\cdot$$

Tri-n-butyltin radicals can also be used to generate radicals from selenium compounds. An example is the formation of acyl radicals from seleno esters.

Boger, D. L.; Mathvink, R. J. *J. Org. Chem.* **1988**, *53*, 3377–3379.

Radicals can also be synthesized by the reduction of alkylmercury salts. For example, in the presence of sodium borohydride, compound **5-2** reacts to form the radical **5-3**.

Giese, B.; Horler, H.; Zwick, W. *Tetrahedron Lett.* **1982**, *23*, 931–934.

5-3 Radical Chain Processes

Most synthetically useful radical reactions occur as chain processes. A radical chain process is one in which many moles of product are formed for every mole of radicals produced. These processes start with one or more initiation step(s) to produce a radical from starting material. This radical then enters a series of steps, called propagation steps, the last of which produces product and a new radical, which can then start the series of propagation steps over again. If the propagation steps are exothermic, a cyclical process occurs in which many product molecules are formed from just one radical molecule, produced in an initiation step. Termination steps involve the removal of radicals from the propagation steps, thus ending a chain. This often occurs by coupling, disproportionation, or abstraction. A compound which stops a chain *via* an abstraction process is called a chain transfer agent. Such compounds remove a radical from the propagation step but generate a new radical in its place. When the new radical is stable, the chain transfer results in termination.

Helpful Hint 5-1. The propagation steps of a chain process must add up to the overall equation for the reaction. The overall equation will not contain any radicals. This means that the radical produced in the last propagation step must be the same as a reactant in the first propagation step.

Radical coupling reactions, reactions in which two radicals react to form a covalent bond, are commonly not good reactions to write as significant product forming steps, because both radicals are reactive intermediates and present in extremely low concentration. This means that the probability for their reacting together is very small. That is, the rate for their reaction, which depends on their concentration, will be very low, and other processes will compete effectively.

Helpful Hint 5-2. Coupling reactions are not energetically efficient ways for generating products in radical reactions.

All of these aspects of radical chain processes are illustrated in the following example.

Example 5-1. Radical chain halogenation by t-butyl hypochlorite.

The overall process is:

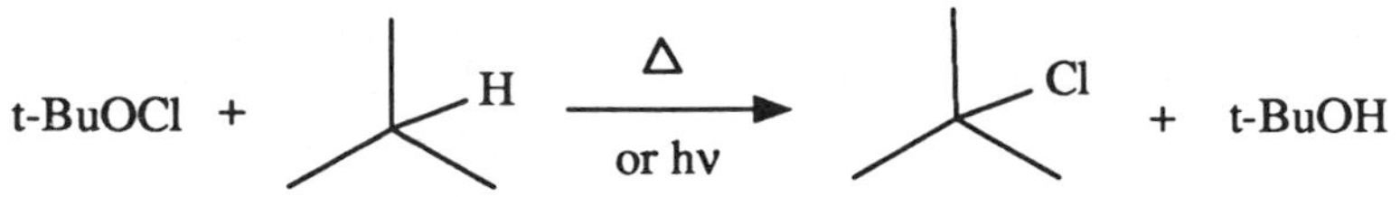

Initiation Step:

Consider t-BuO· to be the chain carrying radical. This means is that t-BuO· will be used up in the first propagation step but regenerated in the last, in this case the second, propagation step.

Propagation Steps:

(2)   + t-BuOCl ⟶ Cl + t-BuO·

Thus, the radical formed in the second step can start another reaction (1). That is why these processes are called chain processes. In other words, starting with just one t-BuO• radical, many sets of propagation steps are carried out. Adding steps (1) and (2) gives the equation for the overall reaction; the radicals cancel when the addition is performed.

Is there another possible reaction pathway? Could the chlorine radical be the chain-carrying radical? If so, equation (3), an exothermic reaction, could represent one of the propagation steps. Now a step must be found which forms the t-butyl chloride product **and** regenerates chlorine radical. This rules

(3) H + ·Cl ⟶ · + HCl

out equation (2) as a product-forming step, because that reaction does not generate the same radical that is used in equation (3). Thus equations (2) and (3) do not represent a chain process and are not feasible energetically, because they require one initiation step for each molecule of product formed.

Another possible product-forming step would be reaction of the t-butyl radical with a chlorine atom. However, this reaction removes radicals without forming new ones, and thus could not be part of the propagation steps of a chain process.

· + ·Cl ⟶ Cl

Termination Steps:
Disproportionation:

The disproportionation process involves the abstraction of hydrogen by one radical from another radical. Abstraction of a hydrogen atom from the carbon adjacent to the radical produces a double bond:

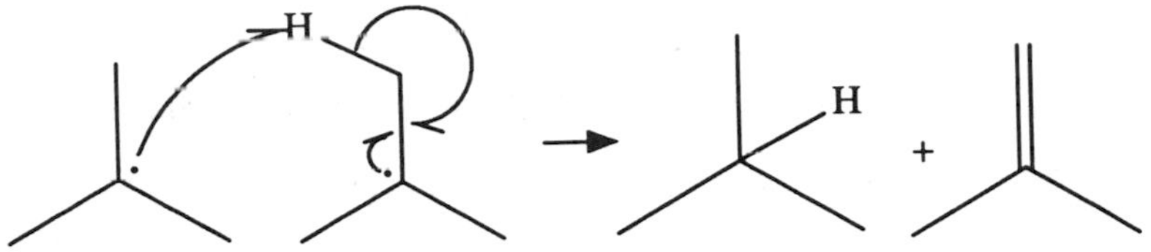

Radical Coupling:
Some possible coupling reactions follow:

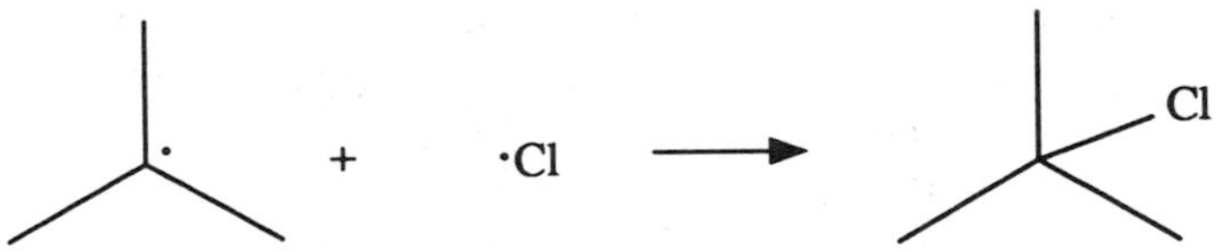

t-BuO· + t-BuO· ⟶ t-BuOOt-Bu

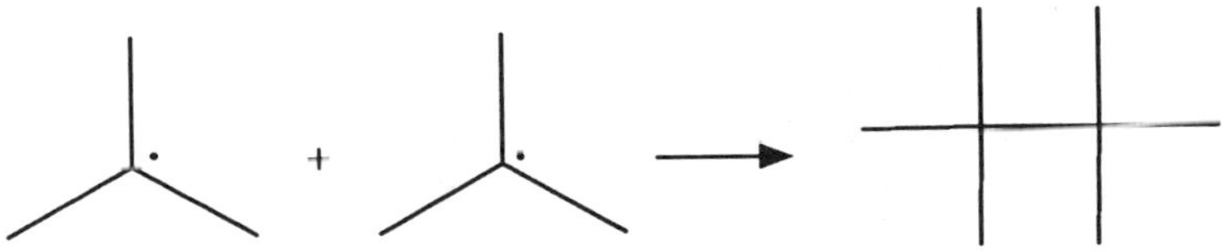

The following radical coupling does not affect the chain process since chlorine atoms are not involved in chain propagation.

·Cl + ·Cl ⟶ Cl—Cl

Problem 5-1. What other radical coupling reactions are possible for the above reaction of t-butyl hypochlorite?

5-4 Radical Inhibitors and Chain Transfer Agents

Radical reactions can be slowed or stopped by the presence of substances called inhibitors. Some common radical inhibitors are shown in Figure 5-1.

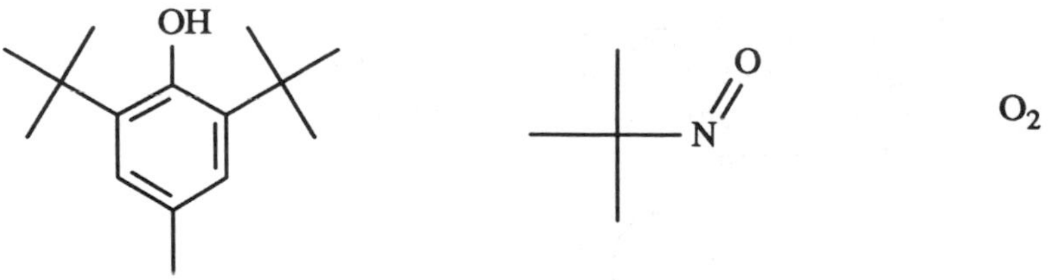

2,6-di-*t*-butyl-4-methylphenol (BHT)   2-nitroso-2-methylpropane   oxygen

Figure 5-1. Common Radical Inhibitors.

BHT is an acronym for butylated hydroxytoluene, an antioxidant commonly used in food. The reaction of oxygen with unsaturated fats gives, after several steps, hydroperoxy radicals which, upon further reaction, give smaller odiferous molecules which can ruin the palatability of foods. BHT can intercept this process by acting as a source for hydrogen abstraction by the hydroperoxy radicals, ROO·. This reaction gives ROOH, which is much less reactive than ROO·. The new radical, **5-4**, which is formed from BHT, is also much less reactive for two reasons: (1) it is stabilized by resonance, and (2) the groups attached to the ring sterically hinder further reaction.

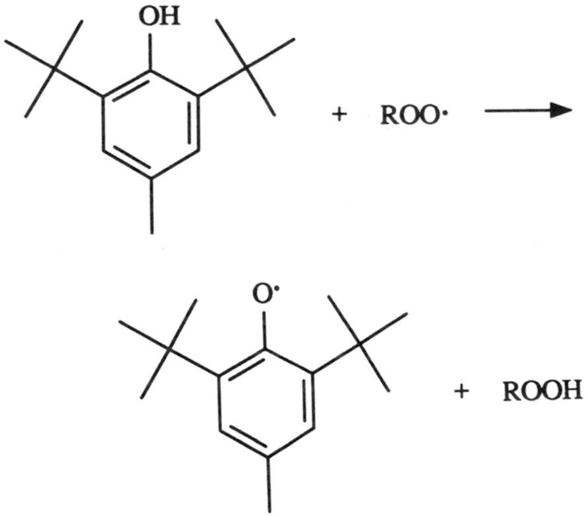

5-4

Radicals can be trapped by their addition to the nitroso group to form nitroxide radicals. In the following equation, diacetyl peroxide is a source of methyl radicals, which are trapped by the nitroso compound. (For further discussion of fragmentation of radicals, see Section 5-7.)

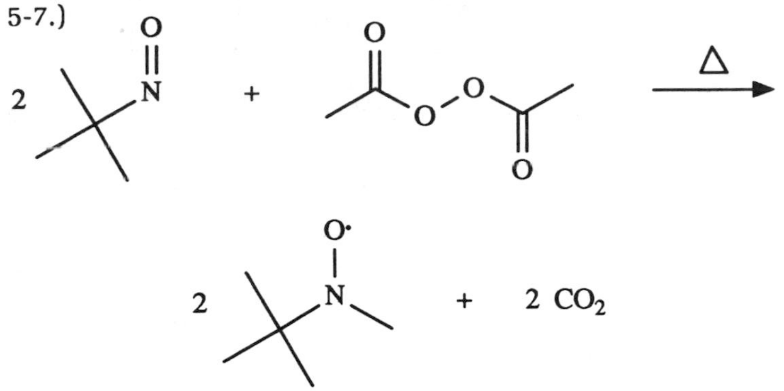

Oxygen can act as an inhibitor by reacting with radicals to produce less reactive hydroperoxy radicals.

$$R\cdot \quad + \quad O_2 \quad \longrightarrow \quad ROO\cdot$$

Figure 5-2 shows two compounds which effectively inhibit electron-transfer reactions.

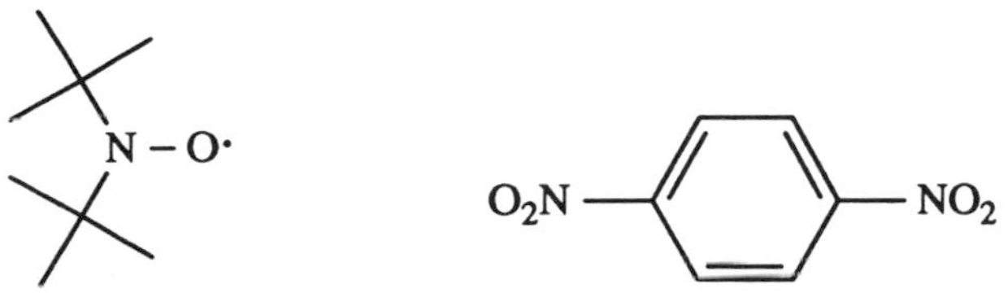

di-*t*-butylnitroxide   1,4-dinitrobenzene

Figure 5-2. Inhibitors of Electron Transfer Reactions.

Transfer of an electron to 1,4-dinitrobenzene gives an anion radical,

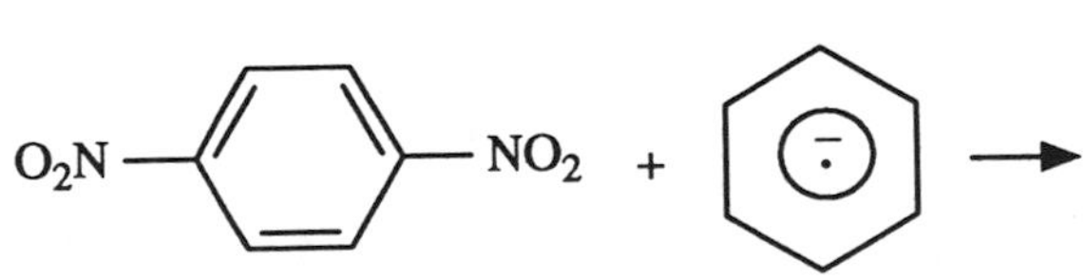

$$O_2N - \overset{\cdot}{\underset{\ominus}{\bigcirc}} - NO_2 \ + \ \bigcirc$$

so stable that it is unlikely to transfer an electron to anything else. The reaction of benzene anion radical with 1,4-dinitrobenzene proceeds to the right, because the anion radical of the 1,4-dinitrobenzene is much more stable than the anion radical of benzene itself. Although 1,3-dinitrobenzene is sometimes used for the inhibition of electron-transfer reactions, it is not as effective as the 1,4-isomer, because its anion radical is not as stable (see Problem 1-5).

Nitroxides are also used to remove anion radicals from a reaction sequence. For example, di-t-butylnitroxide has often been used to inhibit the $S_{RN}1$ reaction, (see Section 5-10). Apparently, nitroxides remove radicals *via* coupling reactions. (Hoffmann, A. K.; Feldman, A. M.; Gelblum, E.; Hodgson, W. G. *J. Am. Chem. Soc.* **1964**, *86*, 639–646.)

A radical reaction, which is initiated only by light, can be prevented by omitting light. Under such circumstances, any reaction which takes place in the dark must proceed by a nonradical pathway.

5-5 Determination of Thermodynamic Feasibility of Radical Reactions

The bond dissociation energies, tabulated below, can be used to determine the effectiveness of radical reactions. In general, a chain will be long enough for a feasible synthetic process, only if all propagation steps are exothermic. Otherwise the chains will be short. For example, the calculations of Example 5-2 show that the radical chlorination of 2-methylpropane by t-butyl hypochlorite is a possible method for the synthesis of t-butyl chloride.

Example 5-2. Determination of the enthalpy change in chlorination by t-butyl hypochlorite.

Consider the reactions shown in Example 5-1. If both propagation steps are exothermic, the chain length will be long enough so that the thermochemistry of the initiation step(s) is unimportant relative to that of the propagation steps.

Table of Bond Dissociation Energies (BDEs) at 298 K in kcal/mole[a]

		Cl—Cl[b] 58 H—Cl[b] 103	Br—Br[b] 46 H—Br[b] 87	I—I[b] 36 H—I[b] 36	F—F[b] 37 H—F[b] 135
CH₃—H	104	CH₃—Cl 73	CH₃—Br 70		
C₂H₅—H	98	C₂H₅—Cl 81	C₂H₅—Br 69	C₂H₅—I 53	C₂H₅—F 106
CH₃CH₂CH₂—H	98	n-C₃H₇—Cl 82	n-C₃H₇—Br 69		
(CH₃)₂CH—H	94.5	(CH₃)₂CH—Cl 81	(CH₃)₂CH—Br 68		
(CH₃)₃C—H	91.0	(CH₃)₃C—Cl 79	(CH₃)₃C—Br 63		
PhCH₂—H	85	PhCH₂—Cl 68	PhCH₂—Br 51		
CH₂=CHCH₂—H	85				
CCl₃—H	95.7	CCl₃—Cl 73	CCl₃—Br 54		
Ph—H	104		Ph—Br 71		
CH₂=CH—H	104				
HOCH₂—H	92				
CH₃O—H	102				
C₂H₅O—H	102				
i-C₃H₇O—H	103				
t-C₄H₉O—H	103	t-C₄H₉O—Cl 44[c]			
CH₃COO—H	112				
CH₃S—H	88				
PhS—H	75				
CH₃—CH₃	88				
PhCO—H	74	PhCO—Cl 74			
HCO—H	88				
CH₃CO—H	88				
CH₃COCH₂—H	92				

References: (a) All values from Kerr, J. A. *Chem. Rev.* 1966, *66*, 465–500, unless otherwise noted. (b) Weast, R. C.; Astle, M. J., Eds *Handbook of Chemistry and Physics*, 63rd ed. 1982-3, Boca Raton: CRC Press, F186ff. (c) Walling, C.; Jacknow, B. B. *J. Am. Chem. Soc.* 1960, *82*, 6108–6112.

Let's calculate the enthalpy change for the reactions in the propagation steps. In equation (1), the C-H bond broken has a BDE of 91.0 and the O-H bond formed has a BDE of 103 kcal/mole. Therefore, this first reaction is exothermic by 91–103 or –12 kcal/mole.

(1)

$$\Delta H = -12 \text{ kcal/mole}$$

For equation (2), the O-Cl bond broken has a BDE of 44 kcal/mole, and the C-Cl bond formed has a BDE of 79 kcal/mole. Thus, the second reaction is exothermic by 44–79 kcal/mole or –35 kcal/mole. Because both reactions are substantially exothermic,

(2)

$$\Delta H = -35 \text{ kcal/mole}$$

this should be a highly favorable process with a long chain length. The enthalpy change for the overall reaction is [(–12) + (–35)] or –47 kcal/mole.

Problem 5-2. Why is the following mechanism for the chlorination of 2-methylpropane by t-butyl hypochlorite not the best mechanism that can be written? You will need to consider the thermochemistry of the processes as well as the fact that a chain process is not involved.

5-6 Addition of Radicals

Radical addition reactions are commonly used in organic synthesis. Such additions range from the simple addition of halocarbons to π bonds to cyclization reactions with demanding stereoelectronic requirements.

5-6-1 ADDITIONS WITHOUT CYCLIZATION

Common types of radicals which add to π bonds are those which can be generated from: RX (where X=halogen), mercaptans, thiophenols, thioacids, aldehydes, and ketones. Like the corresponding electrophilic additions to double bonds, many radical reactions are either regiospecific or highly regioselective.

Example 5-3. Photochemical addition of trifluoroiodomethane to allyl alcohol.

In the initiation step of this reaction, light induces homolytic cleavage of the weak C-I bond.

$$ CF_3I \xrightarrow{\ h\nu\ } \cdot CF_3 \quad + \quad \cdot I $$

The trifluoromethyl radical adds to the double bond, giving the most stable intermediate radical:

$$ F_3C\cdot \quad CH_2 = CHCH_2OH \longrightarrow F_3CCH_2\dot{C}HCH_2OH $$

This radical then abstracts an iodide atom from trifluoroiodomethane to generate product and the chain propagating radical, the trifluoromethyl radical.

$$ F_3CCH_2\dot{C}HCH_2OH + ICF_3 \longrightarrow F_3CCH_2CHICH_2OH + \cdot CF_3 $$

(Note: The paper, presenting this work, reports the product as the isomer shown. However, the evidence cited is only sufficient to claim that this is the major regioisomer produced. That is, the other regioisomer might be a minor component of the reaction mixture. See Park, J. D.; Rogers, F. E.; Lacher, J. R. *J. Org. Chem.* **1961**, *26*, 2089–2095.)

Problem 5-3. Write a step-by-step mechanism for the following reaction:

$$CH_2{=\!=}CHC_6H_{13} \;+\; CBr_4 \xrightarrow[\Delta]{(PhCO_2)_2} Br_3CCH_2CHBrC_6H_{13}$$

Example 5-4. Addition of a radical, formed by reduction of a C-Hg bond.

Addition of radicals, formed from mercury compounds, to alkenes often produce good to excellent yields. The following mechanism illustrates a reaction with a yield of 64%.

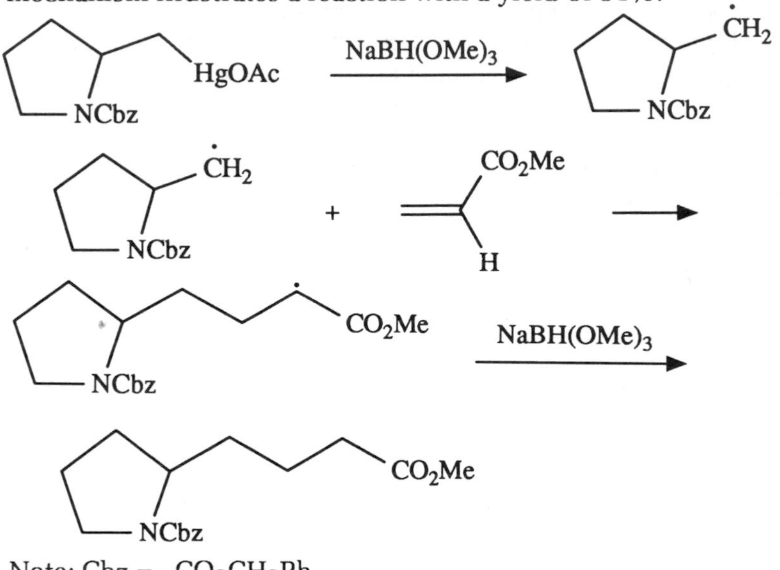

Note: Cbz = $-CO_2CH_2Ph$.

Danishefsky, S.; Taniyama, E.; Webb, II, R. R. *Tetrahedron Lett.* **1983**, *24*, 11–14.

5-6-2 RADICAL CYCLIZATION REACTIONS

In recent years, radical cyclizations have been developed into very useful synthetic reactions. Beckwith, Baldwin, and their co-workers have delineated a number of guidelines. (See Beckwith, A. L. J.; Easton, C. J.; Serelis, A. K. *J. Chem. Soc. Chem. Commun.* **1980**, 482–483 and Baldwin, J. E. *ibid.* **1976**, 734–736.) These papers describe two possible regiochemical outcomes for cyclization, *exo* and *endo*.

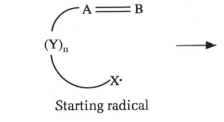

Starting radical

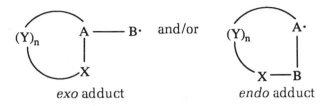

exo adduct   *endo* adduct

For our purposes, the most important guidelines for radical cyclizations are as follows:

1. For ⍵-alkenyl radicals containing up to 8 carbons, when A=B is $CH=CH_2$, the preferred cyclization mode is *exo*. (⍵ means that the alkene is at the terminus away from the radical.) The reaction is controlled by kinetics; the product which forms faster predominates. The rates of formation of the two possible adducts are a consequence of the stereochemical requirements of their transition states. Often the *endo* product is more stable than the *exo* product; but, when this is the case, the more stable product is not the predominant product formed.

2. If the A position has a substituent, (A=B is $CR=CH_2$) reaction to form the *exo* adduct is sterically hindered, and *endo* product predominates. Under these circumstances, it often is the more stable product which predominates.

Example 5-5. Intramolecular cyclization of a vinyl radical.

The overall reaction is shown on the next page.

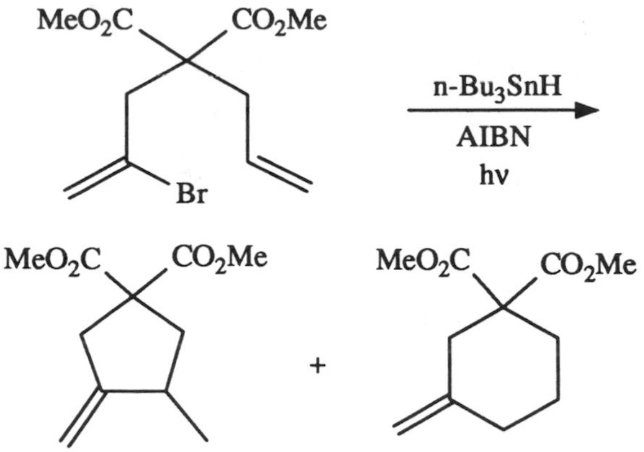

MeO$_2$C   CO$_2$Me

[structure with Br]

$\xrightarrow[\substack{\text{AIBN} \\ \text{h}\nu}]{\text{n-Bu}_3\text{SnH}}$

MeO$_2$C   CO$_2$Me

[cyclopentane structure]

5-5

+

MeO$_2$C   CO$_2$Me

[cyclohexane structure]

5-6

Stork, G.; Baine, N. H. *J. Am. Chem. Soc.* **1982**, *104*, 2321–2323.

By the process shown in Section 5-2, vinyl radical, **5-7**, is formed. This can cyclize to give an *exo* adduct or an *endo* adduct. The resulting radical abstracts hydrogen from n-Bu$_3$SnH to form products **5-5** and **5-6** as well as tri-n-butyltin radical, which continues the chain. As is predicted by the guidelines, the *exo* product, **5-5**, predominates by a factor of 2.

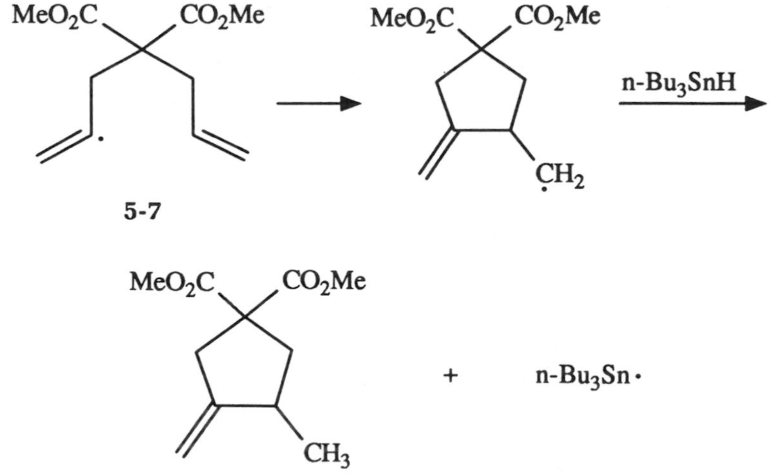

MeO$_2$C   CO$_2$Me

[radical structure]

5-7

$\longrightarrow$

MeO$_2$C   CO$_2$Me

[cyclopentane structure with CH$_2$]
.CH$_2$

$\xrightarrow{\text{n-Bu}_3\text{SnH}}$

MeO$_2$C   CO$_2$Me

[cyclopentane structure with CH$_3$]
CH$_3$

+   n-Bu$_3$Sn ·

Problem 5-4. a. Write the initiation and propagation steps for the following reaction:

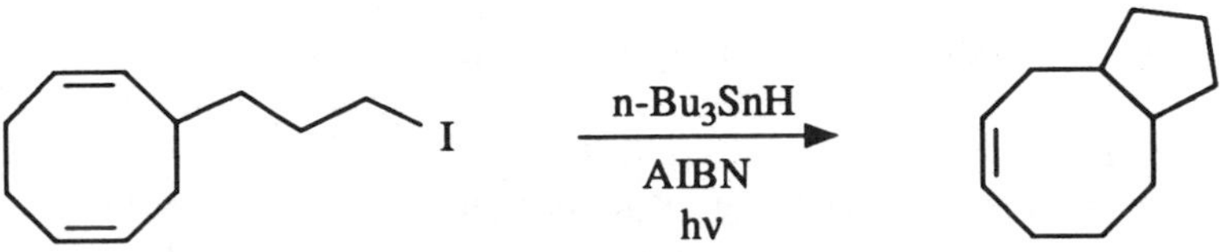

b. The following was a minor product, isolated from the reaction mixture. Show how it might have been formed.

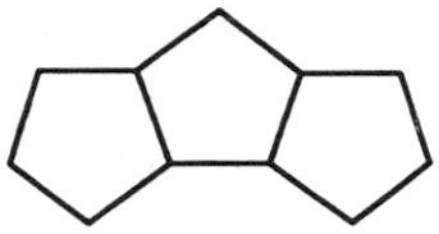

Winkler, J. D.; Sridar, V. *J. Am. Chem. Soc.* **1986,** *108,* 1708–1709.

Problem 5-5. Write the initiation and propagation steps for the following reaction:

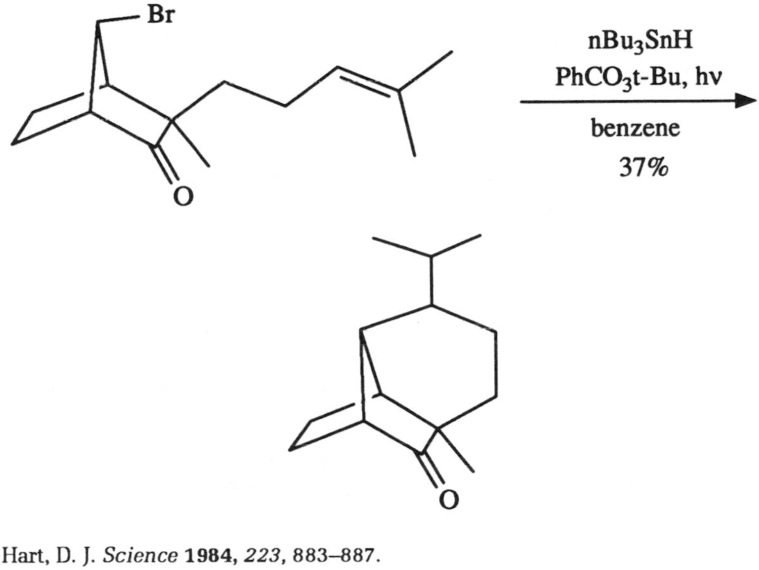

Hart, D. J. *Science* **1984,** *223,* 883–887.

5-7 Fragmentation Reactions

Many radical processes involve the loss of small stable molecules, such as carbon dioxide, nitrogen, or carbon monoxide. Such reactions are called fragmentations. For example, the radical initiator, diacetyl per-oxide, homolyzes at the O-O bond to form carboxy radicals, which then readily lose CO_2 to give methyl radicals. In fact, between 60 and 100 °C, acetyl peroxide can be a convenient source of methyl radicals.

(See Walling, C. Free Radicals in Solution, **1957**, New York: John Wiley, p. 493.) Aryl-substituted carboxy radicals also lose CO_2, but they do so much less readily (e.g., the initial radical formed from benzoyl peroxide, page 298).

The radical, initially produced by homolytic decomposition of a dialkyl peroxide, can undergo further scission. This depends on the temperature and on the stability of the resulting radical(s). For ex-ample, t-butoxy radicals decompose, on heating, to methyl radicals and acetone.

$$t\text{-BuO}\cdot \longrightarrow \ \cdot CH_3 \ + \ CH_3COCH_3$$

Azo compounds often decompose with loss of nitrogen: decomposi-tion of the initiator, AIBN, is an example. (See page 298.)

At elevated temperatures, the radical, **5-8**, generated by hydrogen abstraction from the aldehyde functional group, fragments to give a new carbon radical and carbon monoxide. The fragmentation de-pends on heat: the higher the temperature, the more favorable is the loss of CO from the initial radical.

5-8

When a reagent is present, from which a hydrogen may be readily abstracted, such abstraction may compete with loss of CO.

$$\underset{R}{\overset{O}{\|}}\cdot \quad + \quad R'\!-\!H \quad \longrightarrow \quad \underset{R}{\overset{O}{\|}}H \quad + \quad R'\cdot$$

Another common fragmentation reaction is the loss of CO_2 from RCO_2Br, formed from RCO_2Ag and bromine, the Hunsdiecker reaction:

$$RCO_2Ag \quad + \quad Br_2 \quad \xrightarrow{\Delta} \quad RX \quad + \quad CO_2 \quad + \quad AgBr$$

The mechanism is:

$$RCO_2Ag \quad + \quad Br_2 \quad \longrightarrow \quad RCO_2Br \quad + \quad AgBr$$

Initiation:

$$RCO_2Br \quad \longrightarrow \quad RCO_2\cdot \quad + \quad Br\cdot$$

Propagation 1:

$$RCO_2\cdot \quad \longrightarrow \quad R\cdot \quad + \quad CO_2$$

Propagation 2:

$$R\cdot \quad + \quad RCO_2Br \quad \longrightarrow \quad RBr \quad + \quad RCO_2\cdot$$

Example 5-6. Addition followed by fragmentation.

The radical reaction of carbon tetrachloride with aliphatic double bonds involves addition of the trichloromethyl radical to the double bond, followed by chlorine atom abstraction from carbon tetrachloride by the intermediate radical to give the product. After the addition of the trichloromethyl radical to β-pinene, a fragmentation occurs prior to formation of the product.

The mechanism for this reaction starts with the generation of the trichloromethyl radical, in the usual manner:

$$\text{t-BuOOt-Bu} \xrightarrow{\Delta} 2 \text{ t-BuO} \cdot$$

$$\text{t-BuO} \cdot \quad \text{Cl} - \text{CCl}_3 \longrightarrow \text{t-BuOCl} + \cdot \text{CCl}_3$$

The trichloromethyl radical now adds regiospecifically to the double bond, forming a new carbon radical, **5-9**:

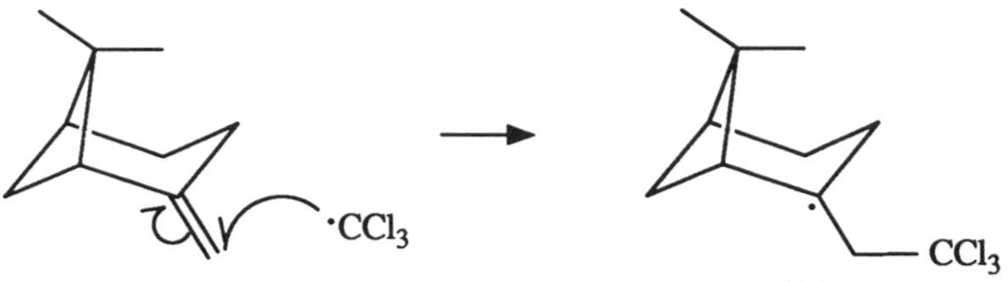

5-9

This radical can now fragment, giving another radical, **5-10** which then abstracts a chlorine atom from another molecule of CCl_4 to give the product:

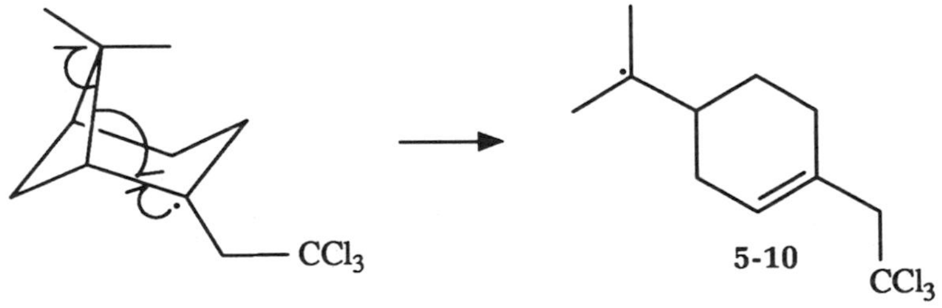

5-10

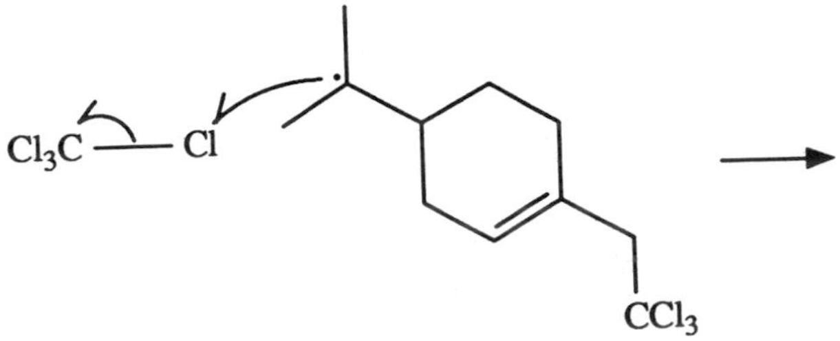

The fact that radical addition of thiolacetic acid to β-pinene gives unrearranged product is evidence for the discrete existence of radical **5-9**. That is, the rate of abstraction of a hydrogen atom from thiolacetic acid by **5-11** is faster than its rate of fragmentation.

(See Claisse, J. A.; Davies, D. I.; Parfitt, L. T. *J. Chem. Soc. (C)* **1970**, 258–262.)

Problem 5-6. Write step-by-step mechanisms for the following reactions:

a.

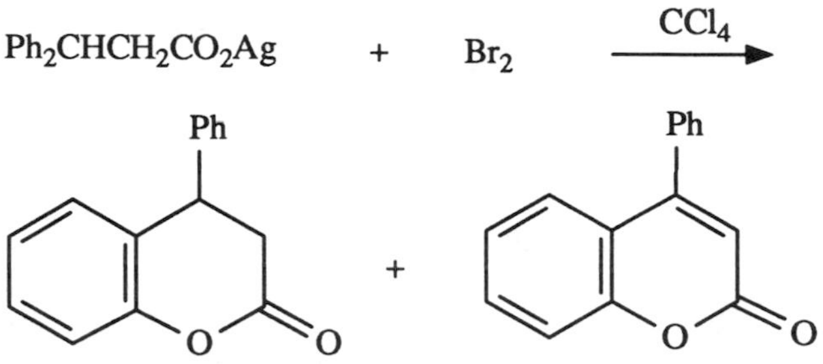

$$Ph_2CHCH_2CO_2Ag \quad + \quad Br_2 \quad \xrightarrow{CCl_4}$$

43%
5-12

+

25%
5-13

Pandet, U. K.; Dirk, I. P. *Tetrahedron Lett.* **1963**, 891–895.

b.

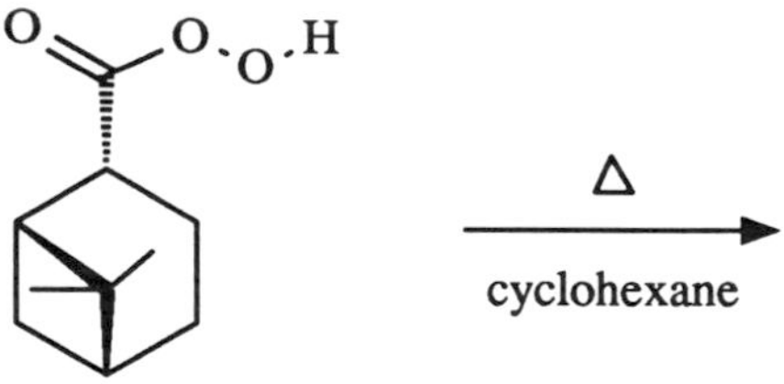

5-14

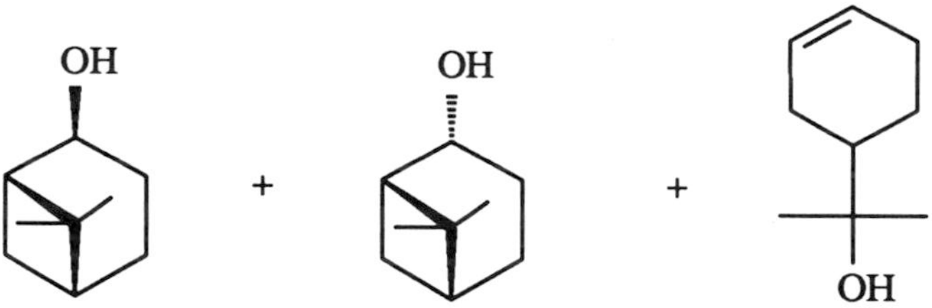

$$\xrightarrow[\text{cyclohexane}]{\Delta}$$

5-15   **5-16**   **5-17**

Your mechanism should take into account the fact that the other isomer (with the peracid group up) of the starting material reacts to give roughly the same ratio of **5-15** and **5-16**, and the fact that the ratio (5–15 + 5–16)/5–17 increases with increasing peracid concentration.

Fossey, J.; Lefort, D.; Sorba, J. *J. Org. Chem.* **1986**, *51*, 3584–3587.

5-8 Rearrangement of Radicals

In radical rearrangements, the groups which migrate are those which can accommodate electrons in a π system (vinyl, aryl, carbonyl) or atoms which can expand their valence shell, i.e. all halogens but fluorine. **Hydrogen and alkyl do not migrate** to radicals. However, an addition-elimination pathway could give the appearance of alkyl migration. (See Example 5-9.)

Example 5-7. Aryl migration.

In the following reaction, **5-19** is a product resulting from migration of an aryl group, and **5-18** is a nonrearranged product. Thus there is competition between rearrangement and hydrogen abstraction by a radical intermediate.

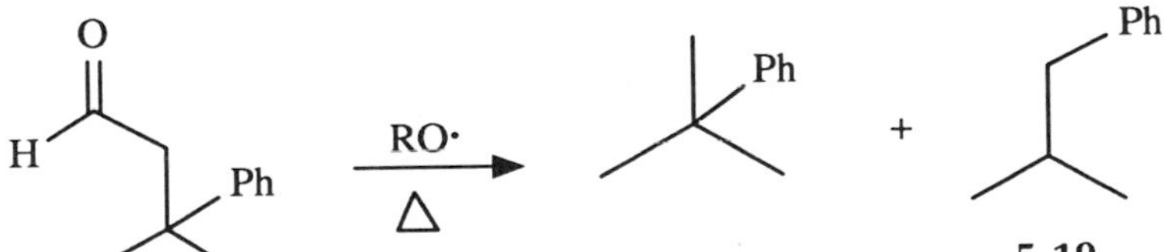

The mechanism for formation of **5-18** and **5-19** involves formation of a hydrocarbon radical from the starting aldehyde. The radical from the initiator abstracts the aldehyde proton to give a carbonyl radical. This loses carbon monoxide to give **5-20**.

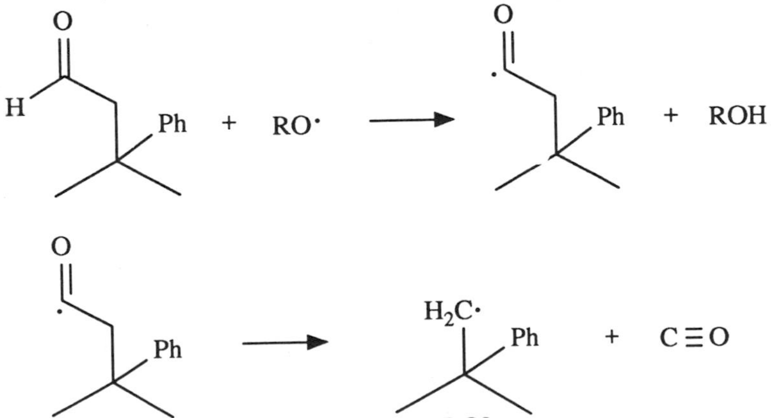

Radical **5-20** can either abstract a hydrogen atom from starting aldehyde to give nonrearranged product, **5-18**, or rearrange *via* phenyl migration to **5-21**, which then abstracts a hydrogen atom to give **5-19**.

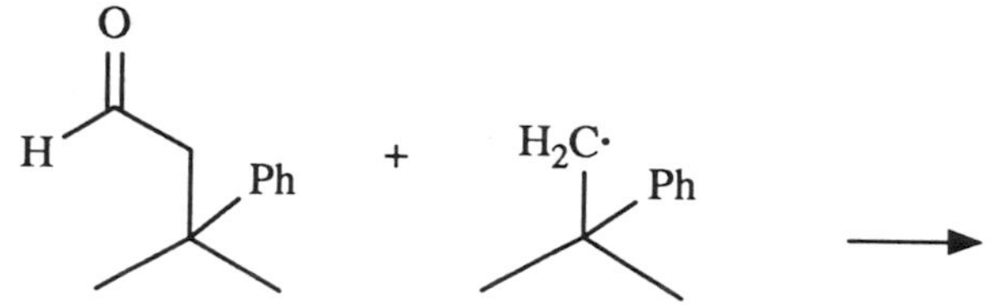

5-20

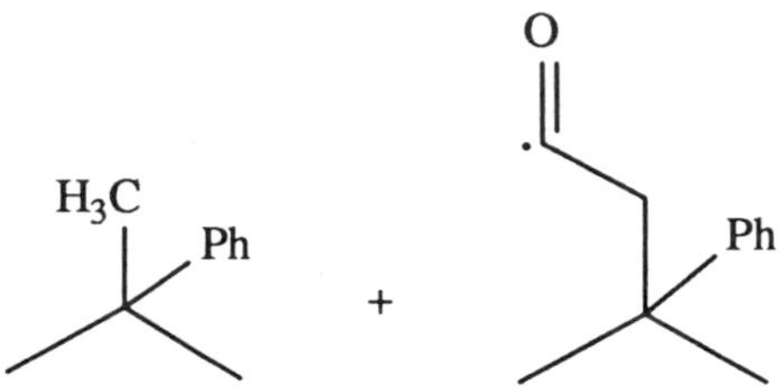

5-18

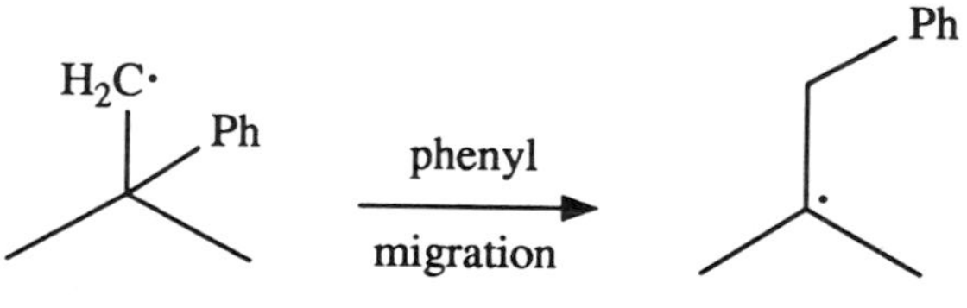

5-20   phenyl migration   5-21

5-21 +

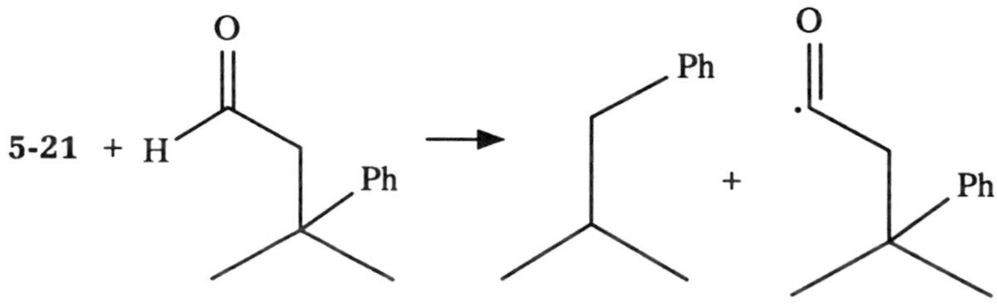

5-19

Example 5-8. Halogen migration.

The radical addition of HBr to 3,3,3-trichloropropene, involves a chlorine atom migration in an intermediate step.

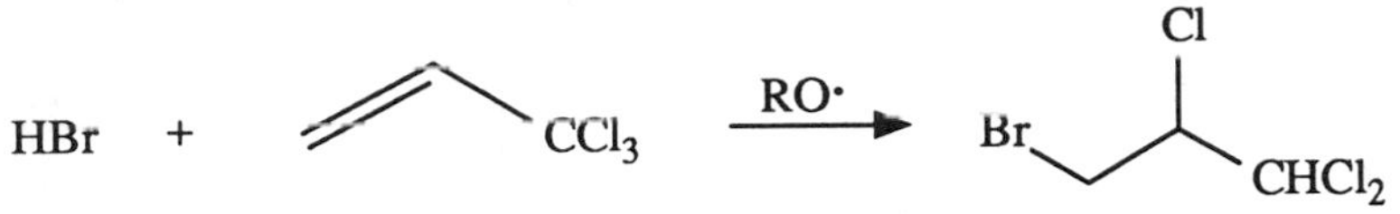

The mechanism for the reaction involves a regiospecific addition of bromine radical to the double bond.

$$RO\cdot \quad + \quad HBr \quad \longrightarrow \quad ROH \quad + \quad Br\cdot$$

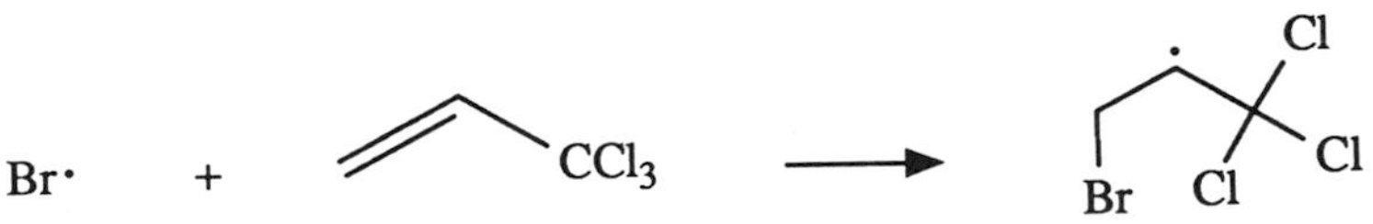

Then a chlorine atom migrates to the adjacent radical giving 5-22. Finally, **5-22** abstracts a hydrogen from HBr to form the product plus another bromine radical to continue the chain.

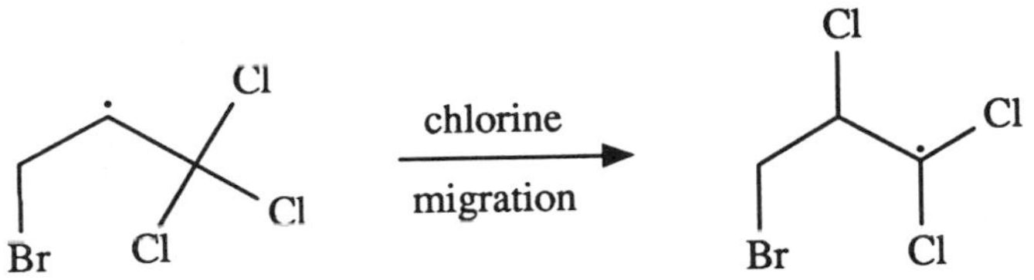

5-22

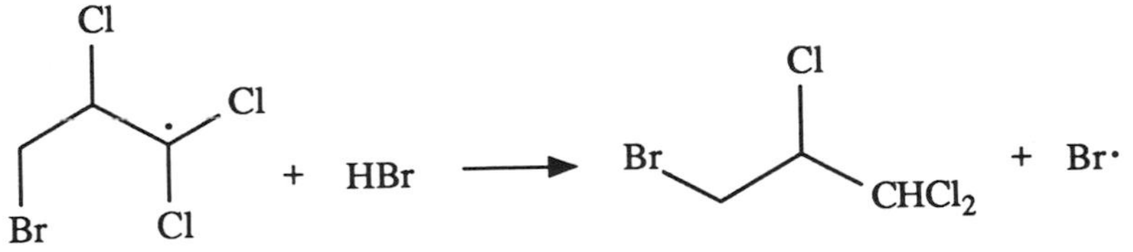

5-22

Example 5-9. An apparent alkyl migration mediated by a carbonyl group.

The overall reaction is as follows:

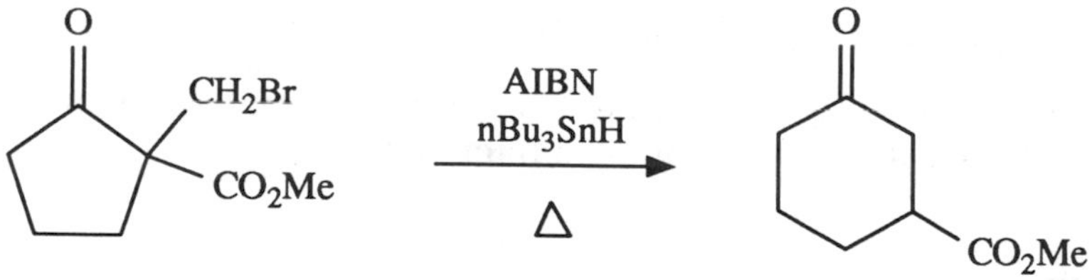

In this reaction, the tri-n-butyltin radical, formed by the usual initiation steps, removes the bromine atom from the starting material to give **5-23**. This radical then adds to the carbonyl carbon, forming a 3-membered ring which then opens at a different bond to give **5-24**.

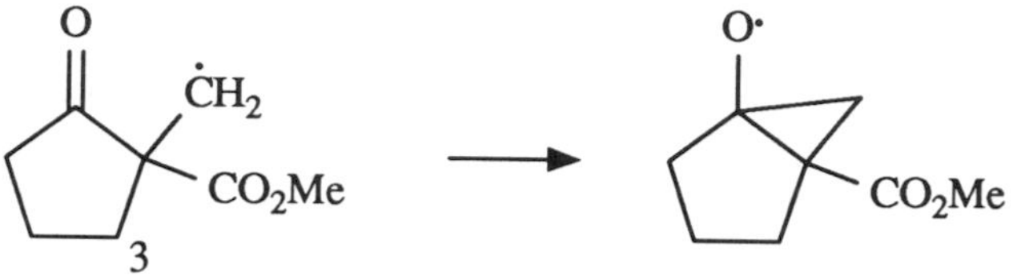

Radical **5-24** can abstract a hydrogen atom from tri-n-butyltin hydride to give the product and a new radical to carry the chain:

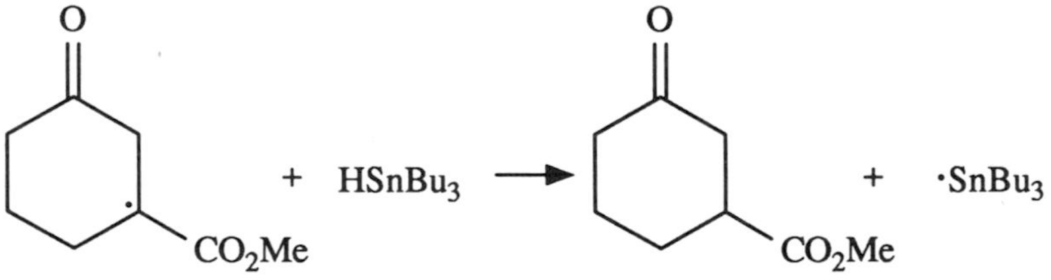

Why can't ring carbon C-3 in **5-23** migrate directly to the radical? This direct rearrangement would be the equivalent of one three center bond containing 3 electrons which means that 1 electron would have to go into an antibonding orbital. This represents such a high energy barrier to reaction that the direct rearrangement does not take place.

Dowd, P.; Choi, S.-C. *J. Am. Chem. Soc.* **1987,** *109,* 3493–3494.

Problem 5-7. Explain why one process gives rearrangement and the other does not. Write step-by-step mechanisms for both processes.

Weinstock, J.; Lewis, S. N. *J. Am. Chem. Soc.* **1957**, *79*, 6243–6247.

Problem 5-8. Write a complete mechanism for the following process:

Beckwith, A. L. J.; O'Shea, D. M.; Westwood, S. W. *J. Am. Chem. Soc.* **1988**, *110*, 2565–2575.

5-9 The $S_{RN}1$ Reaction

In recent years, the $S_{RN}1$ reaction has become a versatile synthetic tool. A general mechanism for its chain propagation steps is given below. The letters of the reaction designation stand for substitution, radical, and nucleophilic. The 1 indicates that the first two steps of the mechanism resemble those of an S_N1 reaction. (See a review by Bunnett, J. F. *Acc. Chem. Res.* **1978**, *11*, 413–420.) $S_{RN}1$ reactions encompass both aliphatic and aromatic compounds.

$$[RX]^{\overline{\cdot}} \longrightarrow R\cdot + X^-$$

$$R\cdot + Y^- \longrightarrow [RY]^{\overline{\cdot}}$$

$$[RY]^{\overline{\cdot}} + RX \longrightarrow RY + [RX]^{\overline{\cdot}}$$

General Mechanism for an $S_{RN}1$ Reaction.

$S_{RN}1$ reactions can be initiated by photochemical excitation, electrochemical reduction, and by solvated electrons (alkali metal in ammonia). In some cases, spontaneous thermal initiation can also take place. The leaving group, X^-, is often a halide, especially bromide or iodide; never fluoride. The nucleophile, Y^-, is commonly an enolate, nitranion, (e.g., **5-25**), or thiolate.

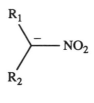

5-25

Since these are free radical reactions, they are sensitive to inhibition; often the inhibitors shown in Figure 5-2 are used.

Adding the steps of the General Mechanism gives the following equation for the overall reaction:

$$RX + Y^- \longrightarrow RY + X^-$$

This equation also describes the overall reaction of either an S_N2 or nucleophilic aromatic substitution process. In some cases, the only way to distinguish an $S_{RN}1$ reaction from these processes is that an $S_{RN}1$ is inhibited by radical inhibitors. Another distinguishing difference is that the relative leaving group abilities of halides are in the opposite order to that found for nucleophilic aromatic substitution by the addition-elimination mechanism. (See Chapter 3.)

Example 5-10. Reaction of an enolate with an aromatic iodide.

The overall reaction is:

70%

The mechanism involves photochemical excitation of the enolate. (An excited state will be indicated by an asterisk, *.) The excited enolate transfers an electron to the aromatic π system. Since a single electron has been added, there must be an unpaired electron present; thus the new intermediate, **5-26**, is a radical. Since an electron has been added to a neutral system, there must be a negative charge. Thus, **5-26** is both an anion and a radical or what is usually called an anion radical.

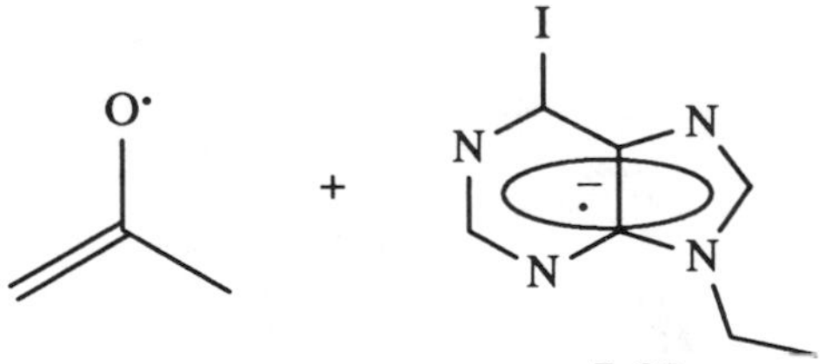

5-26

The anion radical now loses iodide ion to give a radical.

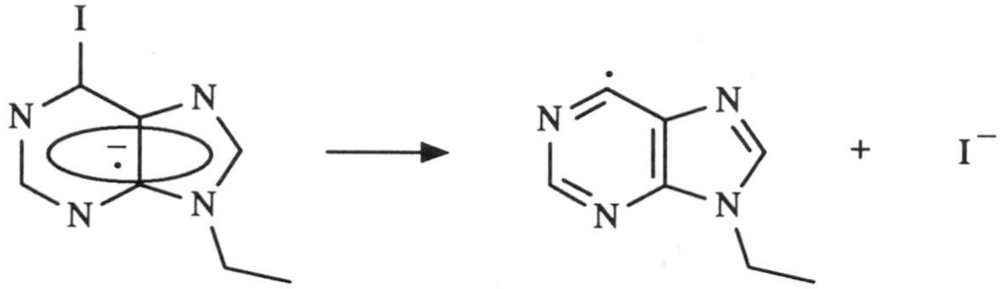

This radical couples with the enolate to give a new anion radical, **5-27**.

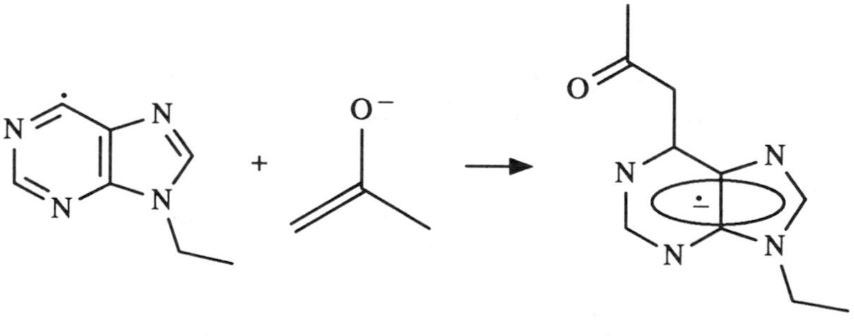

5-27

Intermediate **5-27** transfers an electron to starting material to give a molecule of product and a new molecule of **5-26** to propagate the chain.

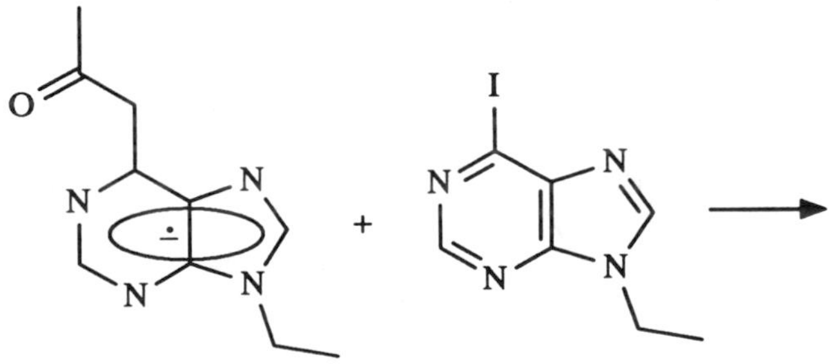

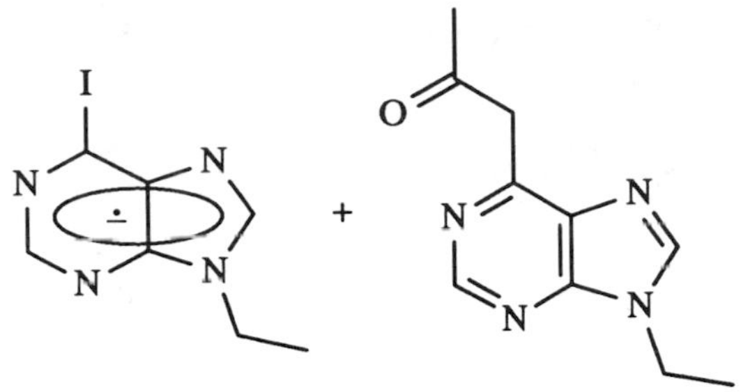

Nair, V.; Chamberlain, S. D. *J. Am. Chem. Soc.* **1985**, *107*, 2183–2185.

Problem 5-9. Write step-by-step mechanisms for the following reactions:

a.

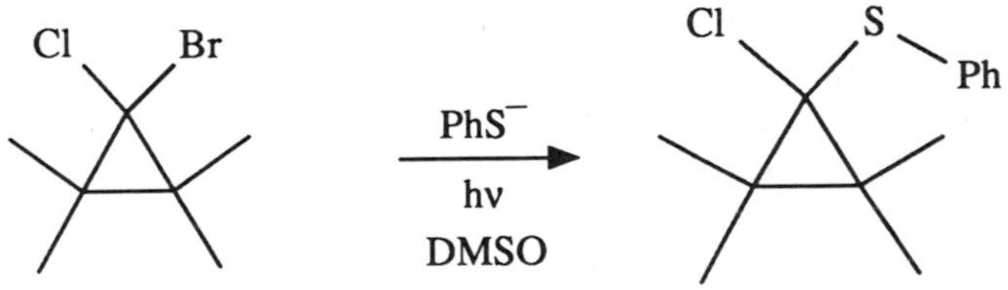

$$\xrightarrow[\text{DMSO}]{\text{PhS}^- \quad h\nu}$$

Meijs, G. F. *J. Org. Chem.* **1986**, *51*, 606–611.

b.

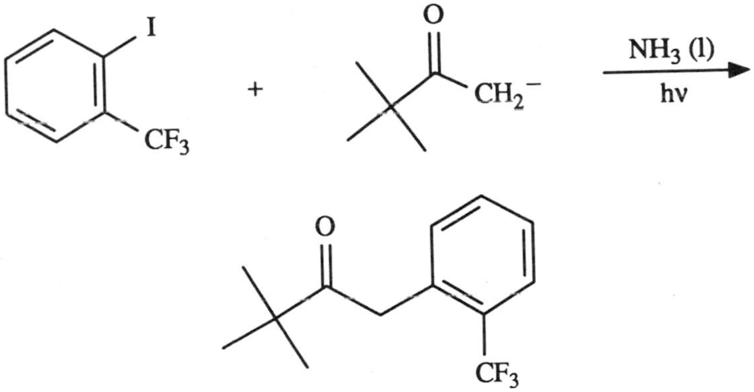

$$\xrightarrow[h\nu]{\text{NH}_3\ (l)}$$

Bunnett, J. F.; Galli, C. *J. Chem. Soc., Perkin Trans. I* **1985**, 2515–2519.

5-10 The Birch Reduction

The usual conditions for the Birch reduction are sodium in liquid ammonia, containing a small amount of ethanol. Workup is generally in acid.

Example 5-11. Birch reduction of benzoic acid.

90%

Under the basic conditions, the carboxylate ion is formed from benzoic acid. Then sodium transfers a single electron to the aromatic system to produce a pentadienyl anion radical, **5-28**. Molecular orbital (MO) calculations indicate that the favorable sites for protonation of the intermediate anion radicals are at the positions *ortho* or *meta* to a π-donating substituent and *para* or *ipso* (*ipso* means at the substituent-containing carbon) to a π-accepting substituent. (See Birch, A.J.; Hinde, A. L.; Radom, L. *J. Am. Chem. Soc.* **1980**, *102*, 3370–3376.) Also pentadienyl anions have more anionic character on the central carbon, and that is where they are protonated in this reaction. Protonation of **5-28** gives the radical, **5-29**, which will be reduced to the corresponding anion, **5-30**, by another sodium atom. Finally, **5-30** is protonated giving the product.

5-28　　　　　　　　　　　　　　　　5-29

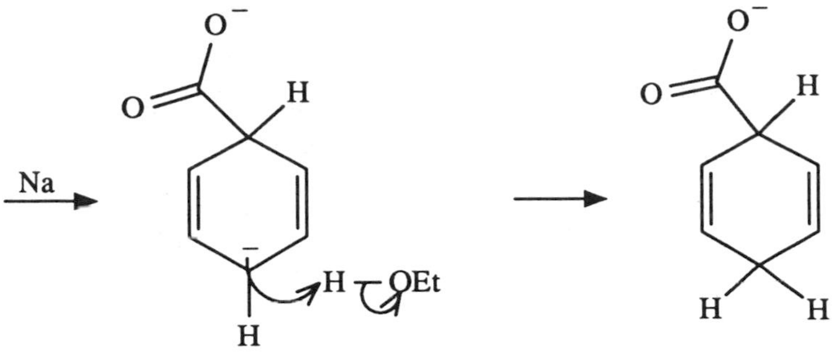

5-30

Problem 5-10. Write a mechanism for the following reaction, including an explanation of the regiochemistry observed.

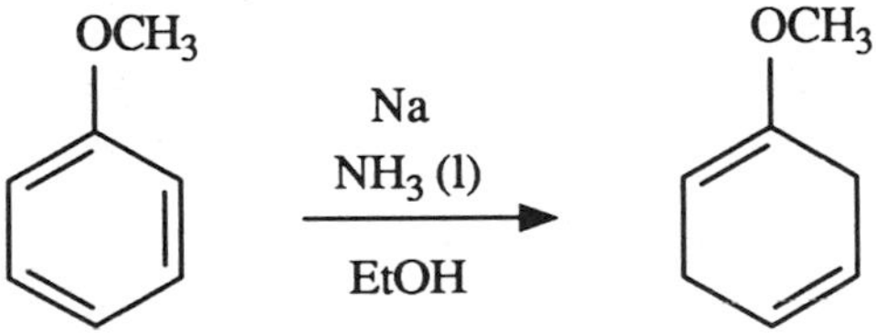

5-11 A Radical Mechanism for the Rearrangement of Some Anions

In the Wittig rearrangement, an anion derived from an ether rearranges to the salt of an alcohol.

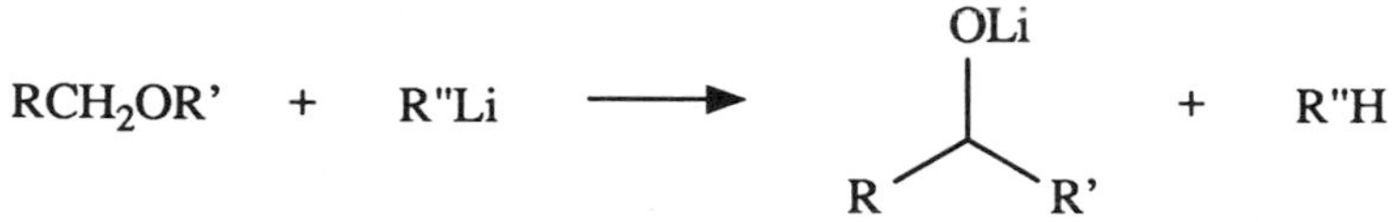

In some cases, the mechanism for the formation of at least some of the product(s) of reaction appears to be radical scission of the initially produced anion.

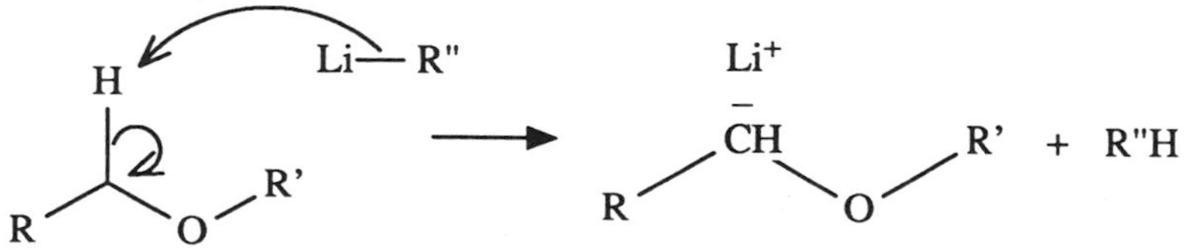

5-31

5-31 →

The radicals can now recombine to form the product:

There is evidence to suggest that some of these reactions go by an ionic mechanism. It is our purpose, in introducing a radical mechanism, to indicate that when there are several possible mechanisms for a reaction, some may be radical in character.

Problem 5-11. Write reasonable mechanisms for the formation of each of the products in the following reaction:

$$(PhCH_2)_2S \xrightarrow[\text{THF}]{\text{BuLi}} \xrightarrow{\text{MeI}}$$

$$PhCH_2CH_2Ph \quad + \qquad\qquad + $$

Biellmann, J. F.; Schmitt, J. L. *Tetrahedron Lett.* **1973**, 4615–4618.

Other reactions, which may follow similar paths, include the rearrangement of anions adjacent to either trivalent or tetravalent nitrogen.

Example 5-12. Unusual rearrangement of an anion α to a nitrogen.

In this example, an initially formed anion undergoes radical scission, recombination, and then anionic rearrangement to the final product.

$$\xrightarrow[\text{2. H}_2\text{O}]{\text{1. NaH, THF}}$$

When Ar = p-tolyl, yield is 57%.

A mechanism for the reaction was written in the cited paper as follows. The hydride removes the proton on the carbon α to the nitrile, leaving a carbanion. Then a benzyl radical cleaves, leaving a resonance stabilized, anion radical, **5-32**.

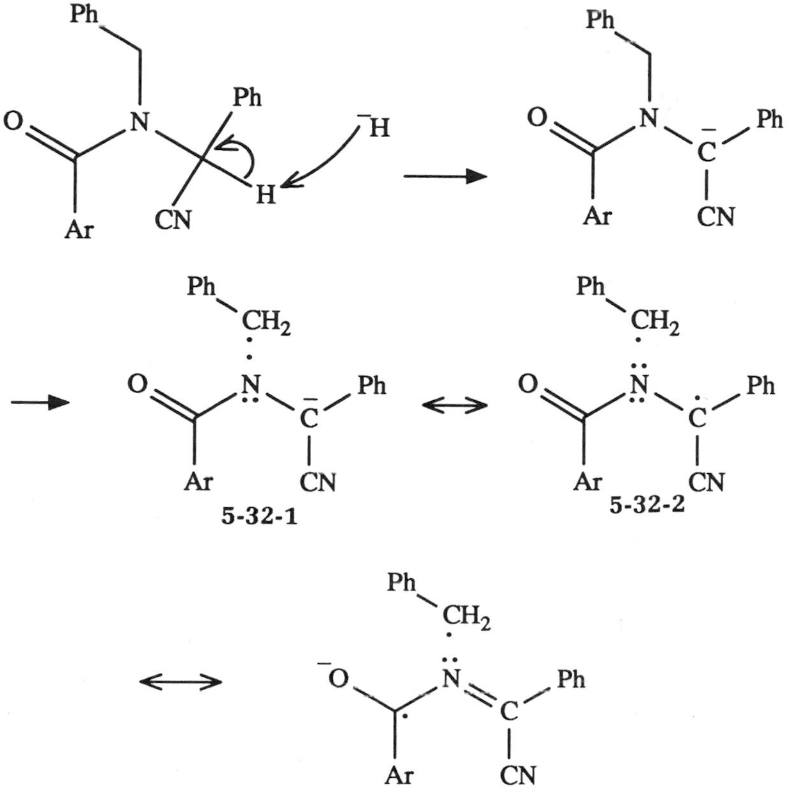

5-32-1   5-32-2

5-32-3

The benzyl radical then recombines at the carbon of the carbonyl group in **5-32**, and subsequent elimination of benzonitrile and cyanide ion gives the product.

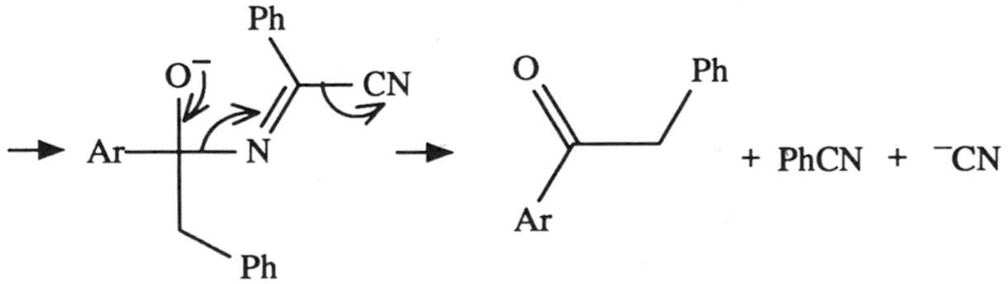

$+ PhCN + {}^-CN$

Stamegna, A. P.; McEwen, W. E. *J. Org. Chem.* **1981**, *46*, 1653–1655.

Problem 5-12. Consider the following data; then write reasonable mechanisms for the formation of 5-33 and 5-34:

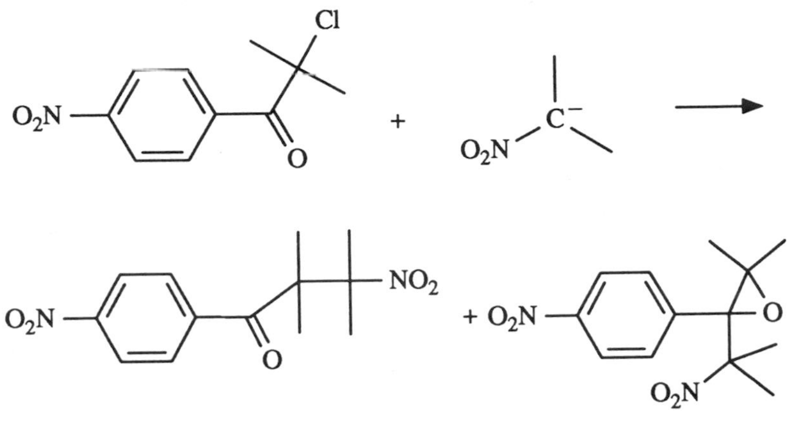

5-33   **5-34**

(a) 10 mol% of di-t-butylnitroxide completely suppressed formation of **5-33**.

(b) The yield of **5-34** was higher in the dark or in the presence of nitroxide.

(c) The yield of **5-33** was lower in the dark than with sunlamp irradiation.

Russell, G. A.; Ros, F. *J. Am. Chem. Soc.* **1985**, *107*, 2506–2511.

Problem 5-13. Write step-by-step mechanisms for the following reactions:

a.

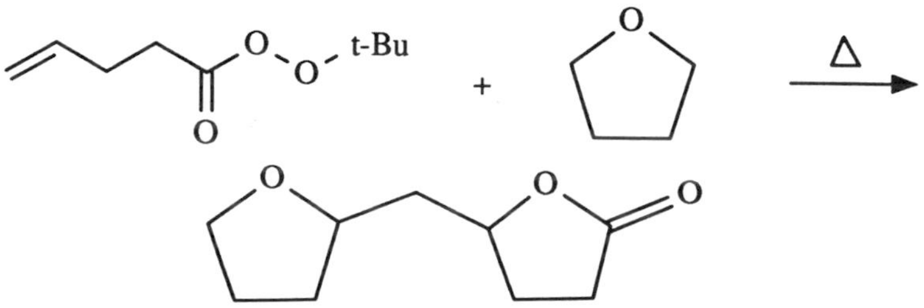

Kharrat, A.; Gardrat, C.; Maillard, B. *Can. J. Chem.* **1984**, *62*, 2385–2390.

b.

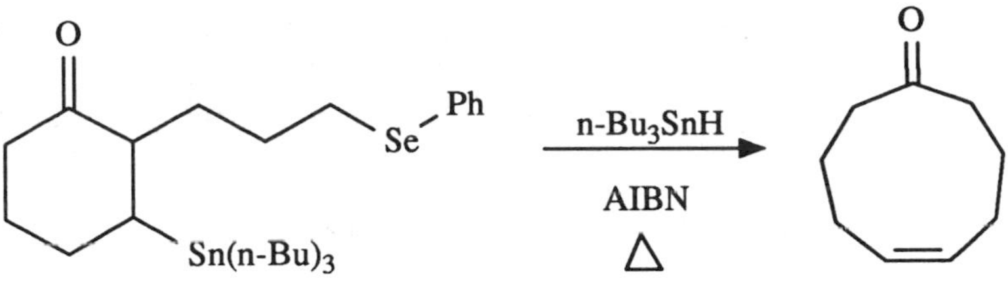

Only 10 mole% of tin hydride is used.

Baldwin, J. E.; Adlington, R. M.; Robertson, J. *J. Chem. Soc. Chem. Commun.* **1988**, 1404–1406.

c.

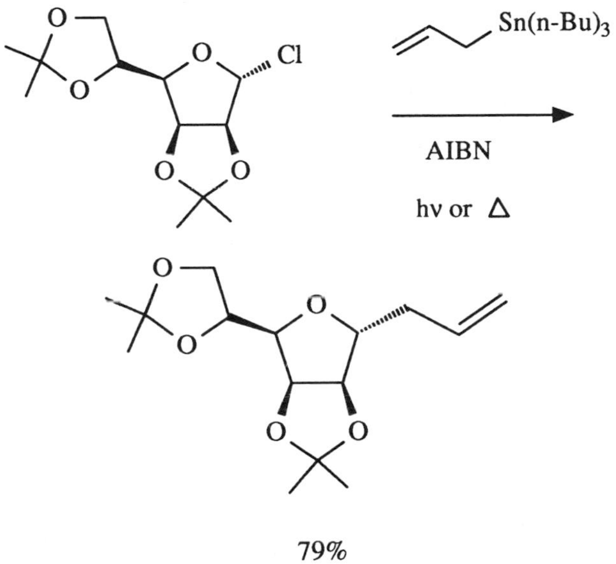

79%

Hart, D. J. *Science* **1984**, *223*, 883–887.

d.

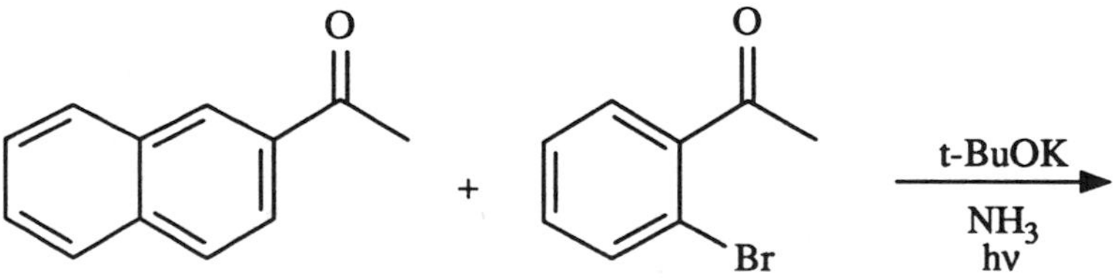

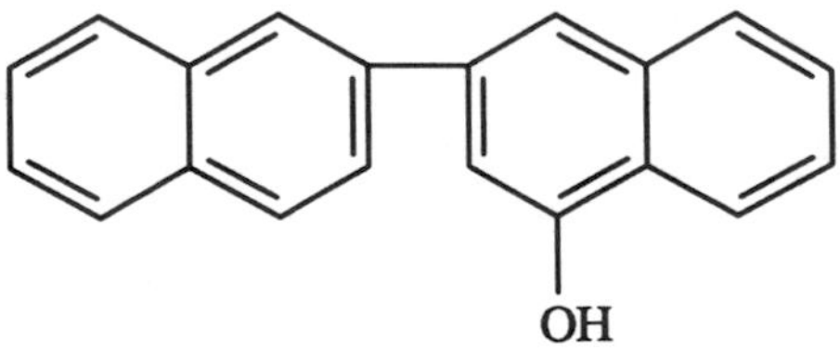

5-35

The yield of **5-35**, when the reaction mixture is irradiated, is 72%; in the dark, none of this product is formed.

Beugelmans, R.; Bois-Choussy, M.; Tang, Q. *J. Org. Chem.* **1987**, *52*, 3880–3883.

e.

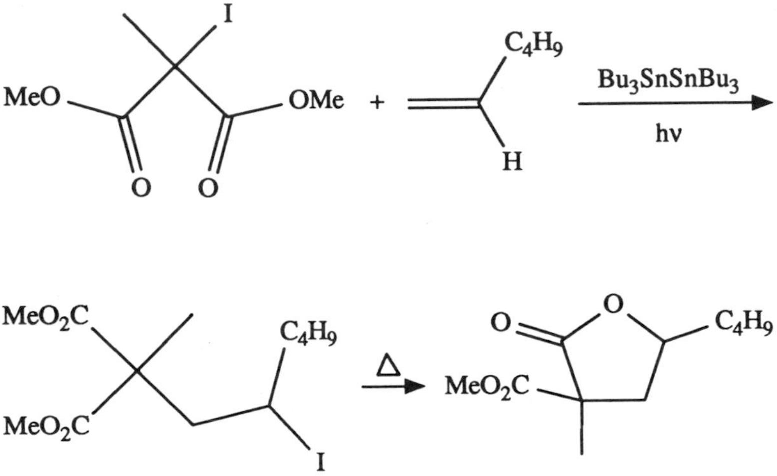

5-36   **5-37**

Hexabutylditin was present in 10 mole%. The yield of **5-37** from 1-hexene and iodomalonate was 69%.

Curran, D. P.; Chen, M.-H.; Spletzer, E.; Seong, C. M.; Chang, C.-T. *J. Am. Chem. Soc.* **1989**, *111*, 8872–8878

Answers to Problems

5-1.

$$t\text{-BuO·} \quad + \quad \underset{\text{(isobutyl radical skeleton)}}{\text{ }} \quad \longrightarrow \quad \underset{\text{ }}{\text{ }}\text{Ot-Bu}$$

$$t\text{-BuO·} \quad + \quad Cl· \quad \longrightarrow \quad t\text{-BuOCl}$$

Even though the last coupling is the reverse of an initiation process, it does remove t-butoxy radical from the chain propagation steps, and thus is a termination process.

5-2. The three equations shown, add up to the same overall reaction as equations (1) and (2) in Example 5-2. Thus, the total enthalpy change, −45 kcal/mole, is the same as the total enthalpy change calculated in the example. Why then do these equations not represent a mechanism for the reaction? There are two reasons. First, since the equations in the problem don't represent a chain process, the thermochemistry of each step, *including the first one,* must be considered. Because this first reaction is highly endothermic, the overall process is not efficient. Second, the third step involves the coupling of two radicals. While the enthalpy for this reaction is very favorable, both radicals will be present in very low concentration, so that the probability for their collision and reaction is very low. In contrast, both propagation steps in Example 5-2 have a radical colliding with a stable molecule whose concentration is much higher than that of any radical intermediate.

5-3. The initiation steps involve homolytic decomposition of the benzoyl peroxide. The resulting carboxy radical abstracts a bromine atom from carbon tetrabromide.

$$(PhCO_2)_2 \xrightarrow{\Delta} 2\ PhCO_2·$$

$$PhCO_2· \quad + \quad CBr_4 \quad \longrightarrow \quad PhCO_2Br \quad + \quad ·CBr_3$$

The propagation steps are:

(1). Regiospecific addition of the tribromomethyl radical to the double bond, leading to the more stable secondary radical.

$$\cdot CBr_3 \;+\; CH_2\!\!=\!\!CHC_6H_{13} \longrightarrow Br_3CCH_2\overset{\displaystyle\cdot}{C}HC_6H_{13}$$

5-38

(2). Abstraction of a bromine atom from carbon tetrabromide by **5-38** to give the product and a new tribromomethyl radical to continue the chain.

$$Br_3CCH_2\overset{\displaystyle\cdot}{C}HC_6H_{13} \;+\; CBr_4 \longrightarrow$$

$$Br_3CCH_2CHBrC_6H_{13} \;+\; \cdot CBr_3$$

5-4.a.   The initiation steps are thermal decomposition of AIBN, then hydrogen abstraction from tri-n-butyltin hydride:

The propagation steps:

(1). The tin radical abstracts an iodine atom.

(2). The resulting radical cyclizes:

5-39

(3). The cyclized radical abstracts a hydrogen atom from tri-n-butyltin hydride to give the product and a new tri-n-butyltin radical to continue the chain.

$$n\text{-Bu}_3\text{Sn} \longrightarrow \text{H} \quad + $$

$$ + \quad n\text{-Bu}_3\text{Sn} \cdot $$

5.4.b.   Radical **5-39** can undergo further cyclization across the ring. Such transannular reactions are common in medium sized rings.

The resulting radical can also abstract a hydrogen from tri-n-butyltin hydride to produce the hydrocarbon and a tin radical to continue the chain.

$$n\text{-Bu}_3\text{Sn} \longrightarrow \text{H} \quad + $$

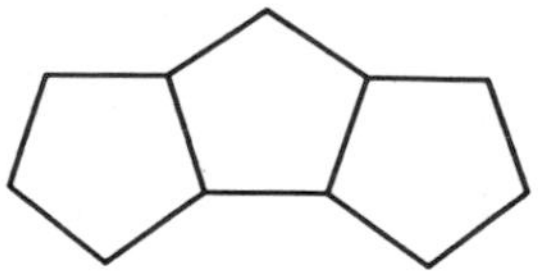

+ $n\text{-}Bu_3Sn\cdot$

5-5. Initiation steps:

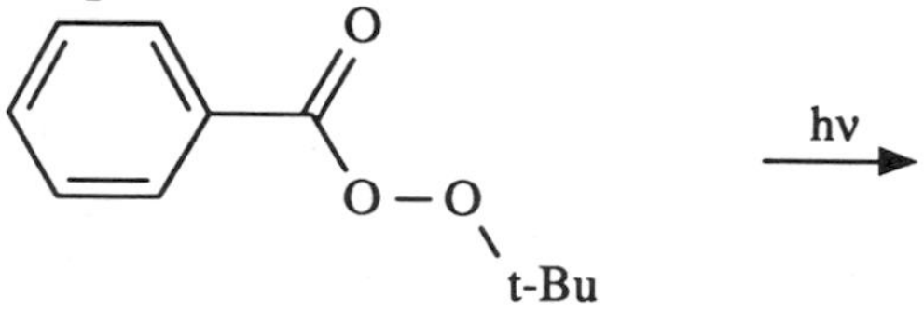

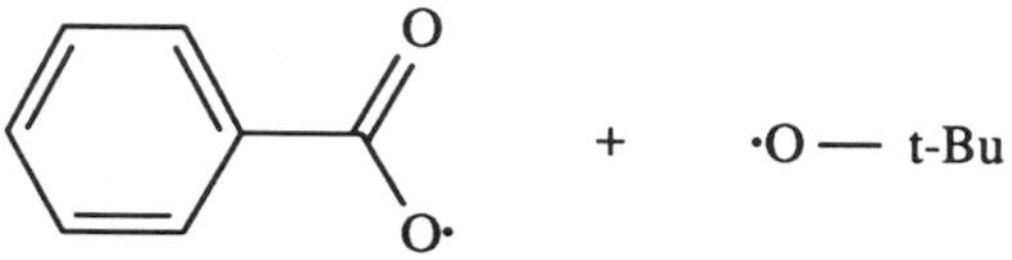

Either of the resulting radicals can remove a hydrogen from tri-n-butyltin hydride.

$$t\text{-}BuO\cdot + H—SnBu_3 \longrightarrow t\text{-}BuOH + \cdot SnBu_3$$

$$PhCO_2\cdot + H—SnBu_3 \longrightarrow PhCO_2H + \cdot SnBu_3$$

Propagation steps:

The tributyltin radical abstracts a bromine atom from the starting material.

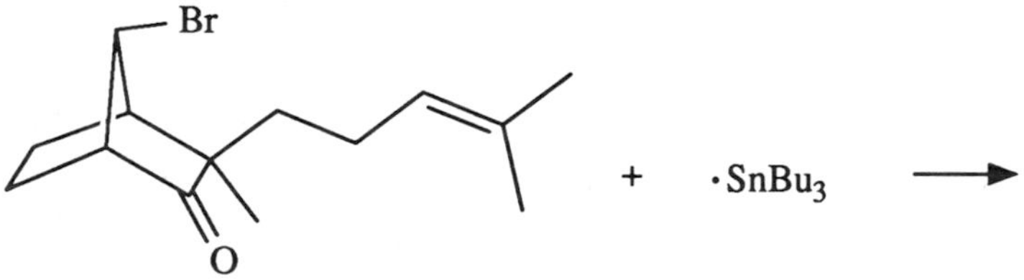

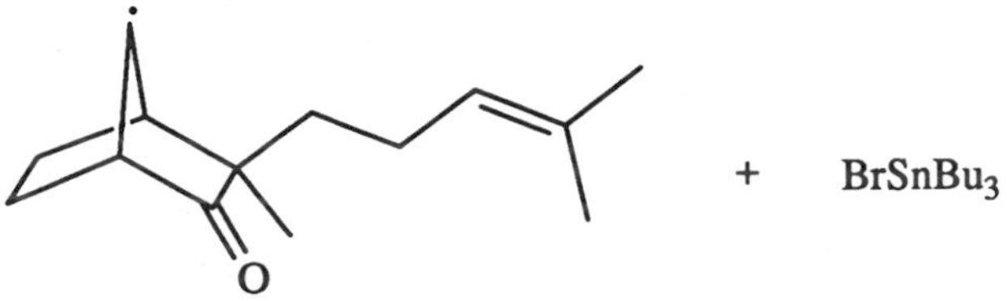

The radical undergoes intramolecular cyclization. This cyclization is regiospecific to form a new radical on the exocyclic carbon. This is expected on the basis of the Baldwin-Beckwith guidelines.

The cyclized radical abstracts a hydrogen from another molecule of tributyltin hydride to generate the product and another tributyltin radical.

$+$ HSnBu$_3$ $\longrightarrow$

$+$ ·SnBu$_3$

5-6.a.   Bromine reacts with the silver salt to form silver bromide and the acyl hypobromite. The latter undergoes homolytic scission to give the carboxy radical, **5-40**.

$$Ph_2CHCH_2CO_2Ag \; + \; Br_2 \longrightarrow Ph_2CHCH_2CO_2Br \; + \; AgBr$$

$$Ph_2CHCH_2CO_2Br \xrightarrow{\Delta} Ph_2CHCH_2CO_2{\cdot} \quad + \quad Br{\cdot}$$

5-40

The carboxy radical can cyclize to form a radical, **5-41**. This radical could be oxidized to the corresponding cation by bromine; loss of a proton gives product **5-12**.

5-40   **5-41**

$$Br{\cdot} \quad + \quad Br^- \quad + \quad \text{(structure)} \quad \longrightarrow \quad \textbf{5-12}$$

An alternative pathway from **5-41** to **5-12** involves abstraction of a bromine atom from bromine to give **5-42**. Product is produced by an elimination mediated by bromide ion.

5-41   **5-42**   **5-12**

Product **5-13** can be produced *via* hydrogen atom abstraction from **5-12** to form radical **5-43**, followed by mechanistic steps the same as those for the transformation of **5-41** to **5-12**.

5-12   + Br· ⟶   **5-43**   + HBr

There is another possible mechanism from **5-40** to **5-12**. From BDE's, abstraction of a hydrogen from one of the *ortho* positions of the phenyl ring by the carboxy radical, **5-40**, is energetically favorable: $\Delta H = -8$ kcal/mol. Cyclization of **5-44** gives a radical which can be oxidized to the corresponding cation by bromine; loss of a proton gives **5-12**.

5-40   **5-44**

5-6.b. Note that the bicyclic structure of the starting material is the same as the bicyclic structure in Example 5-6. It has been written in a different (more old-fashioned) way. Once this is recognized, a mechanism can be written for the reaction which is analogous to Example 5-6.

The first step is homolysis of the weakest bond in **5-14**, the O-O bond. This is followed by loss of CO_2.

+   ·OH

5-45

The fact that the concentration ratio, **(5-15 + 5-16)/5-17**, increases with increasing peracid concentration indicates that **5-15** and **5-16** are formed by a reaction of **5-45** with **5-14** which competes with the fragmentation reaction.

The intermediate radicals **5-45** and **5-46**, symbolized by ·R' in the equation below, can abstract OH from **5-14** to produce the product alcohols.

The fact that roughly the same amounts of **5-15** and **5-16** are produced from either isomer of starting material, **5-14**, shows that the mechanism of the reaction which forms these products is not concerted:

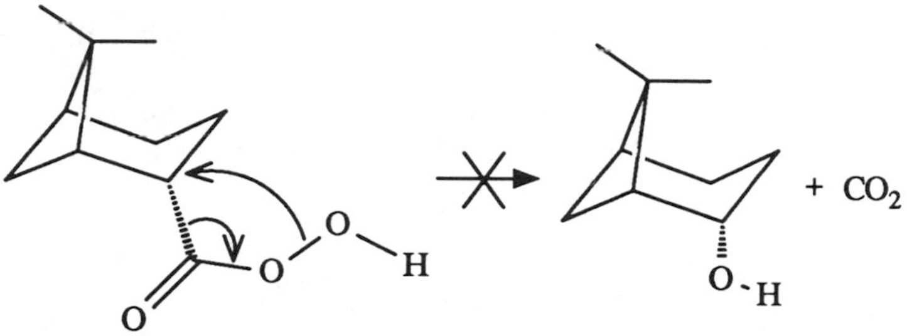

5-7.   n-Butylmercaptan is a better chain transfer agent than the aldehyde, because hydrogen abstraction from the mercaptan is much easier. Thus the radical, formed by addition of n-butylthio radical to the aliphatic double bond, is captured by mercaptan before it can rearrange. This is not the case for reaction with aldehyde.

t-BuOOH $\longrightarrow$ t-BuO· + ·OH

t-BuO· + n-BuSH $\longrightarrow$ t-BuOH + n-BuS·

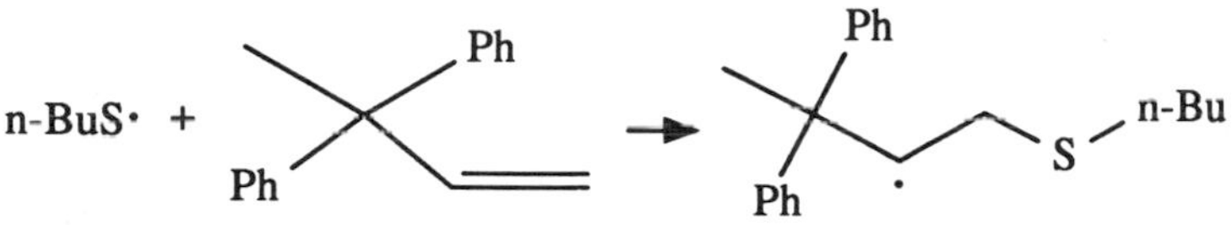

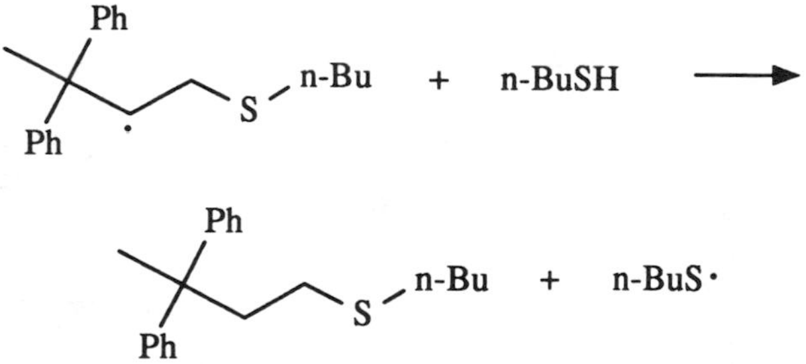

The reaction with aldehyde is as follows:

t-BuO· + n-PrCHO $\longrightarrow$ t-BuOH + n-PrC=O

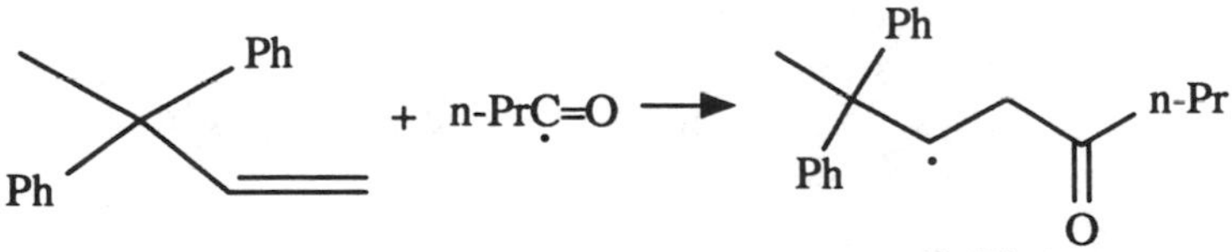

5-47

In this case, rearrangement of **5-47** competes effectively with its abstraction of hydrogen from starting material.

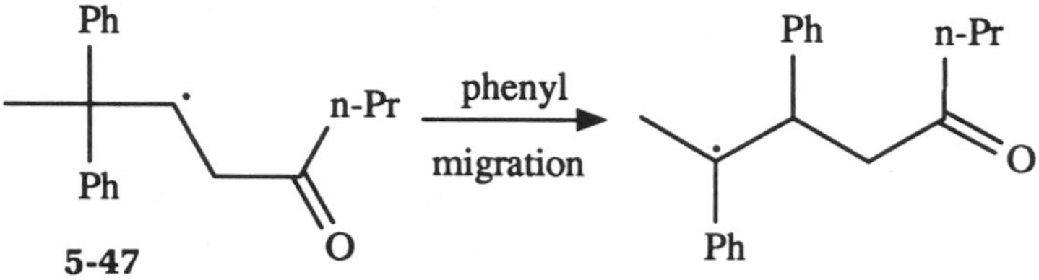

5-47

The rearranged radical abstracts a hydrogen from aldehyde to continue the chain.

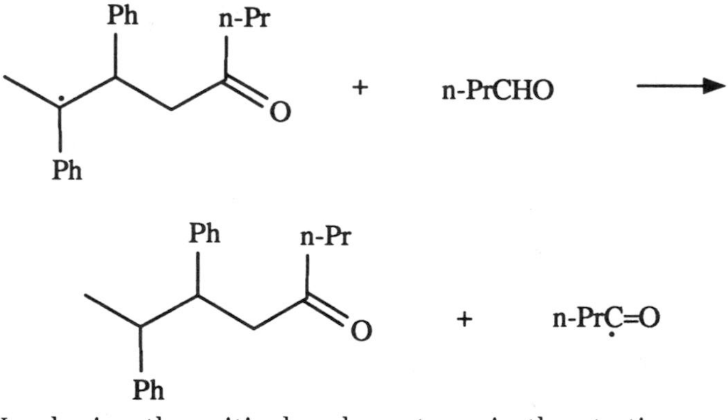

5-8.   Numbering the critical carbon atoms in the starting material and product gives a good indication of the course of the reaction.

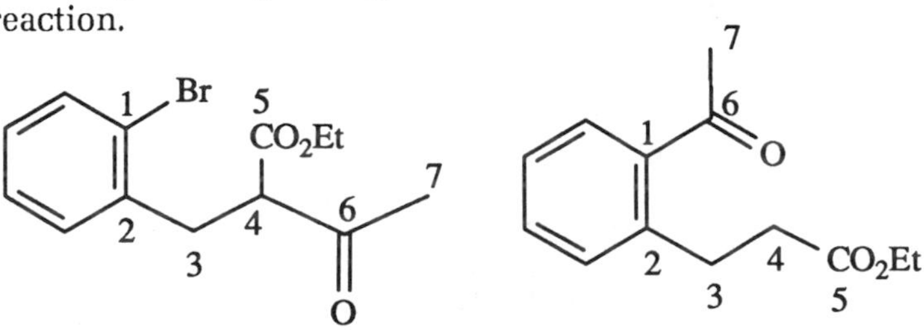

The acetyl group has "migrated" from C-4 in the starting material to C-1 in the product. Thus, the following sequence of events can be anticipated: (1) removal of the bromine atom from C-1, (2) addition of the resulting radical to C-6, (3) cleavage of the C-6 to C-4 bond, and (4) reaction with tri-n-butyltin hydride to give the product and a radical to continue the chain.

The initiation steps, to form the tri-n-butyltin radical, are the same as those in Problem 5-4. This radical then abstracts bromine from the starting material.

The aryl radical adds to the carbonyl group, a bond cleaves, and the resulting radical abstracts hydrogen from tri-n-butyltin hydride to form the product and a new chain-carrying radical.

5-9. The fact that hv is shown over the arrows for the reactions of parts a and b indicates that light is required for the reactions to take place. This rules out a simple S_N2 substitution for part a and nucleophilic aromatic substitution, by an addition-elimination reaction, in part b. Thus, the most likely mechanism for each of these reactions is an $S_{RN}1$.

5-9.a. The thiolate anion is excited by the light.

$$PhS^- \xrightarrow{\text{hv}} PhS^{*-}$$

The excited thiolate anion transfers an electron to cyclopropane.

The better leaving group, bromide, leaves.

Thiolate reacts, as a nucleophile, with the electrophilic radical, to produce the anion radical of the product.

Electron transfer to another molecule of starting cyclopropane gives the product and another cyclopropane anion radical to continue the chain.

The following is not a proper product forming step, because it does not also produce a radical to continue the chain. It is actually a termination step, involving the coupling of two radicals (see above). Formation of a major product by the coupling of two radicals is unlikely because the concentration of radical intermediates is usually quite low.

5-9.b. Each step in the mechanism for this reaction has a direct correspondence to a step in the mechanism for part a.

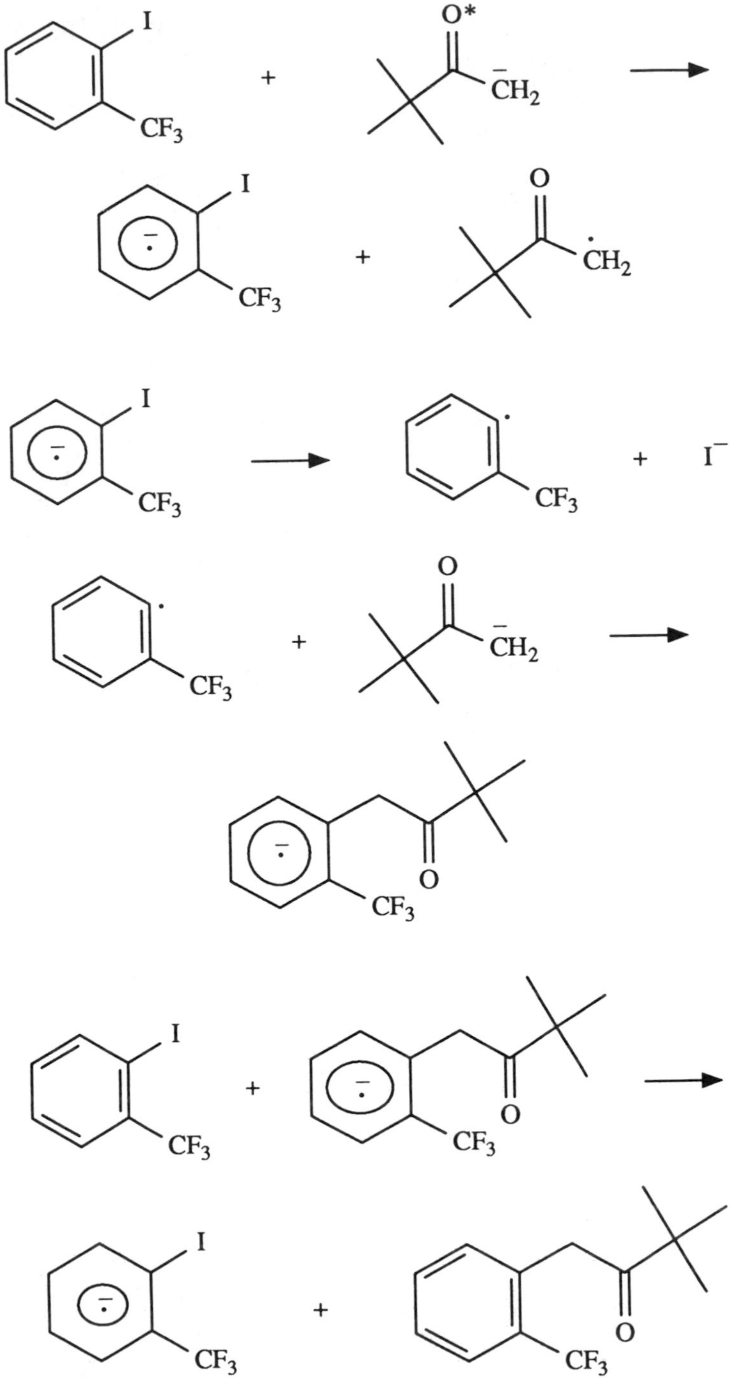
I
O*
CH2
CF3
+
CF3
CH2

5-10. Sodium transfers a single electron to the aromatic system to produce a pentadienyl anion radical, **5-48**. This anion is protonated at either the position *ortho* or *meta* to the π-donating methoxy group to give an intermediate radical. When a second electron is added to this radical, the pentadienyl anion is again protonated at its central carbon, *ortho* or *meta* to the methoxy substituent.

5-48-1   **5-48-2**

5-11. Comparison of the starting material with products shows that methyl iodide is present in order to methylate sulfur. Since the reaction takes place in base, the most likely mechanism for the methylation is an S_N2 reaction with thiolate ion:

$$RS^- \quad CH_3 - I \longrightarrow RSCH_3 + I^-$$

The fact that the first two products have centers of symmetry suggests that they might be products of radical coupling. Butyllithium can remove a proton from the carbon α to the sulfur. The resulting anion could then fragment into a radical and a radical anion by scission of the S-C bond.

$$PhCH_2 \cdot \quad \underline{S}CHPh$$

Two benzyl radicals can couple to give the first product.

$$PhCH_2 \cdot \quad \cdot CH_2Ph \longrightarrow PhCH_2CH_2Ph$$

Two thiolate anion radicals can couple to give the dithiolate precursor, **5-49**, of the second product.

5-49

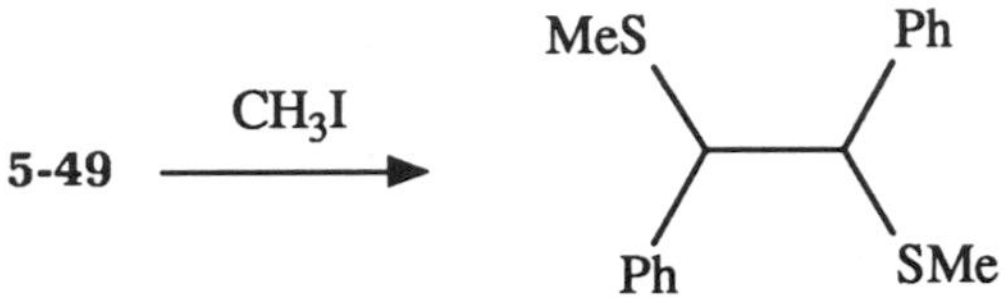

And finally, the two initially formed radicals can couple, at carbon, to give the thiolate precursor, **5-50**, of the third product.

5-50

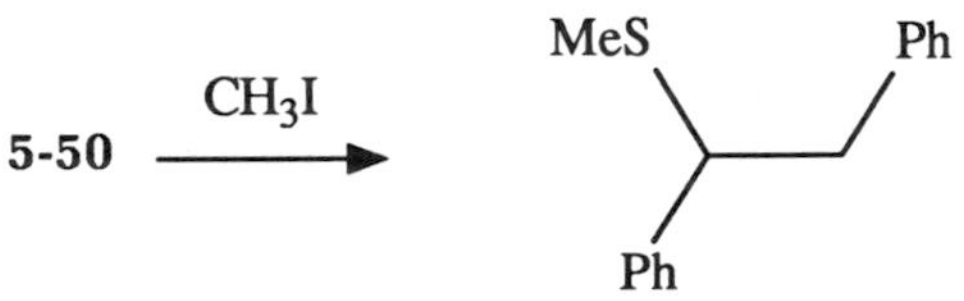

This is an example of a reaction where radical coupling is significantly involved in product formation. Thus, this is an exception to the rule that radicals don't often react with other radicals.

5-12. Formation of **5-33** is completely suppressed by addition of a free-radical inhibitor, but is stimulated by light. These facts strongly support a radical pathway to **5-33**. Other factors also make ionic S_N pathways unlikely. An S_N2 reaction is ruled out, because **5-33** is formed by substitution at a 3° carbon. An S_N1 ionization is not a favorable process, because formation of a carbocation next to the partially positive carbonyl would be required. Therefore, we are left with the strong possibility that **5-33** is formed by an $S_{RN}1$ mechanism. On the other hand, **5-34** forms, in better yield, in the dark or in the presence of a radical inhibitor, data which support a non-radical reaction. The regiochemistry is also different for the two reactions.

An $S_{RN}1$ mechanism for the formation of **5-33** can be written as follows:

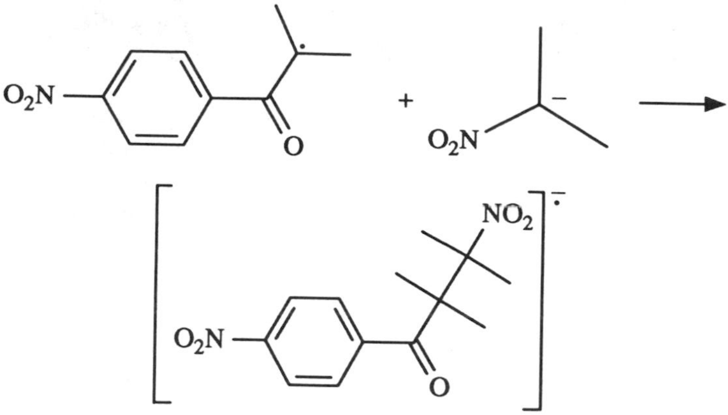

5-52

5-52 +

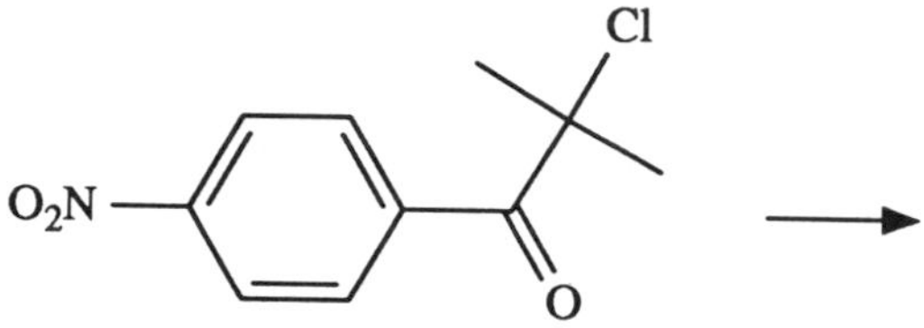

5-33 +

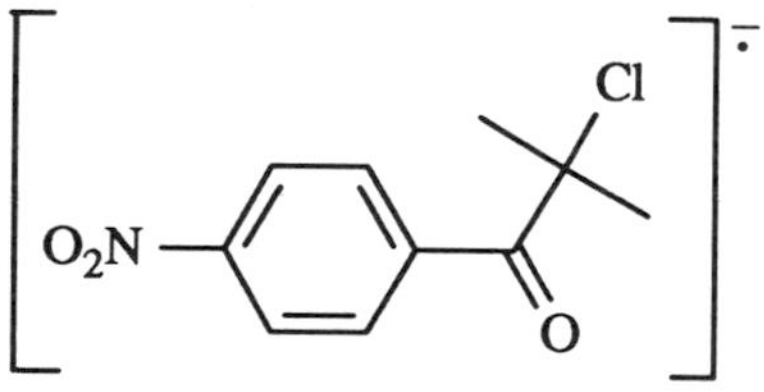

It is tempting to write a radical coupling to form product **5-33**, because it does not involve as many steps. However, this mechanism would be energetically inefficient, because it does not continue the chain process. Thus, the following **IS NOT** the major product-forming step.

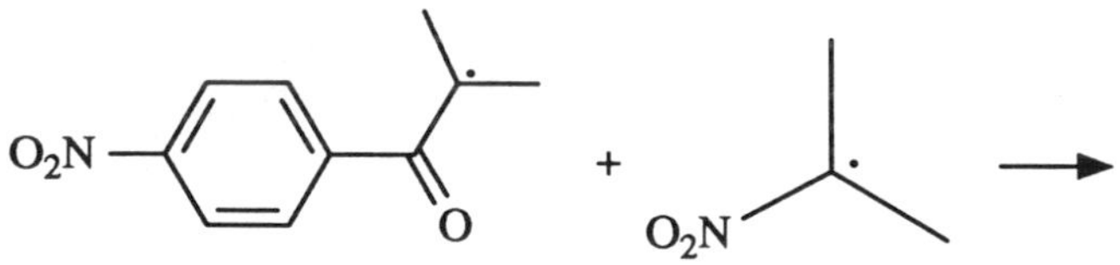

5-33

Numbering of atoms in the starting materials and in **5-34** gives a good indication of how **5-34** must be formed.

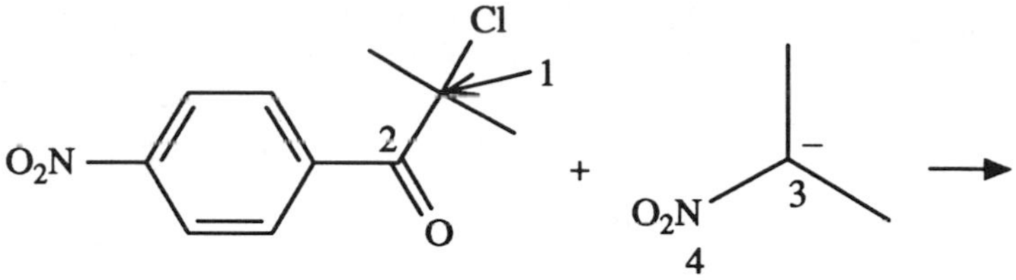

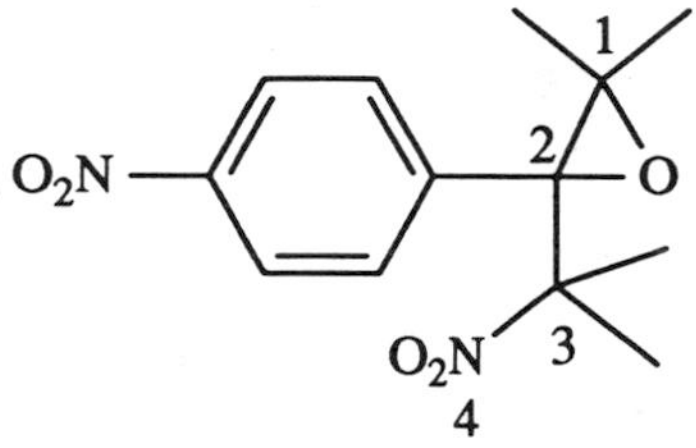

5-34

Since C-3 becomes attached to C-2, a nucleophilic reaction of the nitranion at the electrophilic carbonyl carbon must take place. Sterically this carbon is the most accessible. Then an intramolecular nucleophilic reaction of the alcoholate with the adjacent carbon (numbered 1 in the equation above) occurs.

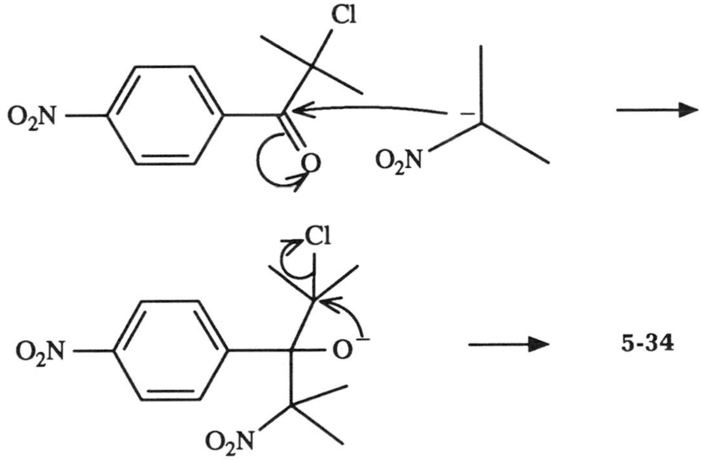

5-34

5-13.a. The perester could serve as a free radical initiator. At first glance, it appears that two tetrahydrofuran molecules form the two rings of the product, but the product contains one additional carbon atom. Therefore, the perester must be the source of the atoms of one ring, as well as a source of radicals. Since the perester contains a carbonyl group, the simplest explanation is that the lactone ring is derived from the perester. These considerations suggest the following chain mechanism:

Initiation:

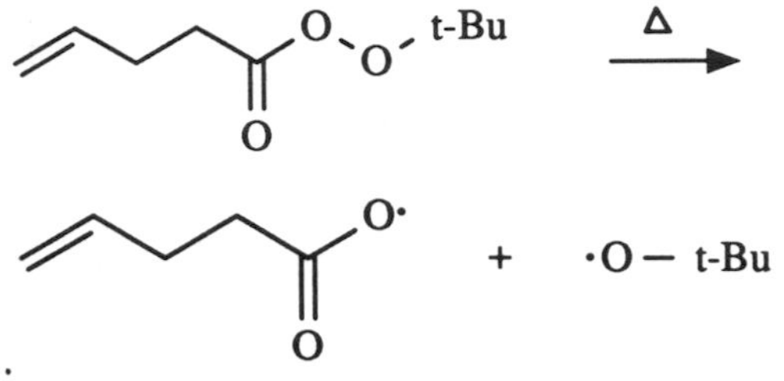

Propagation:

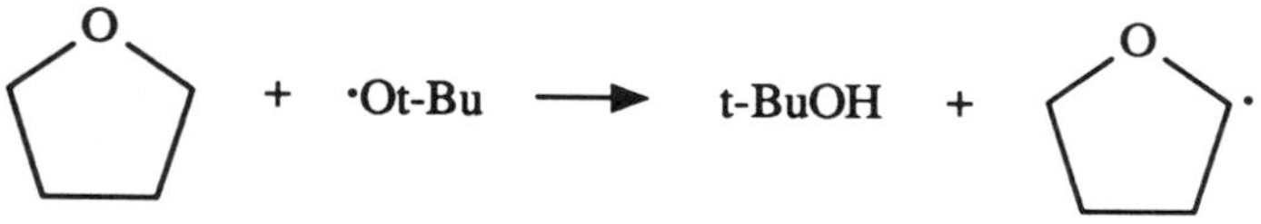

5-53

The tetrahydrofuranyl radical, **5-53**, adds to the carbon-carbon double bond of the perester.

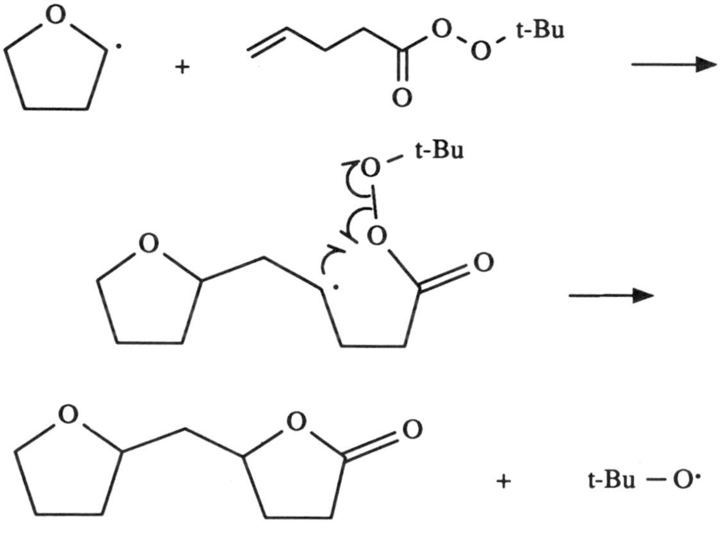

The following process, which is not a radical chain reaction, would not be as efficient, because two radicals must collide in order to form the product:

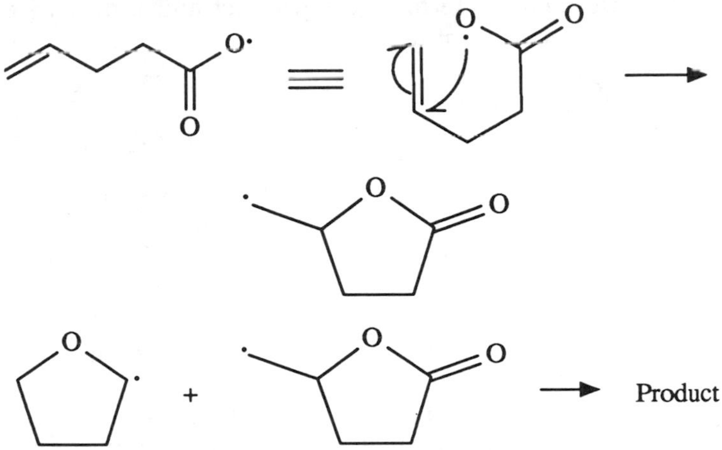

5-13.b. From the reaction conditions, we can make the following speculations and/or conclusions: (1) The tri-n-butyltin radical will be the chain carrier. (2) Tri-n-butyltin will be lost from the starting material. (3) A bond in the 6-membered ring breaks at some point in the mechanism.

In Section 5-2, we learned that tri-n-butyltin radicals react with selenium compounds to produce radical intermediates. Therefore, this might be a logical first step, after the initiation reaction between AIBN and tri-n-butyltin hydride has produced tri-n-butyltin radicals.

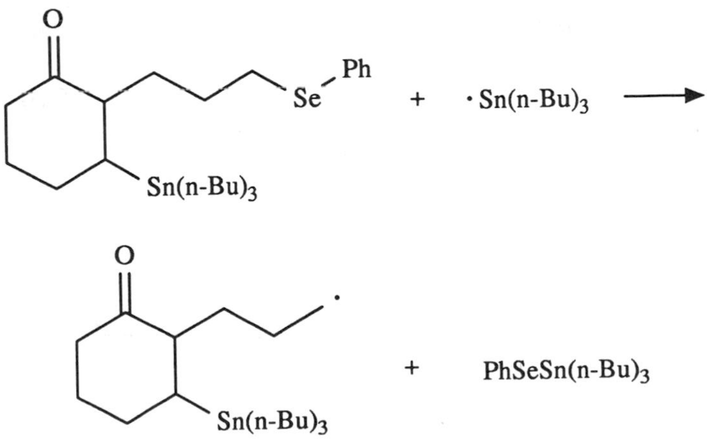

5-54

The carbonyl group, of radical **5-54**, can undergo an intramolecular cyclization reaction with the alkyl radical. The shared bond, between the 5- and 6-membered rings, then breaks to form the product and a new tri-n-butyltin radical, to continue the chain.

Product   +   · Sn(n-Bu)$_3$

The big difference between this reaction and that of Example 5-9, is that only 10 mole% of tin hydride is present. This is enough to initiate the reaction, but not enough to significantly reduce intermediate radicals.

5-13.c. Clues to the reaction mechanism are (1) chlorine is missing in the product, (2) AIBN is present as an initiator and (3) the tributyltin group is present. Thus it appears that tributyltin radicals abstract chlorine from the starting material. In the initiation steps, AIBN forms 2-cyano-2-propyl radicals in the usual way and these radicals react with allyltri-n-butyltin to produce tri-n-butyltin radicals by an addition-elimination mechanism.

Initiation:

Propagation:

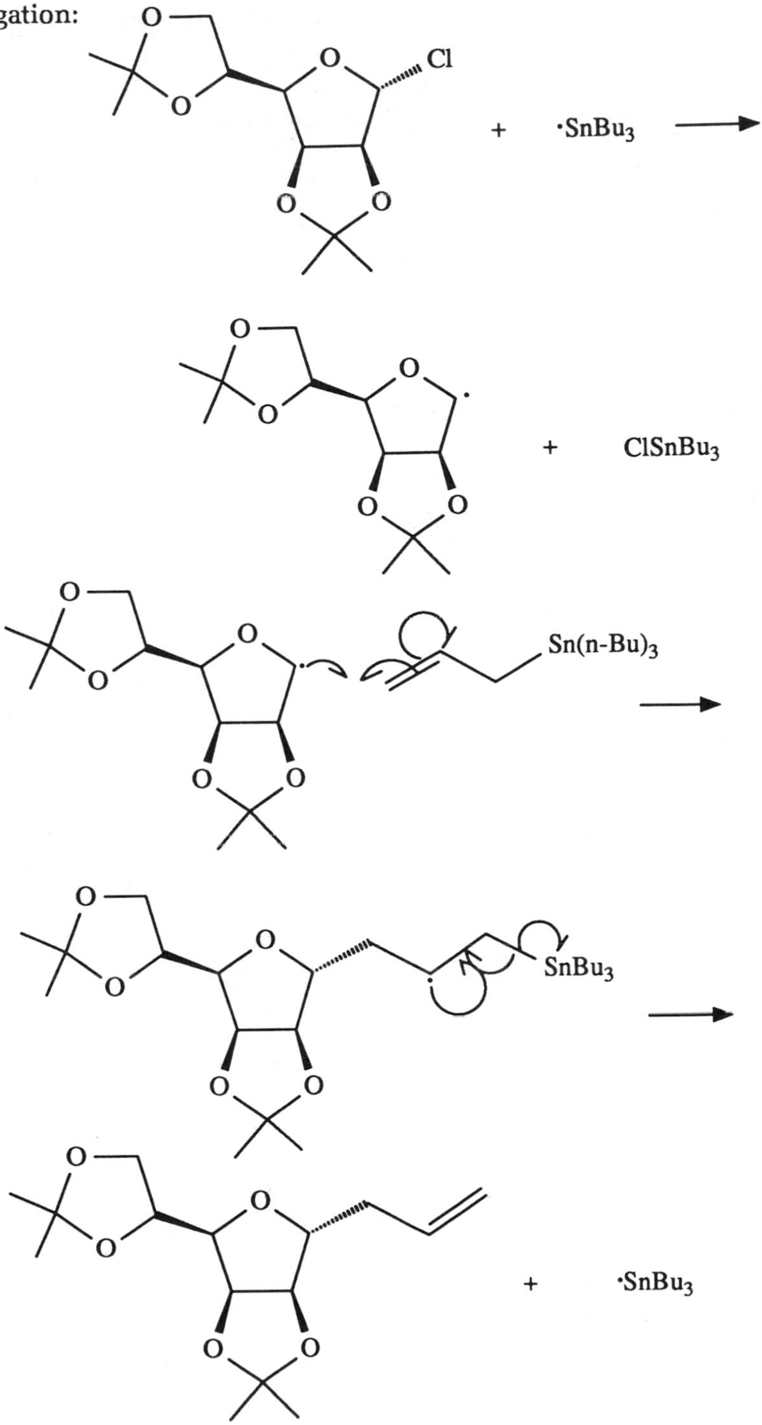

$+ \cdot SnBu_3 \longrightarrow$

$+ \quad ClSnBu_3$

$Sn(n\text{-}Bu)_3$

$\longrightarrow$

$SnBu_3$

$\longrightarrow$

$+ \quad \cdot SnBu_3$

5-13.d.   The fact that this reaction doesn't go in the dark strongly suggests a radical mechanism for at least part of the reaction pathway. An $S_{RN}1$ reaction looks like a promising possibility. In the initiation steps, the anion of the naphthyl ketone, excited by light, donates an electron to the bromo ketone, to form an anion radical. The product of a $S_{RN}1$ reaction would be **5-55**.

5-55

If the six highlighted carbon atoms in **5-55** form a six-membered ring, the correct carbon skeleton for the final product is obtained. In fact, an intramolecular aldol condensation, followed by elimination of water, gives the final product.

     The aldol condensation of **5-55** involves several steps. First, a proton is removed from the methyl group by the t-butoxide ion. (If the proton were removed from the methylene group, α to the other carbonyl, condensation would be at the carbonyl attached to the methyl. This would produce a 5-membered ring.) The resulting enolate adds to the other carbonyl group.

Protonation of the resulting alkoxide ion gives alcohol **5-56**.

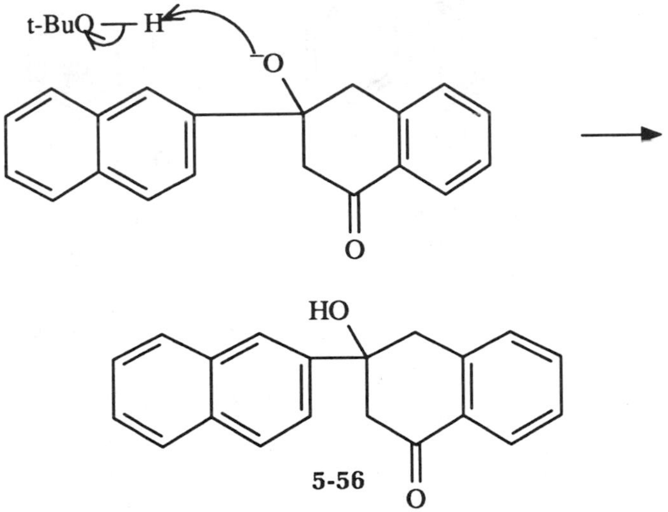

5-56

The alcohol can undergo base-promoted elimination to give two possible products. **5-57** or **5-58**. The new double bond in each of these products is stabilized by conjugation with the carbonyl group and at least one of the aromatic rings.

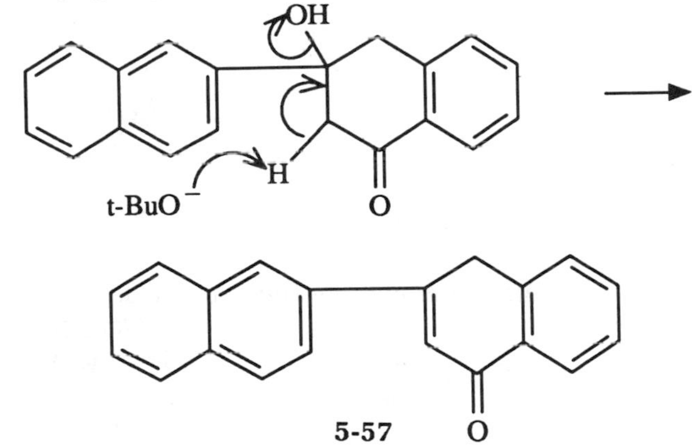

5-57

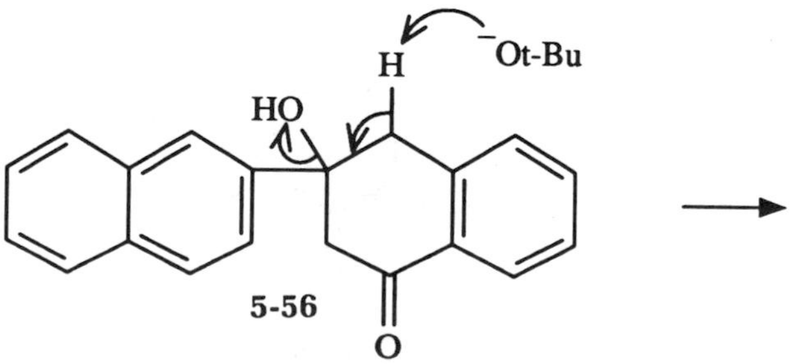

5-56

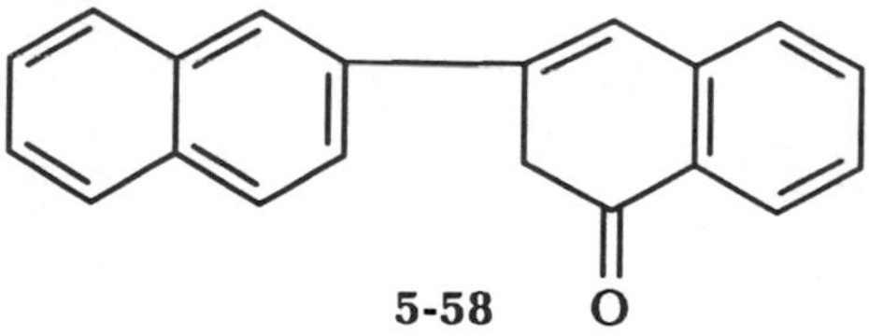

5-58   O

Either **5-57** or **5-58** will readily tautomerize to give the final product. Since both mechanisms are so similar, only one is shown.

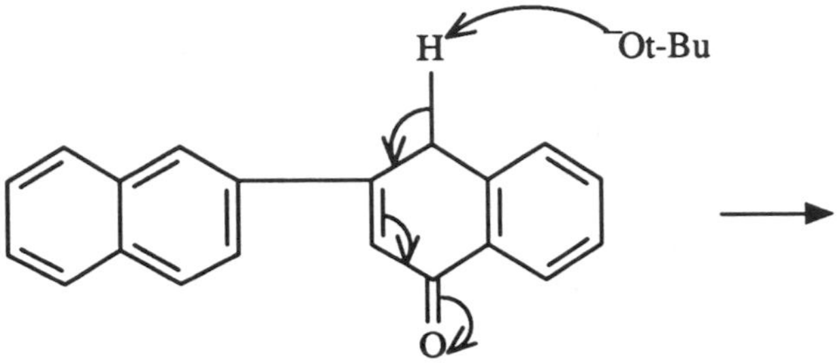

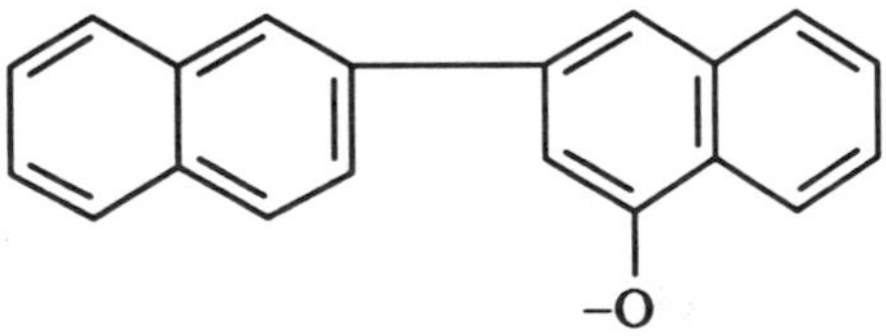

5-59

Upon acidic workup the naphthoxide ion **5-59** will be protonated to give **5-35**.

Another plausible mechanism for the final stages of the reaction is tautomerization of **5-56** to **5-60**, followed by elimination, directly to the phenol. (**5-61** would not be produced, because the aromaticity of the righthand ring is interrupted.) However, the above mechanism is preferable, because alcohol **5-60** would not be as stable as **5-57** or **5-58**.

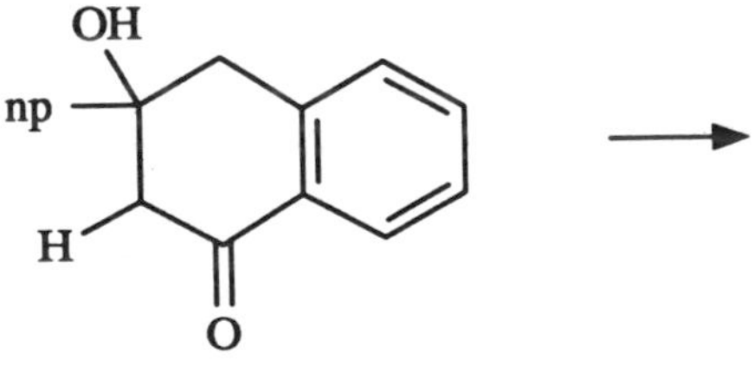

5-56

5-60   **5-61**

What remains is to write the mechanism of the initial $S_{RN}1$ reaction.

Initiation Steps:

Propagation Steps:

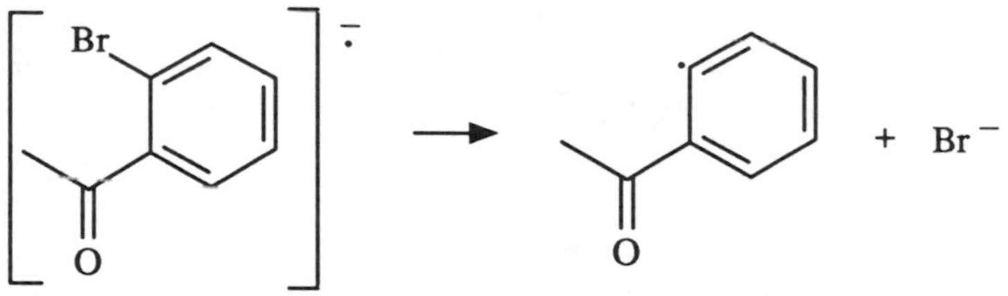

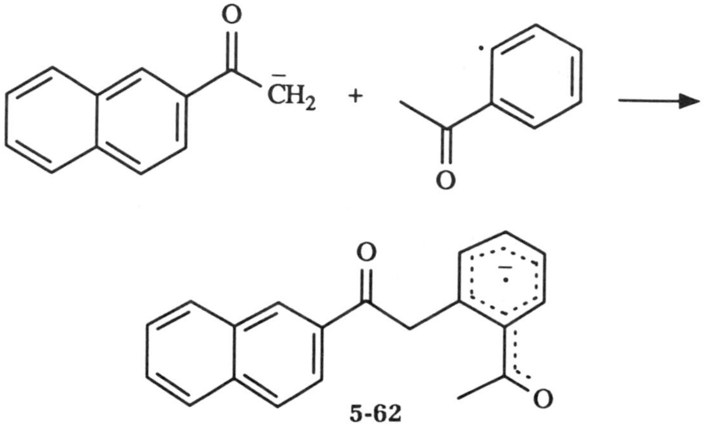

5-62

In the last step, a new anion radical of the starting bromo compound is formed, which continues the chain.

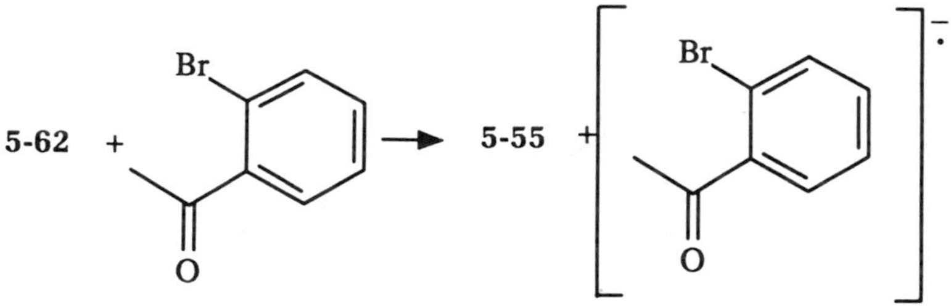

5-13.e. The fact that the overall reaction goes in 69% yield, with only 10 mole% of hexabutylditin present, indicates that this must be a chain process. Since iodide is present, probable initiation steps would be: photochemical decomposition of the ditin compound to tri-n-butyltin radicals, followed by abstraction of iodine from the starting material.

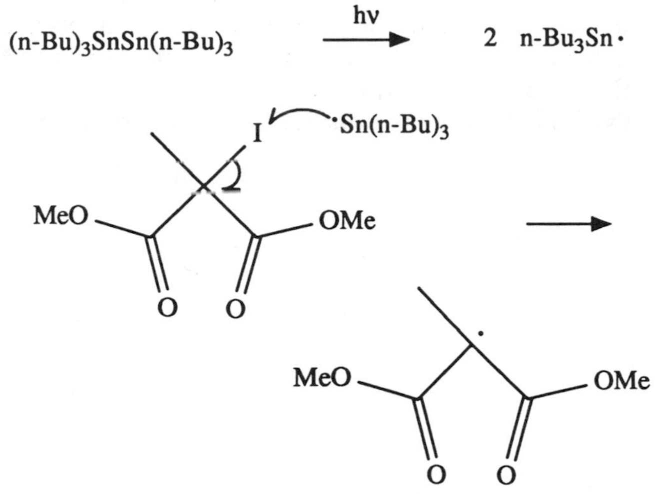

$$(n\text{-}Bu)_3SnSn(n\text{-}Bu)_3 \xrightarrow{\ h\nu\ } 2 \ n\text{-}Bu_3Sn\cdot$$

5-63

Radical **5-63** can add regiospecifically to the 1-alkene to produce an intermediate radical, which can then abstract iodine from the starting ester to produce **5-36** and a new radical to continue the chain.

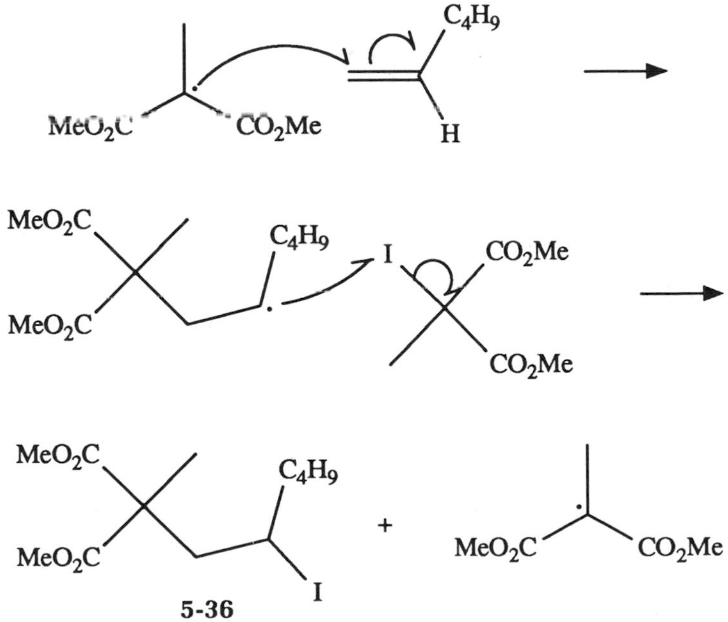

5-36

Because C-I bonds break easily, both homolytically and heterolytically, and the authors of the paper didn't specify the reaction solvent, either a radical or ionic mechanism can be written for the formation of **5-37** from **5-36**.

$$\longrightarrow \quad \textbf{5-37} \quad + \quad CH_3I$$

$$\longrightarrow \quad \textbf{5-37} \quad + \quad CH_3I$$

Chapter 6

Pericyclic Reactions

6-1 Introduction

Pericyclic reactions are reactions which involve a cyclic transition state. Examples which will be discussed in this book are electrocyclic transformations, cycloadditions, sigmatropic reactions, and the ene reaction. **Electrocyclic transformations** are: intramolecular ring closing reactions, where the bond formation occurs at the ends of a conjugated π electron system; or the reverse, ring opening reactions of cyclic polyenes. **Cycloadditions** are reactions of two π systems, at their termini, to produce a new ring. Common examples are the Diels-Alder reaction, and many 1,3-dipolar cycloadditions. **Sigmatropic reactions** involve the movement of a σ bond, mediated by a π system. An example is the Cope rearrangement of a 1,5-diene. The **ene reaction** combines aspects of cycloadditions and sigmatropic reactions. Separate sections of this Chapter are devoted to each of these reaction types.

Many pericyclic transformations are concerted. Almost always, such reactions turn out to be "symmetry allowed". That is, certain symmetry characteristics of the molecular orbitals involved in a transformation are necessary in order for a concerted reaction to occur. Conversely, some helpful hints can be applied to the molecular orbitals involved in a reaction which in many cases allow us to predict how readily a particular transformation with a particular stereochemistry will occur. Of prime importance for π systems are the amplitudes of the wave function at the ends of the system. We will represent the sign of the amplitude by a + or − within an orbital lobe. (An alternate representation uses shaded or unshaded lobes.) When a node, defined as the transition point between + and − amplitude (or zero amplitude), falls at a nuclear position, it will be indicated by lobes with no signs. The qualitative methods for generation of the amplitudes, presented in this Chapter, suffice for predicting or understanding many reactions. However, regiochemical predictions require calculations which are outside the scope of this book.

Of special importance are the molecular orbital of highest energy which contains one or two electrons (HOMO—Highest Occupied Molecular Orbital) and the molecular orbital of lowest energy which does not contain any electrons (LUMO—Lowest Unoccupied Molecular Orbital).

6-2 Representations of Molecular Orbitals

In this section, we show how to generate molecular orbitals for conjugated π systems, starting with the individual p orbitals of the carbons. Orbitals of individual atoms are called atomic orbitals (AO's); overlap of AO's leads to the molecular orbitals (MO's) of molecules. The shape of the orbitals is the shape of the wave character associated with particular electrons. Most critical to this discussion are the SIGNS of the amplitude of the wave function associated with a particular MO. For example, in order to form a bond between two different MO's, the signs of the amplitude of the wave functions must be the same (both + or both −), in the region where the bond is formed. Thus the waves add (form a bond) if the signs are the same, or subtract (antibond) if the signs are different.

The probability density for finding an electron within a particular space is equal to the square of the wave function.

6-2-1 THE MOLECULAR ORBITALS FOR ETHYLENE

Molecular orbitals, generated by addition and subtraction of atomic orbitals, are said to be formed by a linear combination of AO's. The number of MO's, that can be generated from n AO's, is n. Thus the two π MO's for ethylene are generated from overlap of the two p AO's. The relative amplitudes (signs) of MO's are characterized by: (1) the number of nodes; (2) the symmetry relative to a mirror plane, m_2; and (3) the symmetry relative to a two-fold axis, C_2. For each MO, these three characteristics will be unique. In other words, in any molecule these three descriptors cannot be identical for two different MO's.

To produce the MO's for ethylene, the p atomic orbital from the second atom is added to the p atomic orbital of the first atom to give a bonding MO and subtracted to give an antibonding MO. The process of subtraction of an AO simply means that the sign of the amplitude of the wave function is changed. Figure 1 shows this addition and subtraction. The MO's are shown with the bonding orbital (the orbital of lower energy) at the bottom and the antibonding orbital (the orbital of higher energy) at the top. A π bond between two carbons exists only when the amplitudes of the p AO wave functions (+ or −) are the same on both sides of the internuclear line, which represents the σ bond between the atoms. This is true only for the MO of lower energy in Figure 1.

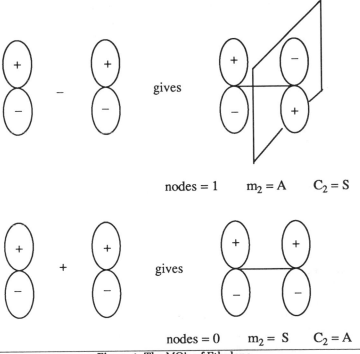

nodes = 1 m_2 = A C_2 = S

nodes = 0 m_2 = S C_2 = A

Figure 1. The MO's of Ethylene.

The bonding orbital is called a π MO and the antibonding orbital is called a π* MO.

The symmetries of the MO's, especially the HOMO's and LUMO's, are essential to understanding the concertedness (or lack thereof) and stereochemistry of pericyclic reactions. The symmetry elements which must be considered are a two-fold axis, C_2, and a mirror plane, m_2. Both are perpendicular to the plane of the paper, and both pass through the geometric center of the molecule. Thus, the C_2 axis lies in the m_2 plane, shown in the upper MO in Figure 1. (This is also the location of the node in this MO.) The MO's (π wave functions) for ethylene are either antisymmetric (A) or symmetric (S) relative to these symmetry elements. Symmetric means that the wave function amplitude (+ or −) is not changed by the symmetry operation. Antisymmetric means that the amplitude is reversed (+ to −, − to +) at all locations by the symmetry operation.

Since each original p atomic orbital donates an electron, the MO's shown must accommodate two electrons. Just as electrons are distributed in atomic orbitals, both electrons are placed in the MO of lower energy with their spins paired. In Figure 2, and subsequent figures, each electron is designated by either an up or a down arrow. Since the bonding level is filled, and there are no electrons in the antibonding level, this represents a very stable electron distribution. This also means that, for ethylene, the bonding MO is the HOMO and the antibonding MO is the LUMO.

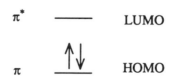

π^* ——   LUMO

π ⇅   HOMO

Figure 2. Relative Energies of MO's in Ethylene.

6-2-2 HELPFUL HINTS FOR GENERATING MOLECULAR ORBITALS

The MO's for conjugated π systems can be built up by linear combinations of p atomic orbitals, one on each carbon of the system. The number of MO's will be equal to the number of p atomic orbitals used. By noting some general characteristics of MO's, a set of helpful hints can be formulated, from which the MO's for simple π systems can be written.

Helpful Hint 6-1. All MO's must be either totally symmetric or totally antisymmetric, relative to m_2 and C_2.

Helpful Hint 6-2. The properties of the MO's for a particular π system alternate with respect to symmetry or antisymmetry in regard to C_2 and m_2. That is, the MO of lowest energy is always symmetric (S) relative to m_2 and antisymmetric (A) relative to C_2. The MO next higher in energy, is A with respect to m_2 and S with respect to C_2. The next higher will be S with respect to m_2 and A with respect to C_2, and so on.

Helpful Hint 6-3. The MO of lowest energy always has no nodes. The next higher in energy has one node, the next has two nodes, and so on. These nodes must be symmetrically located on either side of the geometric center of the molecule. Therefore, a single node is always located at the center, on m_2; two nodes are always placed symmetrically on either side of m_2; and so on. When MO's contain an odd number of atoms, it is conventional to place nodes on nuclei (except those at either end). However, nodes are never placed on adjacent carbons: for example, see the MO for pentadienyl, with 3 nodes, in Problem 6-2.

Example 6-1. What is wrong with the following MO, generated for the 3-carbon system?

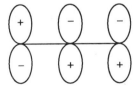

This MO is partly antisymmetric and partly symmetric with respect to both C_2 and m_2. Also, the single node is not located at the geometric center.

By applying Helpful Hints 6-1, 6-2, and 6-3, all MO's necessary for consideration of the stereochemistry and concertedness of pericyclic reactions can be generated.

Problem 6-1. Explain why the following would not be the two lowest energy MO's for 1,3-butadiene:

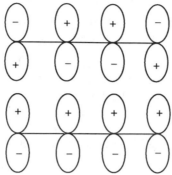

Problem 6-2. Write all the MO's for the 4-carbon and 5-carbon π systems. Give the symmetries of each, with respect to m_2 and C_2, and indicate the number of nodes.

Example 6-2. The π electron configurations for allyl cation, radical, and anion.

The allyl system starts out with p orbitals on each of the three atoms. Thus, three MO's can be generated, and the electrons present in the π system are distributed in the usual way, filling the lowest energy orbital first, next highest next, and so on. Since the allyl cation, $CH_2=CHCH_2^+$, contains two π electrons, only the MO of lowest energy is filled.

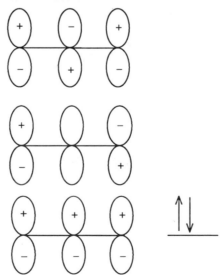

The allyl radical contains one more electron, which must go into the MO next highest in energy.

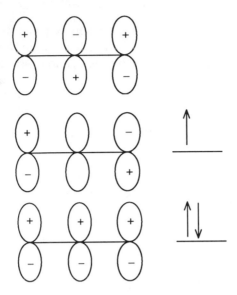

Finally, the allyl anion contains a total of 4 π electrons, and will have the following electron configuration:

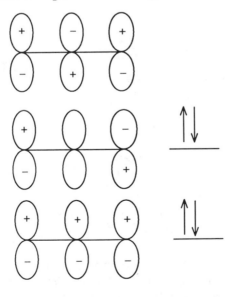

Problem 6-3. Using the orbitals drawn in Problem 6-2, indicate the electronic configuration for the pentadienyl cation, radical and anion. Label the HOMO and LUMO for each case.

In a photochemical process, an electron is promoted into a higher MO, producing an excited state. This changes the HOMO and LUMO for a π-system in the excited molecule. Such a process is shown below for butadiene.

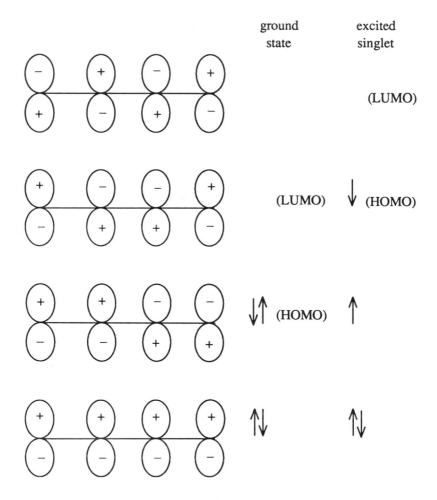

Because an electron has moved to the next higher MO in the excited state, the HOMO and LUMO are also moved up by one MO, relative to the ground state for the molecule.

6-3 A General Method to Predict Whether a Reaction is Symmetry Allowed or Symmetry Forbidden

The symmetries of the orbitals involved in a concerted reaction are crucially important to the prediction of how readily the reaction will occur. Symmetry allowed processes usually take place much more readily than those which are symmetry forbidden.

Several different approaches have been developed to predict whether a reaction is symmetry allowed or forbidden. A general helpful hint, 6-4, which covers all types of pericyclic changes, uses the total number of electrons involved in each component of the process under consideration as well as the fact that only orbitals (or lobes of orbitals) which have the same sign of the wave function may produce new bonds. In several kinds of reactions this leads to the prediction of a particular stereochemical outcome for the concerted process.

By each component of the process, we mean each π system, either an isolated double bond or a conjugated π system, and each σ bond involved in the pericyclic change. Each of these components is categorized by the number of its electrons involved, that is, as either [4q + 2] or [4r], where q and r are integers. If a component has 2, 6, 10, 14 ... electrons, it is a [4q + 2] component; if a component has 4, 8, 12, 16 ... electrons, it is a [4r] component.

The stereochemical course of reaction of a component is designated as either suprafacial (abbreviated as s or *supra*) or antarafacial (abbreviated as a or *antara*). For a π component, suprafacial means that both reacting lobes, one on each end of the π system, are on the same side of the plane defined by the carbons of the π system and their adjoining atoms. In contrast, antarafacial means that both reacting lobes are on opposite sides of the plane defined by the carbons of the π system and their adjoining atoms.

For example, reaction of the HOMO of butadiene at the shaded lobes, on the same side of the plane, is called a suprafacial process:

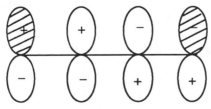

Reaction of the lobes on opposite sides of the plane is called an antarafacial process:

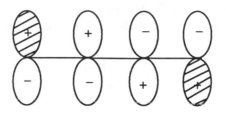

For a σ bond, reaction of either the two front lobes or the two back lobes is considered to be a suprafacial process:

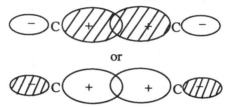

Reaction of the front lobe of one atom and the back lobe of the other is considered to be an antarafacial process:

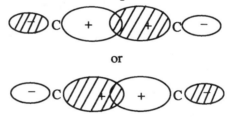

Helpful Hint 6-4. A thermal pericyclic change is symmetry allowed when the number of [4q + 2] suprafacial ($[4q +2]_s$) components, plus the number of [4r] antarafacial ($[4r]_a$) components, equals an odd total number. A photochemical pericyclic change is symmetry allowed when this total number is even.

If Helpful Hint 6-4 is not obeyed, the reaction is said to be symmetry forbidden. Symmetry forbidden reactions can take place, although ordinarily such reactions occur with difficulty, if at all. When a thermal reaction is forbidden, by symmetry considerations, it is often said that it is not thermally allowed. Thus, "symmetry forbidden" and "not thermally allowed" mean the same thing, and apply when the number of counted components is even. Similarly, "symmetry forbidden" and "not photochemically allowed" mean the same thing, and apply when the number of counted components is odd.

Often, components of a pericyclic reaction are described by groups of three symbols: first, a subscript σ or π, to indicate the type of MO that is reacting; then, the number of electrons involved in reaction of the component; then, a subscript a or s for antarafacial or suprafacial reaction of that component. The overall reaction is then described as a sum of these symbol groups enclosed in square brackets. For example, the designation for the Diels-Alder reaction would be $[_\pi4_s + _\pi2_s]$.

Example 6-3. Are the following processes symmetry allowed, thermally or photochemically?

(1) $[_\pi8_a + _\sigma2_s]$ The $_\pi8_a$ antarafacial component does contain [4r] electrons (r=2), so this component is counted. The $_\sigma2_s$ suprafacial component contains [4q + 2] electrons (q=0), so this component is counted. With an even number, 2, of counted components, the process is symmetry allowed for a photochemical reaction.

(2) $[_\pi4_s + _\pi2_s]$ The $_\pi4_s$ suprafacial component does not contain [4q + 2] electrons, so this component is not counted. The $_\pi2_s$ suprafacial component does contain [4q + 2] electrons, so this component is counted. Thus, there is an odd number, 1, of counted processes, and the symmetry allowed reaction is predicted to be a thermal process.

Problem 6-4. On the basis of Helpful Hint 6-4, which of the following concerted processes are thermally allowed and which are photochemically allowed?

a. $[_\pi4_a + _\pi2_a]$ b. $[_\pi8_s + _\pi2_s]$ c. $[_\pi10_a + _\sigma0_s]$

6-4 Electrocyclic Reactions

6-4-1 GENERAL PRINCIPLES WITH EXAMPLES

Electrocyclic reactions are intramolecular ring openings or ring closings. Reactions of substituted cyclobutenes to form butadienes illustrate many of the principles which must be considered for prediction of the stereochemical outcome of concerted electrocyclic reactions. While in theory, allowed electrocyclic reactions can proceed in either direction, in practice, one side of the equation is usually favored over the other.

Because of the large strain energy of the four-membered ring, cyclobutene rings usually thermally open readily to give butadienes, and the reverse reactions are not thermodynamically favored. In cyclobutene above, since there are no polar substituents, the direction of the electron flow arrows was chosen arbitrarily.

Helpful Hint 6-5. In the application of Helpful Hint 6-4 to electrocyclic reactions, it is more convenient to consider the symmetry aspects of the polyene than those of the cyclized product.

Example 6-4. The thermal ring opening of *cis*-3-chloro-4-methylcyclobutene—an antarafacial, conrotatory process.

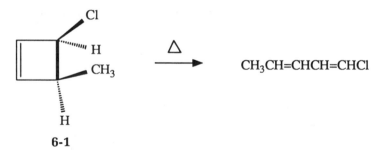

6-1

Orbital symmetry considerations can be used to predict the stereochemistry of the chlorine and methyl in the product butadiene. First, count the total number of electrons in the π system. According to Helpful Hint 6-5, this count, made in the open chain product, butadiene is 4. According to Helpful Hint 6-4, a 4 electron system will react antarafacially in a thermal reaction. That is, this system is [4r] (r=1) with just one component (an odd number). The designation for the process is [$_\pi 4_a$].

As reaction takes place, the lobes for the highlighted C-C σ bond of **6-1**, the bond which is breaking, rotate into the plane defined by the π bond. Because the process is antarafacial, these lobes, which are converted to p orbitals on the terminal carbons of the product butadiene, must end up on opposite sides of the plane in the product.

In general there are two modes of ring opening which can produce this antarafacial orientation. These modes are considered as rotations taking place at each end of the breaking σ bond. In the diagrams below, the dashed lines, extending to the right between the constituents on the ring are drawn only to clarify the possible modes of rotation.

The rotation, shown in Equation 1, leads to an antarafacial orientation of the lobes which comprised the σ bond.

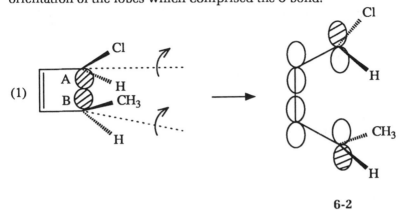

(1)

6-2

[Note: In this example and in Example 6-5 the signs of the amplitude of the lobes of the p orbitals are omitted for clarity. In each case the product is the HOMO of butadiene. All the changes in the orbitals as the reaction proceeds are not depicted, but have no bearing on the stereochemical outcome at this level of analysis. For more detail see Woodward, R. B.; Hoffmann, R. The Conservation of Orbital Symmetry **1970**, Weinheim: Verlag Chemie and Academic Press, p. 38 ff. and the references cited later in this example.]

In Equation 1 lobe A of the σ bond rotates upward and lobe B of this bond rotates downward. Therefore, these lobes (both are shaded) end up on opposite sides of the plane defined by the carbons of the butadiene product. By definition this is an antarafacial process. Note that the indicated rotations determine a specific stereochemistry for the product.

There is another rotation, shown in Equation 2, which also leads to an antarafacial orientation of the lobes which originally comprised the σ bond.

(2)

6-3

In this process, the top lobe of the σ bond rotates downward, and the bottom lobe rotates upward. These lobes are on opposite faces of the butadiene **6-3**; thus, this is also an antarafacial process. Note that only a single stereoisomer is formed as the product of this particular rotation.

In this thermal reaction only E,Z-1-chloro-1,3-pentadiene, **6-2**, is formed, the isomer in which the Cl has rotated outward. An explanation for this preference, of one allowed process over the other, is beyond the scope of this book; however, possible explanations can be found in the following references: Dolbier, W. R., Jr.; Koroniak, H.; Burton, D. J.; Bailey, A. R.; Shaw, G. S.; Hansen, S. W. *J. Am. Chem. Soc.* **1984,** *106,* 1871–1872; Rondan, N. G.; Houk, K. N. *J. Am. Chem. Soc.* **1985,** *107,* 2099–2111; and Kirmse, W.; Rondan, N. G.; Houk, K. N. *J. Am. Chem. Soc.* **1984,** *106,* 7989–7991.

For either mode of ring opening in this example, the direction of both rotations in the molecule is the same; in other words, both arrows depicting rotation point in the same direction. Such rotations are said to be conrotatory.

Example 6-5. The suprafacial ring opening of cis-3-chloro-4-methylcyclobutene—a thermally forbidden process.

Let's look at the necessary rotations and stereochemical outcome of the suprafacial ring opening of **6-1**. In this reaction both lobes of the σ bond end up on the same side of the plane defined by the carbons in the product.

Note that the rotations at the two ends of the σ bond, depicted by the arrows, occur in opposite directions; such rotations are called disrotatory.

This suprafacial ring opening is not thermally allowed. According to Helpful Hint 6-4, a suprafacial process is counted only when it has [4q + 2] electrons. The cyclobutene-butadiene conversion is a [4r] process with r=1. Such 4r processes are counted only if they are antarafacial. Thus, since no processes are counted, the total number is 0 (an even number), and according to Helpful Hint 6-4, the reaction is not thermally allowed. The designation for this reaction would be [$_\pi 4_s$].

To summarize the results in Examples 6-4 and 6-5: Each of the four rotational modes of ring opening leads to a particular stereoisomer of the product. When such ring openings are concerted processes, only products from allowed modes, either conrotatory (antarafacial) or disrotatory (suprafacial) are observed. In actual practice, sometimes only one of the two possible stereoisomers from the "allowed" processes is found as the reaction product.

6-4-2 THE FRONTIER ORBITAL APPROACH

The frontier orbital method, an alternative to Helpful Hint 6-4, can also be used to predict whether a reaction is symmetry allowed. In the application of this method to electrocyclic transformations, the HOMO of the open chain polyene is considered.

Helpful Hint 6-6. In electrocyclic transformations, the lobes at the end of the HOMO of the polyene will rotate in such a manner as to produce a bonding interaction between the ends of the chain. In other words, the lobe of + amplitude will interact with the lobe of + amplitude or the lobe of – amplitude will interact with the lobe of – amplitude to produce a bonding interaction.

Example 6-6. The frontier orbital approach to thermal ring-opening of cyclobutenes.

First draw the MO for the HOMO for the ring-opened π system, butadiene. Then consider the rotation of the lobes at the end of the chain so that either the + lobes or the – lobes overlap in the resulting cyclobutene. (At the same time the π bond of the product becomes bonding. See note and references in Example 6-4.)

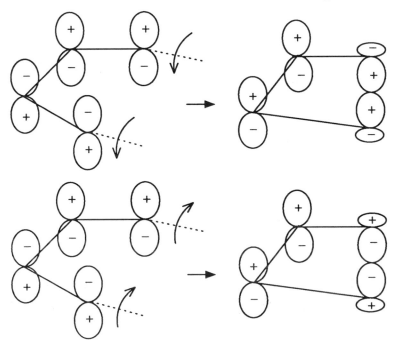

Thus the conrotatory process in the HOMO allows lobes of like amplitude to interact, producing a new bond. A disrotatory process, on the other hand, would lead to an antibonding interaction between the ends of the chain and would not lead to reaction.

*Problem 6-5. Applying either Helpful Hints 6-4 or 6-5 or Helpful Hint 6-6 decide whether the following **thermal** reactions are "symmetry allowed" or "symmetry forbidden".*

a.

Spellmeyer, D. C.; Houk, K. N.; Rondan, N. G.; Miller, R. D.; Franz, L.; Fickes, G. N. *J. Am. Chem. Soc.* **1989**, *111*, 5356–5367.

b.

X= (CH₂)₃

Vos, G. J. M.; Reinhoudt, D. N.; Benders, P. H.; Harkema, S.; Van Hummel, G. J. *J. Chem. Soc. Chem. Commun.* **1985**, 661–662.

6-4-3 OTHER ASPECTS OF ELECTROCYCLIC TRANSFORMATIONS

Helpful Hint 6-7. Only the π electrons involved in bond changes are counted.

Example 6-7. The thermal cyclization of cyclooctatetraene to bicyclo[4.2.0]octatriene.

Only 6 π electrons are necessary to effect this cyclization, and only these 6 electrons control the stereochemical course of the reaction. Thus, the thermal reaction follows a disrotatory mode, which gives a *cis* orientation at the ring junction. If all eight electrons had been counted, a conrotatory process would have been predicted, leading to a *trans* ring junction which is unlikely for such a ring fusion. In practice, cyclooctatetraene is the more stable isomer and the presence of the bicyclic compound has been demonstrated only through trapping experiments.

Problem 6-6. What is the relative stereochemistry of the ambiguous groups in each of the following concerted processes?

a.

b. (1) The different products formed when the starting material is heated or irradiated with uv light:

(2) The product of photochemical ring opening of the thermal product of the reaction in (1).

Darcy, P. J.; Heller, H. G.; Strydon, P. J.; Whittall, J. *J. Chem. Soc. Perkin Trans. I* **1981**, 202–205.

c.

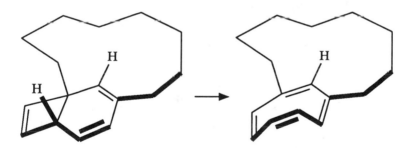

Huisgen, R.; Dahmen, A.; Huber, H. *J. Am. Chem. Soc.* **1967**, *89*, 7130–7131.

d.

Sauter, H.; Gallenkamp, B.; Prinzbach, H. *Chem. Ber.* **1977**, *110*, 1382–1402.

Problem 6-7. Is the following reaction "symmetry allowed" or symmetry forbidden? Explain.

Paquette, L. A.; Want, T.-Z. *J. Am. Chem. Soc.* **1988**, *110*, 3663–3665.

The relative rates of concerted ring opening of substituted cyclopropyl tosylates can be explained on the basis of electrocyclic principles. The best orientation for the lobes developing as the ring bond breaks is opposite the leaving group (similar to neighboring group participation). Orbital symmetry considerations dictate a disrotatory process, because this is a 2π electron system. Moreover, the disrotatory mode, which places the methyl groups on the outside, is much faster.

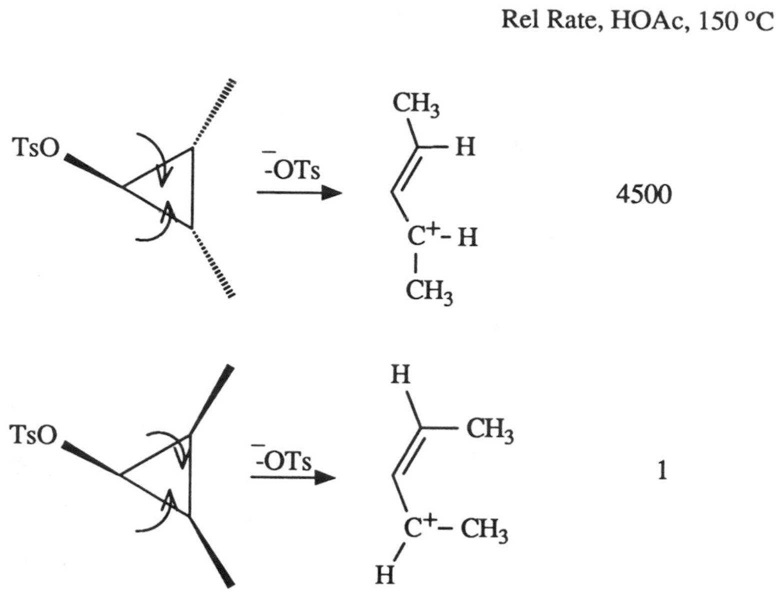

Rel Rate, HOAc, 150 °C

4500

1

von R. Schleyer, P.; Van Dine, G. W.; Schollkopf, U.; Paust, J. *J. Am. Chem. Soc.* **1966**, *88*, 2868–2869.

In small fused cyclopropyl systems, only the disrotatory mode, which moves the bridgehead hydrogens outward, is geometrically feasible.

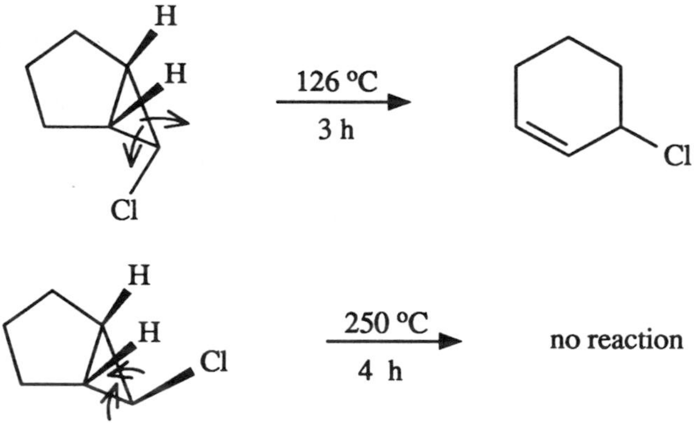

no reaction

Baird, M. S.; Lindsay, D. G.; Reese, C. B. *J. Chem. Soc. (C)* **1969**, 1173–1178.

Problem 6-8. Explain why the reactivity is so different in the two following reactions:

Jefford, C. W.; Hill, D. T. *Tetrahedron Lett.* **1969**, 1957–1960.

6-5 Cycloadditions

6-5-1 TERMINOLOGY

A cycloaddition is the reaction of two (occasionally more) separate components, at each of their termini, to produce a ring. Cycloadditions may be intermolecular or intramolecular, and may involve reactions of cyclic compounds to produce one or more new rings. One way to describe a cycloaddition is to separately record the number of electrons in each component which is involved in the reaction.

Example 6-8. Terminology and stereochemistry of the Diels-Alder reaction.

This is a [4 + 2] cycloaddition, because the diene contains 4π electrons and the dienophile contains 2π electrons in the double bond undergoing the cycloaddition. In other words, the 4 and the 2 within the brackets are the number of electrons in the π systems whose termini are reacting to give the cycload-duct. Note that the carbonyl groups, in the dimethyl maleate starting material, are conjugated with the π bond undergoing reaction. However, since the π electrons of these carbonyl groups are not forming new bonds in the course of the reaction, they are not counted.

Each component reacts suprafacially, so a more complete designation would be $[_\pi 4_s + _\pi 2_s]$. The fact that the reaction is suprafacial in both components also means that the stereo-chemical relationships among substituents in the starting ma-terials are maintained in the product. Thus, the two carbomethoxy groups in the dienophile component which are *cis* in the starting material, are also *cis* in the product. Suprafacial reaction at the diene component leads to a *cis* ori-entation of the two methyl groups in the product.

Example 6-9. A cycloaddition reaction with three π components.

Cookson, R. C.; Dance, J.; Hudec, J. *J. Chem. Soc.* **1964**, 5417–5422.

Because the two π bonds in the bicycloheptadiene are uncon-jugated, each is designated separately in the description of the reaction. Only the carbons, at each end of the C=C bond of ma-leic anhydride, are forming new bonds; so only the 2 π elec-trons of this bond are counted. Thus this is a $[_\pi 2_s + _\pi 2_s + _\pi 2_s]$ cycloaddition.

Problem 6-9. Designate the following cycloadditions according to the number of electrons, in each component, involved in the process.

a.

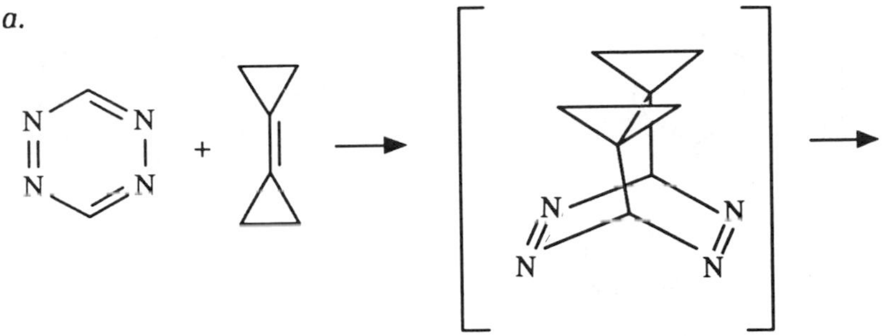

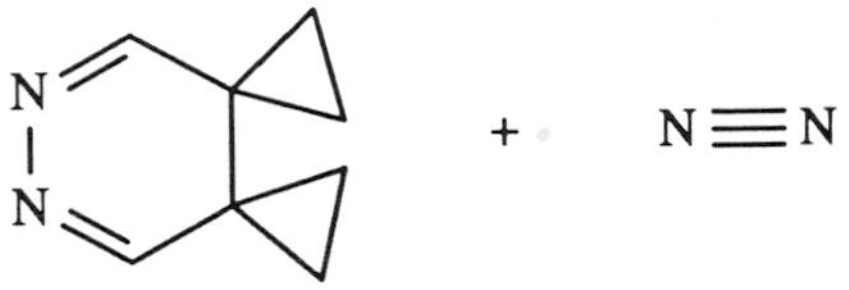

de Meijere, A.; Erden, I.; Weber, W.; Kaufmann, D. *J. Org. Chem.* **1988**, *53*, 152–161.

The reaction of the unstable intermediate is called a cycloreversion or retro Diels-Alder reaction. It is ordinarily designated by considering the reverse cycloaddition of the products of the reaction.

b.

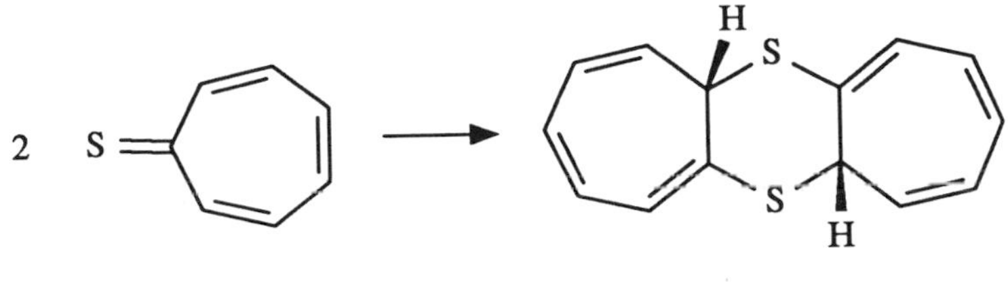

Machiguchi, T.; Hasegawa, T.; Itoh, S.; Mizuno, H. *J. Am. Chem. Soc.* **1989**, *111*, 1920–1921.

c.

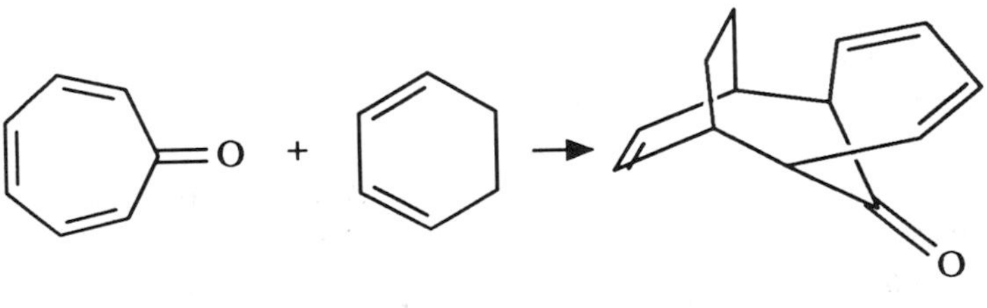

Takeshita, H.; Sugiyama, S.; Hatsui, T. *J. Chem. Soc. Perkin Trans. II* **1986**, 1491–1493.

d.

$+ \ {}^1O_2 \ \longrightarrow$

Carte, B.; Kernan, M. R.; Barrabee, E. B.; Faulkner, D. J.; Matsumoto, G. K.; Clardy, J. *J. Org. Chem.* **1986**, *51*, 3528–3532.

Problem 6-10. Give the complete formulation [$_\pi 2_S$, etc.] for each of the following cycloadditions.

a.

Minami, T.; Harui, N.; Taniguchi, Y. *J. Org. Chem.* **1986**, *51*, 3572–3576.

b.

Xu, S. L.; Moore, H. W. *J. Org. Chem.* **1989**, *54*, 6018–6021.

Unfortunately, another kind of terminology has come into use in the last few years. This is simply to designate the total number of atoms in each reacting component **counting both termini and all atoms in between.** In these cases, the **number of atoms, not the number of electrons,** is given in the description for the reaction. The two methods for naming cycloadditions do not give the same numbers, which can be a source of confusion.

Example 6-10. Counting the atoms involved in a cycloaddition.

Consider the following reaction:

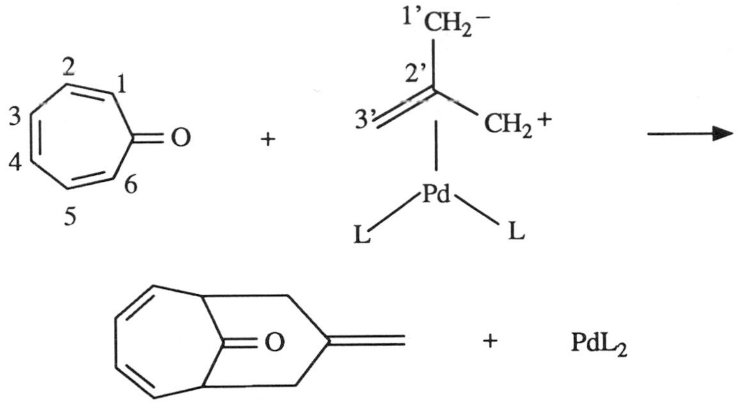

Trost, B. M.; Seoane, P. R. *J. Am. Chem. Soc.* **1987,** *109,* 615–617.

The authors call this a [6 + 3] cycloaddition, because the reacting portion of tropone (the ketone) contains 6 atoms and the alkene component contains 3 atoms. The mechanism could be shown with arrows as follows:

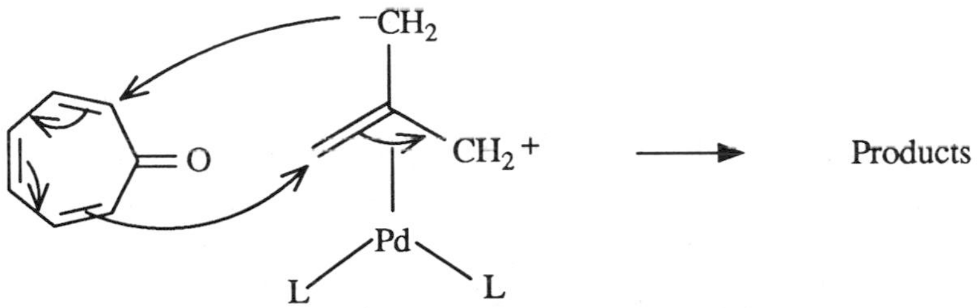

In tropone, 6π electrons are reacting, and in the alkene 4π electrons are reacting. Therefore, if this is a concerted reaction, it could also be called a [6 + 4] cycloaddition. If the distinction is made that [6 + 3] refers to the cycloadduct and [6 + 4] refers to the cycloaddition, the two methods of nomenclature are compatible. However, this distinction is not always made in the literature and, unless one thinks about the details of the particular reaction, the descriptions may be confusing.

Problem 6-11. If it were concerted, how would the following reaction be described by both systems of nomenclature?

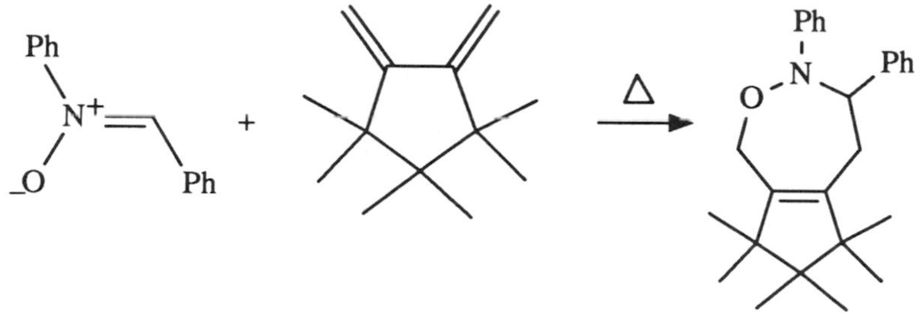

Baran, J.; Mayr, H. *J. Am. Chem. Soc.* **1987**, *109*, 6519–6521.

Problem 6-12. What is the product from an intramolecular [4 + 2] cycloaddition of the following molecule?

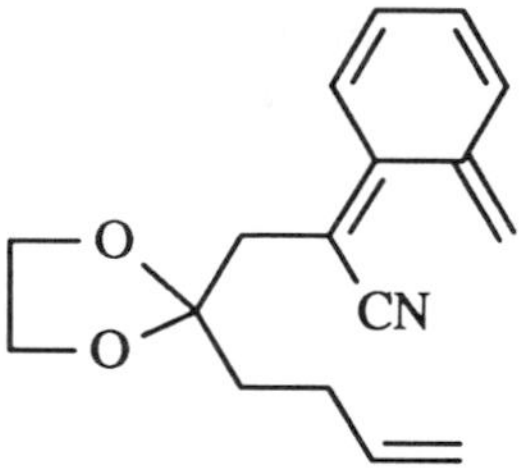

Kametani, T.; Suzuki, Y.; Honda, T. *J. Chem. Soc. Perkin Trans. I* **1986**, 1373–1377.

6-5-2 DETERMINING WHETHER A CONCERTED CYCLO-ADDITION IS THERMALLY OR PHOTOCHEMICALLY "ALLOWED"

As stated earlier, Helpful Hint 6-4 is general and thus can be applied to cycloaddition reactions.

Example 6-11. A [$_\pi 4_s$ + $_\pi 4_s$] reaction.

A [$^{\cdot}_\pi 4_s$ + $_\pi 4_s$] reaction consists of two suprafacial processes of 4π electron systems. Since [4r] processes (r=1 in this example) are not counted if they are suprafacial, there is an even number (zero) of counted processes. Thus, this would be a photochemically allowed reaction.

Example 6-12. A [$_\pi$12$_s$ + $_\pi$2$_s$] process.

A [$_\pi$12$_s$ + $_\pi$2$_s$] reaction consists of one [4q + 2] suprafacial process (the $_\pi$2$_s$ component) and no [4r] antarafacial processes. Thus, the number of counted processes is odd, and this reaction would be thermally allowed.

6-5-3 PREDICTION OF ALLOWED OR FORBIDDEN BY FRONTIER ORBITAL THEORY

Helpful Hint 6-8. According to frontier orbital theory, two-component cycloadditions are symmetry-allowed, if the symmetry of the HOMO of one component is such that it can overlap, in a bonding way, with the LUMO of the other component.

Example 6-13. Frontier orbitals applied to the dimerization of ethylene.

Consideration of the HOMO and LUMO of ethylene shows that a *supra-supra* cycloaddition cannot be concerted. That is, the amplitude of the wave functions does not allow simultaneous bond formation between both ends of each molecule when the HOMO and LUMO react suprafacially.

Ethylene LUMO:

Ethylene HOMO:

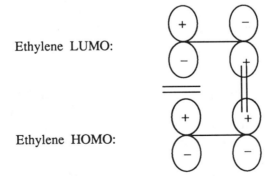

In other words a [$_\pi$2$_s$ + $_\pi$2$_s$] cycloaddition of ethylene is not a thermally allowed reaction. Note that this reaction has two [4q + 2] suprafacial processes and would be predicted to be photochemically allowed.

On the other hand, if the HOMO were to react suprafacially and the LUMO antarafacially, the symmetries would match.

Ethylene LUMO:

Ethylene HOMO:

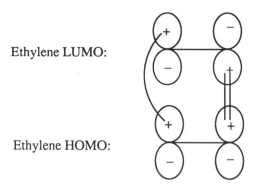

This would also be true if the HOMO were to react antarafacially and the LUMO suprafacially.

Ethylene LUMO:

Ethylene HOMO:

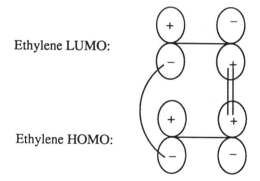

Nonetheless, such [$_\pi2_s$ + $_\pi2_a$] thermal cycloadditions do not occur as concerted reactions (though "allowed" by Helpful Hint 6-4 or Helpful Hint 6-8), because the necessary twist of the LUMO double bond is not geometrically feasible.

Example 6-14. Application to a [4+2] cycloaddition.

For [4 + 2] cycloadditions the diene is often the HOMO, and the dienophile is the LUMO. So, in the general illustration below, the LUMO is the 2π electron component, and the HOMO is the 4π electron component.

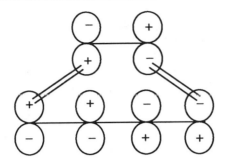

In this case, both ends of each termini can overlap suprafacially in a bonding way, and a concerted reaction is possible. Application of Helpful Hint 6-4 would count only the 2 electron process, and the reaction would be thermally allowed.

Example 6-15. Analysis of a [2 + 2] Reaction.

An example of a reaction, in which a σ bond and a π bond are involved, is shown in the following intramolecular cycloaddition. The bonds reacting are highlighted.

The lobes of the involved frontier orbitals, in the ground state configuration, are shown in the next equation. The π bond is the HOMO and the σ bond, the LUMO. Note that there is no geometrically feasible way for both termini of each component to interact concertedly in a bonding way. The appropriate lobes do not match in one of the necessary interactions. Thus, this reaction is not thermally allowed.

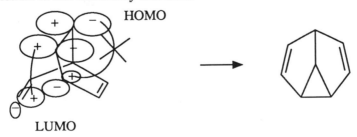

If the HOMO is the σ bond and the LUMO is the π bond, there is the same problem:

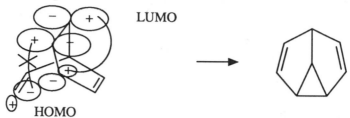

Now, it is the other end of the termini where the reaction is not feasible. Note: the front lobe of the π bond does not extend far enough to the left and downward to overlap effectively with the necessary + lobe of the HOMO.

The next thing to consider is a photochemical process. The HOMO is an antibonding π* orbital, because upon absorption of light, an electron is promoted from the bonding to the antibonding orbital of the double bond. (It is easier to photochemically excite a π bond than a σ bond.) The symmetries now match for the transformation. Thus, this is a photochemically allowed reaction.

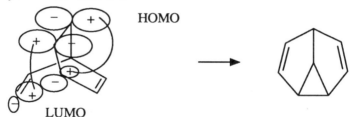

Applying Helpful Hint 6-4 to this reaction, gives the same prediction as frontier orbital theory, because there are no $[4q + 2]_s$ or $[4r]_a$ processes, an even number.

For a useful discussion of frontier orbital theory, see Fukui, K. *Acct. Chem. Res.* **1971**, 57–64.

6-5-4 SECONDARY ORBITAL INTERACTIONS

In the reaction of cyclopentadiene with dimethyl fumarate, the two carbomethoxy groups end up on the *endo* side of the ring in the product, **6-4**. None of the *exo* isomer, **6-5**, is formed.

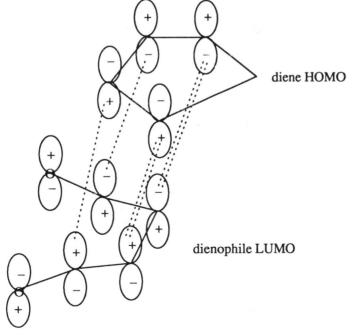

6-4   + no   **6-5**

This is a result of secondary orbital interactions, between orbitals which are not involved in the bonding changes taking place in the cycloaddition. Thus, in this case, interactions of the **entire** π system of the dienophile with the π system of the diene must be considered.

Consider the HOMO of the diene and the LUMO of the dicarbonyl-conjugated double bond (total π electrons = 6):

diene HOMO

dienophile LUMO

. . . . represents secondary orbital interaction
: : : : : : : : : : : represents primary orbital interaction

Doubly dashed lines indicate bond forming overlap between the ends of the diene and the ends of the carbon-carbon double bond of the dienophile to produce the cycloadduct. Singly dashed lines indicate attractive interactions between orbitals which do not overlap in the final product. However, these attractive interactions control the stereochemistry of the product by holding the carbonyl groups underneath the 4π electron system of cyclopentadiene as the reaction proceeds. This produces the *endo* stereochemistry.

6-5-5 OTHER [4 + 2] CYCLOADDITION REACTIONS

In this section cycloadditions of allyl cations and cycloadditions of 1,3-dipoles will be discussed. Some 1,3-dipolar additions were introduced in Example 6-10 and Problem 6-11.

Example 6-16. The allyl cation as dienophile.

A mechanism with several steps can be envisioned for the following reaction:

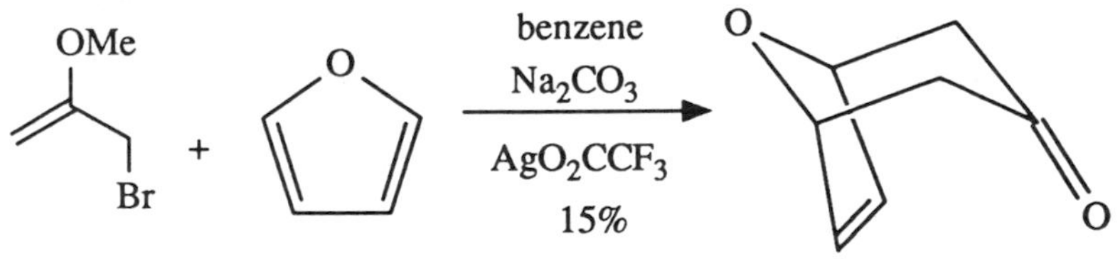

The first step, ionization of the bromide to the 2-methoxyallyl cation, **6-6**, is assisted by the silver ion:

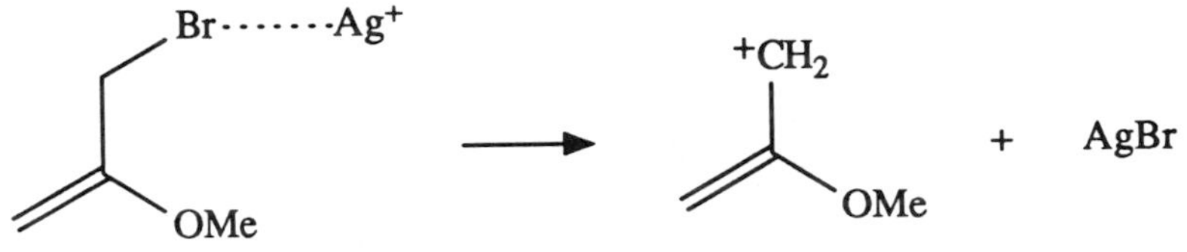

Cation **6-6**, which contains 2π electrons, can cycloadd to furan, which acts as the 4π electron component, to give cyclized cation, **6-7**:

6-6 **6-7**

Under the workup conditions, dilute nitric acid, **6-7** will hydrolyze to the product ketone.

Hill, A. E.; Greenwood, G.; Hoffmann, H. M. R. *J. Am. Chem. Soc.* **1973**, *95*, 1338–1340.

From a synthetic standpoint, the 1,3-dipolar cycloaddition is a very important reaction. In this [4 + 2] cycloaddition, the four electron component is dipolar in nature, and the two electron component is usually referred to as the dipolarophile. When the thermal reaction is concerted, such reactions are suprafacial in both components; i.e., they are $[_\pi 4_s + _\pi 2_s]$ cycloadditions.

Example 6-17. 1,3-Dipolar cycloaddition of a nitrile oxide and an alkyne.

In the reaction shown, the nitrile oxide is a 4π electron system, and one of the acetylenic π bonds is a 2π electron system.

Huisgen, R. *Angew. Chem. Int. Ed. Engl.* **1963**, *2*, 565–598.

According to the newer terminology, they react to form what can be called a [3 + 2] adduct. If the mechanism is written with arrows, the usual rules apply. That is, the flow of electrons is away from negative charge and toward positive charge.

The regiochemistry of 1,3-dipolar additions can be explained on the basis of frontier orbital theory. However, coefficients of the wave functions are required, and their derivation is beyond the scope of this book. In the examples and problems, the major products of the reactions are given. For further information, see Houk, K. N. *Acct. Chem. Res.* **1975**, 361–369.

Example 6-18. The intramolecular 1,3-dipolar cycloaddition of an azide.

Tsai, C.-Y.; Sha, C.-K. *Tetrahedron Lett.* **1987**, *28*, 1419–1420.

This is a concerted reaction; 4 of the π electrons of the azide group undergo cycloaddition with the 2π electrons of the carbon-carbon double bond of the α,β-unsaturated ester:

Such intramolecular cycloadditions are very powerful synthetic tools for the synthesis of fused ring systems.

Problem 6-13. Write the possible products of the 1,3-dipolar cycloaddition of 6-8 with acrylonitrile. Because of their instability, many 1,3-dipoles are formed in situ when needed. Such a formation of a nitrile ylide, 6-8, is shown in the first step.

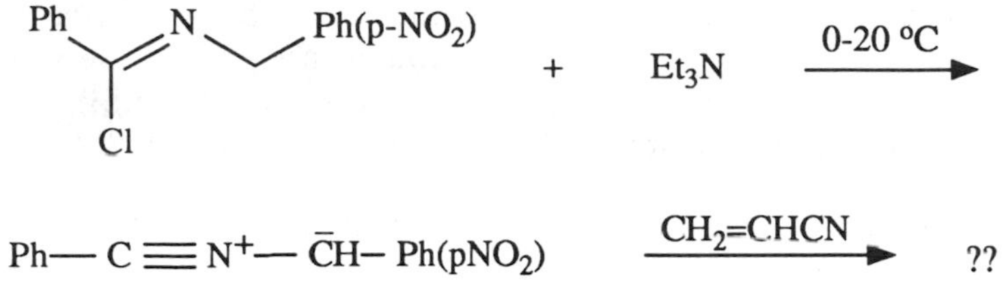

Ph—C≡N⁺—C̄H—Ph(pNO₂) $\xrightarrow{\text{CH}_2=\text{CHCN}}$??

6-8

Huisgen, R. *Angew. Chem. Int. Ed. Eng.* **1963**, *2*, 565–598.

Problem 6-14. Show how the following reaction might occur:

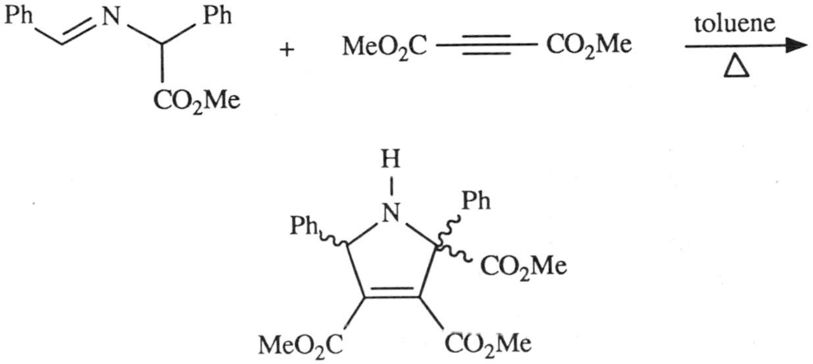

Grigg, R.; Kemp, J.; Sheldrick, G.; Trotter, J. *J. Chem. Soc. Chem. Commun.* **1978**, 109–111.

6-6 Sigmatropic Reactions

6-6-1 TERMINOLOGY

In a sigmatropic reaction, movement (tropic is from the Greek word "tropos," to turn) of a σ (sigma) bond takes place, producing rearrangement. Common types of sigmatropic reactions are the familiar 1,2-hydride, or alkyl, shift in carbocations; and the Cope rearrangement.

Sigmatropic reactions can be designated by the same scheme we have used for π electron reactions; i.e., [$_\pi 4_s + _\sigma 2_s$], etc. However, sigmatropic reactions are more often designated in a different way: First, label both atoms of the original (breaking) σ bond "1." Then, count the atoms along the chains on both sides until you reach the atoms which form the new σ bond. The numbers assigned to these atoms are then given in brackets, separated by a comma; e.g., [1,5], or [3,3]. This nomenclature is illustrated in Examples 6-19 through 6-21.

Example 6-19. A [1,2] sigmatropic shift in a carbocation.

The σ bond which is broken in the starting material and the σ bond formed in the product are highlighted. Hydrogen (1') has moved from carbon 1 to carbon 2, so the reaction is designated as a [1,2] sigmatropic shift. The "1", of this designation, does not refer to the number on carbon, but to the number on hydrogen. This indicates that the same atom (hydrogen) is at one end of the σ bond in both starting material and product. The "2" of the designation, [1,2], is the number of the carbon where the new σ bond is formed, relative to the number 1 for the carbon where the old σ bond was broken.

This reaction can also be designated as $[\pi 0_s + \sigma 2_s]$. This is more appropriate than labelling the reaction as $[\pi 2_s + \sigma 0_s]$, because this is considered to be a hydride shift, not a proton shift.

Example 6-20. A [3,3] sigmatropic shift.

Consider the classical Cope rearrangement:

The atoms at the ends of the σ bond being broken are numbered "1" and "1'". Atoms are then numbered along each chain until the atoms at the ends of the new σ bond are reached. The atoms of the new σ bond are numbered "3" and "3'". The σ bond, originally at the 1,1' position, has formally moved to the 3,3' position. Thus, this is called a [3,3] sigmatropic shift. It could also be designated as a $[_\pi 2_s + _\pi 2_s + _\sigma 2_s]$ reaction.

Example 6-21. A [1,5] sigmatropic shift.

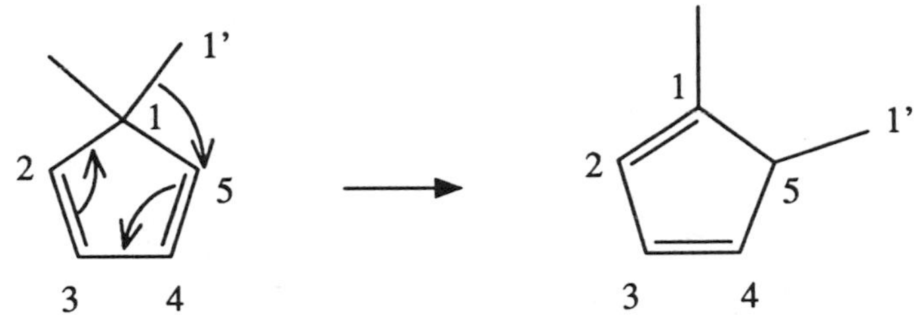

All the carbons in the ring must be counted as part of the process. That is, this reaction is **not** a simple [1,2] shift, because the π electrons in the ring must also rearrange. Thus, this is a [1,5] shift or $[_\pi 4_s + _\sigma 2_s]$ reaction.

Problem 6-15. Designate the type of sigmatropic shift which occurs in each of the following reactions (e. g. [1,3]). Also give the appropriate reaction designation ($[_\pi 2_s$, etc.]).

a.

b.

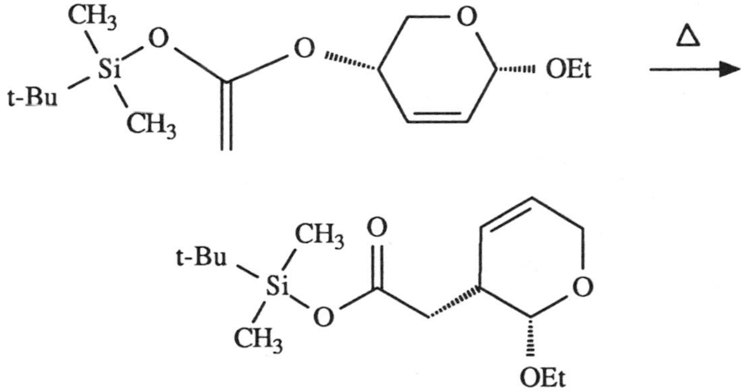

Curran, D. P.; Jacobs, P. B.; Elliott, R. L.; Kim, B. H. *J. Am. Chem. Soc.* **1987,** *109,* 5280–5282.

c.

Vedejs, E. *Acc. Chem. Res.* **1984**, *17*, 358–364.

d.

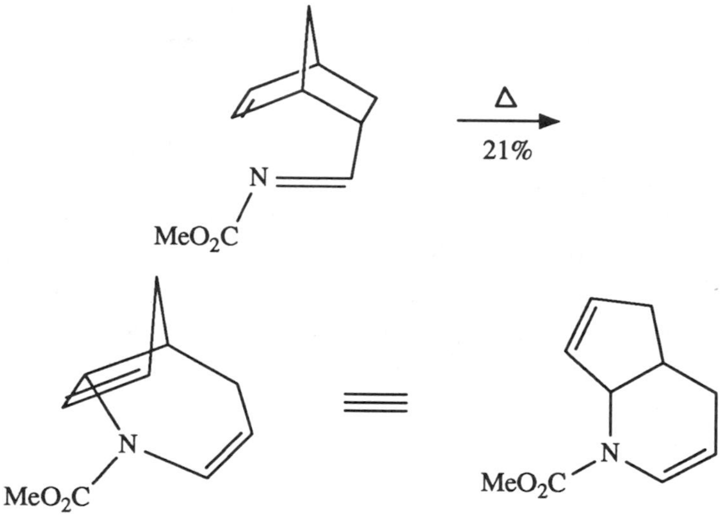

Wu, P.-L.; Chu, M.; Fowler, F. W. *J. Org. Chem.* **1988**, *53*, 963–972.

6-6-2 WHICH REACTIONS ARE THERMALLY ALLOWED AND WHICH ARE PHOTOCHEMICALLY ALLOWED?

One way to decide whether a reaction is thermally or photo-chemically allowed is to apply the general Helpful Hint, 6-4. First, determine the number of involved electrons in each component. Then determine the suprafacial or antarafacial character of its reaction. Helpful Hint 6-4 predicts that the $[\pi 0_s + \sigma 2_s]$ reaction of Example 6-19 is thermally allowed. Since only the two electron process is counted, there is an odd number, one, of counted processes.

Example 6-20 illustrates a $[\pi 2_s + \pi 2_s + \sigma 2_s]$ reaction. Since 2-electron suprafacial processes are counted, there is an odd number, three, and the reaction is thermally allowed. Example 6-21, $[\pi 4_s + \sigma 2_s]$, is also thermally allowed, because only the two electron component is counted.

Example 6-22. Is a [1,3] sigmatropic shift thermally allowed?

If this were a $[\pi 2_s + \sigma 2_s]$ concerted rearrangement, it would not be thermally allowed, because both 2-electron processes would be counted. On the other hand, if the π system were to react antarafacially, the rearrangement would be the thermally allowed, $[\pi 2_a + \sigma 2_s]$, because a 2-electron, antarafacial process is not counted. However, because it is difficult to maintain a concerted reaction with the geometry required for the antarafacial component, thermal [1,3] sigmatropic shifts are not common. Note that a hydrogen atom, with a spherical 1s orbital, cannot react antarafacially. However, a carbon, bonding with hybrid orbitals, has the capability of reacting antarafacially, as illustrated by Example 6-23.

Example 6-23. The stereochemical consequences of a concerted, thermally-allowed [1,5] sigmatropic shift.

An elegant demonstration of the stereochemistry of a thermally allowed [1,5] sigmatropic shift was reported by Roth and co-workers in 1970. They studied the stereochemistry of the reaction of the optically active starting material, **6-9**. Consider the products from each of the two possible suprafacial reactions, and each of the two possible antarafacial reactions.

Suprafacial rearrangement of hydrogen across the top face of the π system gives **6-10**.

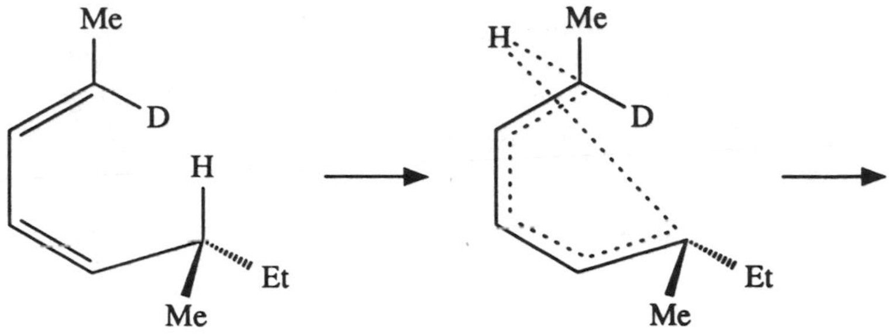

6-9

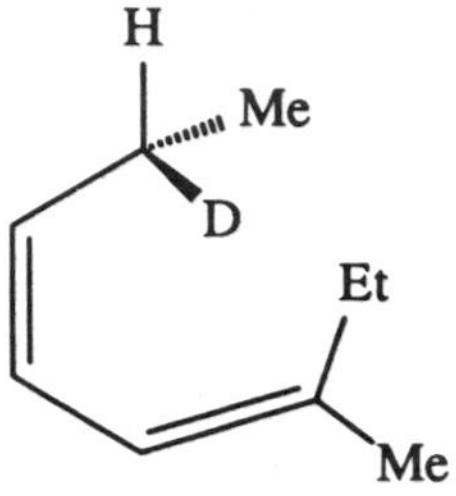

6-10

Rotation about the carbon-carbon single bond axis between the 2-butyl group and the π bond gives a new conformation of **6-9**. In a suprafacial movement of hydrogen across the bottom face of the π system, the isomeric **6-11** is produced.

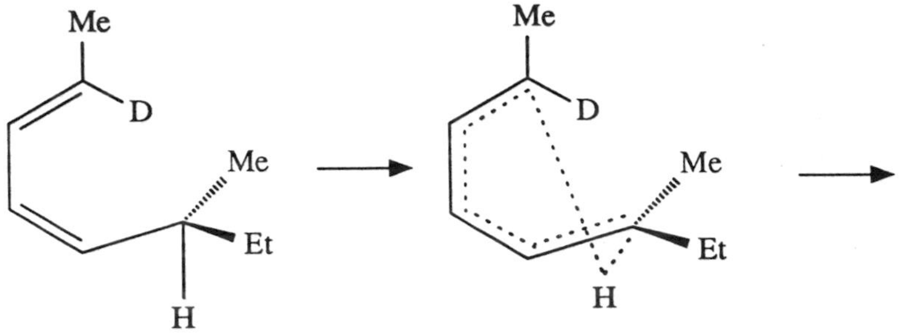

6-9

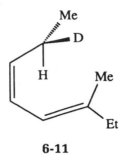

6-11

Antarafacial movement of hydrogen also gives two possible products, **6-12** and **6-13**. In the reaction to give **6-12**, hydrogen moves from the top face to the bottom face, and in the reaction to give **6-13**, hydrogen moves from the bottom face to the top face.

6-9

6-12

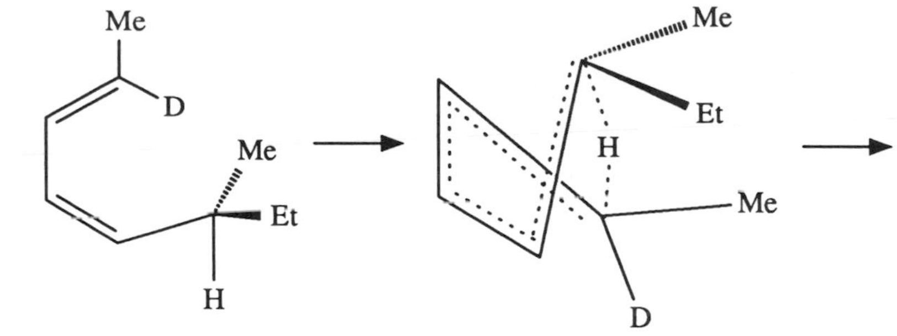

6-9

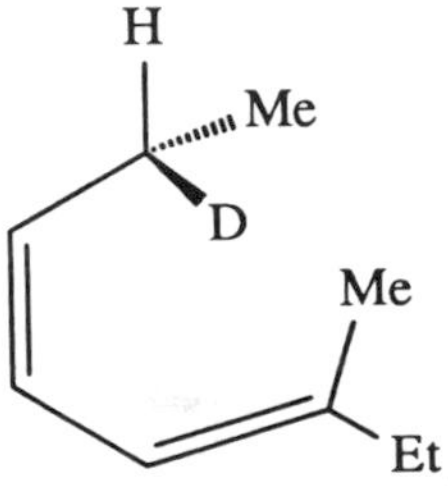

6-13

In agreement with theory, compounds **6-10** and **6-11**, and not **6-12** or **6-13**, were the products of the reaction. Because of their suprafacial character, thermal [1,5] sigmatropic shifts occur with facility. That is, the [1,5] sigmatropic shift is a $[\pi 4_s + \sigma 2_s]$ process which is thermally allowed.

Roth, W. R.; Konig, J.; Stein, K. *Chem. Ber.* **1970**, *103*, 426–439.

Another interesting example of the stereochemical consequences of orbital symmetry is the result of sigmatropic shifts in the 6-methyl-bicyclo[3.1.0]hexenyl cation. In this cation, the *exo*-6-methyl group will remain *exo* as the migration of carbon 6 proceeds around the ring.

Example 6-24. Sigmatropic shifts in the *exo*-6-methyl-bicyclo[3.1.0]hexenyl cation.

(Note that the numbers shown in the equation are not those used to name the compound. Thus, the 1' carbon is numbered 6 for nomenclature purposes, and the methyl on this carbon is referred to as the exo-6-methyl.) Analysis of the reaction, in the usual way, shows that this is a [1,4] sigmatropic shift of the carbon labelled 1' in the equation. If this were a $[_\sigma2_s + _\pi2_s]$, reaction, it would not be thermally allowed. On the other hand, if it were a $[_\sigma2_a + _\pi2_s]$ reaction, it would be allowed. (The other allowed reaction, $[_\sigma2_s + _\pi2_a]$, is not a feasible concerted reaction, because of the high distortion necessary for a twist about the π system.)

Consideration of these restrictions indicates that the π system reacts on the same side at both termini, and that when the carbon migrates, it is the back lobe of the broken σ bond which interacts with the other end of the π system to give the product. As a consequence, the methyl group always remains *exo*. If the σ bond were to react suprafacially, the same lobe that formed the original σ bond would interact with the other end of the π system, and the methyl group would be *endo*.

The frontier orbital approach to this reaction is shown in the following equation. The allyl system is the LUMO and the σ bond is the HOMO:

For some, it is difficult to see that such a transformation leads to the methyl group remaining *exo*. Molecular models are very useful in visualizing such reactions.

For the actual structures studied see Hart, H.; Rodgers, T. R.; Griffiths, J. J. *Am. Chem. Soc.* **1969**, *91*, 754–756.

Problem 6-16. Show how the following concerted processes can be explained on the basis of one or more sigmatropic shifts. What are the designations for the shift(s) involved?

a.

Miller, B.; Baghdadchi, J. *J. Org. Chem.* **1987**, *52*, 3390–3394.

b.

Barrack, S. A.; Okamura, W. H. *J. Org. Chem.* **1986**, *51*, 3201–3206.

6-7 The Ene Reaction

The ene reaction appears to combine the characteristics of cycloadditions and sigmatropic reactions.

If concerted, it is described as [$_\sigma 2_s + _\pi 2_s + _\pi 2_s$], a thermally allowed reaction. This looks similar to the Diels-Alder reaction, in which a C-H bond replaces the double bond of the diene component. In some cases, the ene reaction actually competes with the Diels-Alder reaction. Because of the similarities, the allyl component is often called the enophile, and the other component, the ene.

The ene reaction can also take place intramolecularly, and thus lead to new rings.

Example 6-25. An intramolecular ene reaction.

The unconjugated diene, **6-14**, reacts to form a cyclic molecule.

X=CO$_2$Me

6-14

EtAlCl$_2$

82%

6-14

Since the proton is transferred to the top of the double bond, the carbomethoxy group, X, is forced down. Cautionary note: there is no proof that this is a concerted reaction.

Snider, B. B.; Phillips, G. B. *J. Org. Chem.* **1984**, *49*, 183–185.

Problem 6-17. Draw the product for the ene reaction of the following components:

+

Problem 6-18. Write appropriate mechanisms for each of the following reactions. For any possibly concerted steps, include the formulation [$_\pi 2_s$, etc.] as part of your answers.

a.

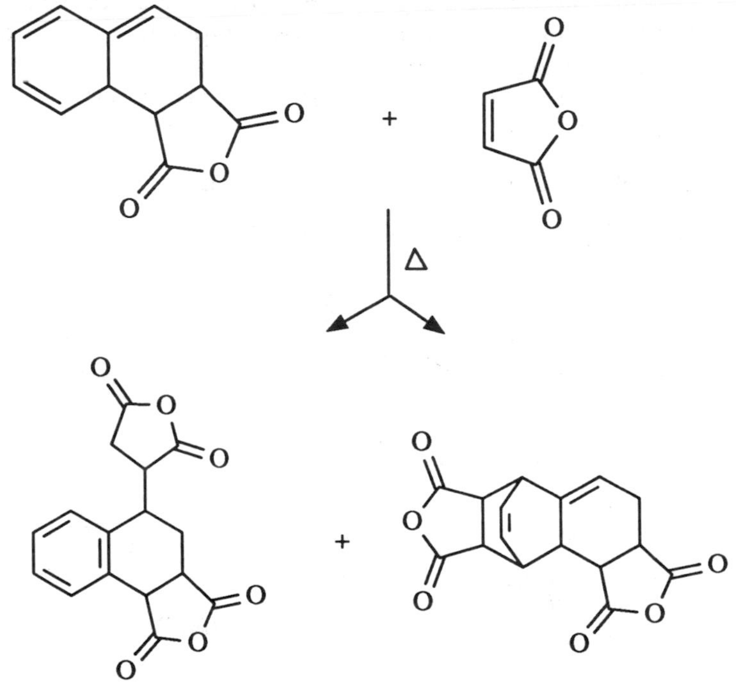

+

Δ

+

89% of product   11% of product

Alder, K.; Schmitz-Johnson, R. *Ann.* **1955**, *595*, 1–37; Hoffmann, H. M. R. *Angew. Chem. Int. Ed. Engl.* **1969**, *8*, 556–577.

b.

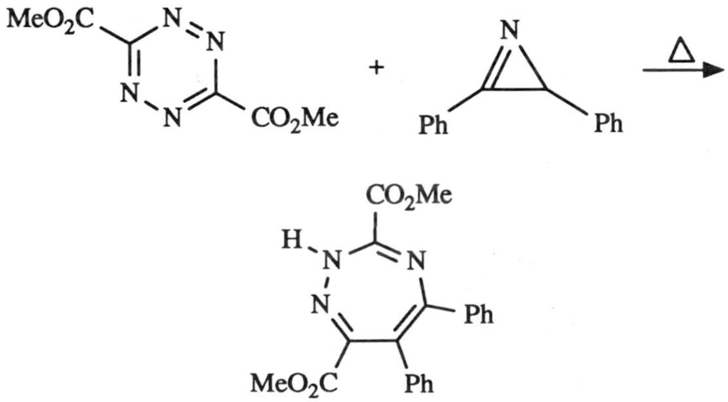

Δ

Johnson, G. C.; Levin, R. H. *Tetrahedron Lett.* **1974**, 2303–2307.

c.

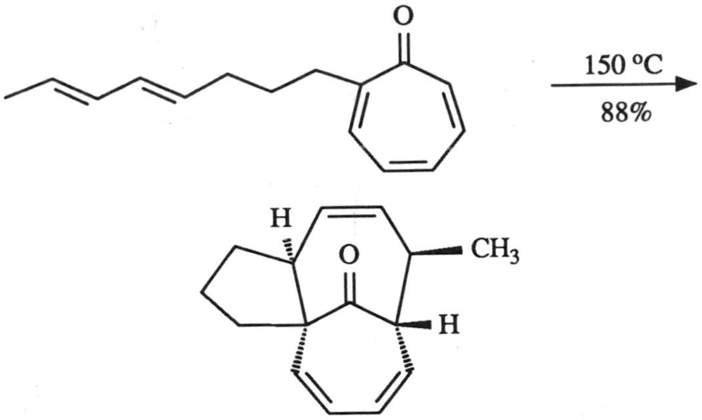

Funk, R. L.; Bolton, G. L. *J. Am. Chem. Soc.* **1986**, *108*, 4655–4657.

d.

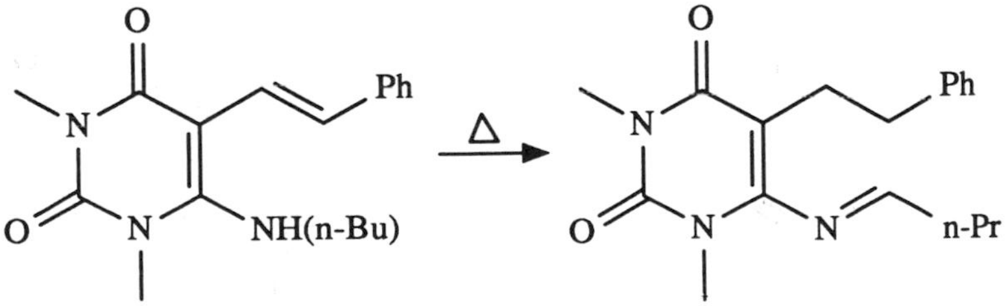

Matyus, P.; Zolyomi, G.; Eckhardt, G.; Wamhoff, H. *Chem. Ber.* **1986**, *119*, 943–949.

e.

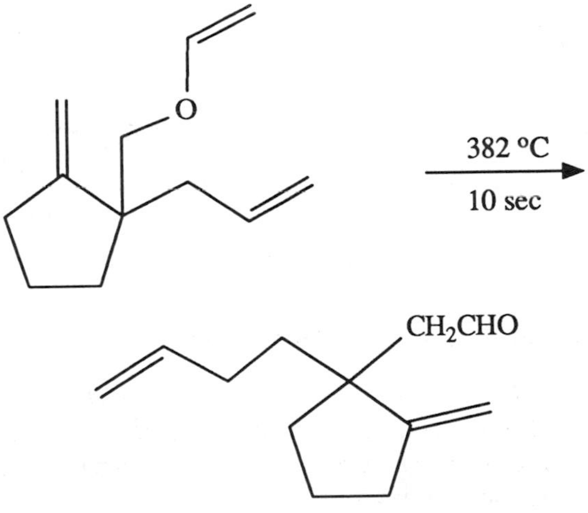

Ziegler, F. E.; Piwinski, J. J. *J. Am. Chem. Soc.* **1982**, *104*, 7181–7190.

f.

Alder, K.; Dortmann, H. A. *Chem. Ber.* **1952**, *85*, 556–565.

Problem 6-19. The following thermal transformation involves 3 pericyclic changes; the first two are electrocyclic, and the third is a sigmatropic rearrangement. Give structures for the two intermediates in the reaction.

R_1 = alkyl

Shishido, K.; Shitara, E.; Fukumoto, K.; Kametani, T. *J. Am. Chem. Soc.* **1985**, *107*, 5810–5812.

Problem 6-20. *The equation below represents a thermal [4 + 4] cycloaddition, which occurs readily. Discuss orbital symmetry considerations in detail, and propose a mechanism for the reaction.*

Heine, H. W.; Suriano, J. A.; Winkel, C.; Burik, A.; Taylor, C. M.; Williams, E. A. *J. Org. Chem.* **1989**, *54*, 5926–5930.

Problem 6-21. *Transformation of starting material to 6-15 involves a concerted electrocyclic transformation, followed by anionic-promoted ring opening of the epoxide. The other transformations are one step concerted processes. Explain the stereochemical preferences of each reaction.*

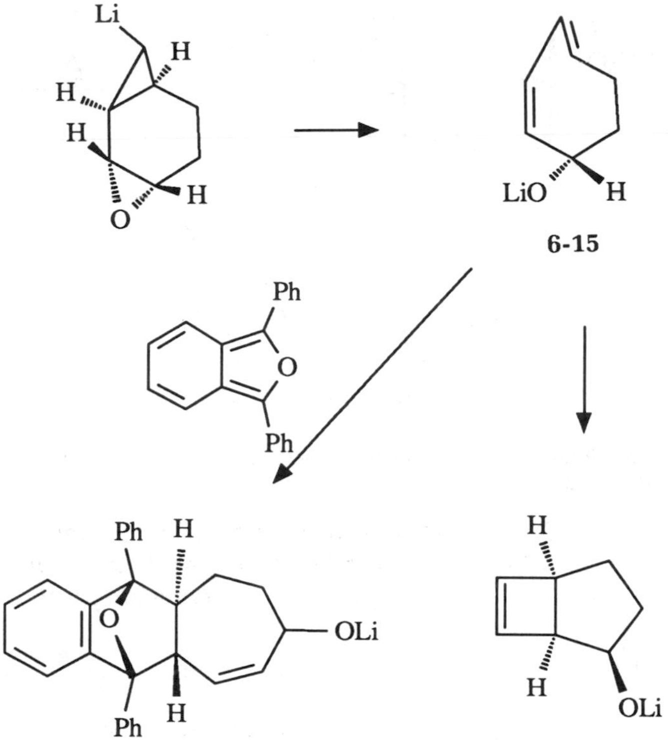

6-15

6-16

6-17

Also explain why the electrocyclic transformation of **6-15** to **6-17** is favored, while such reactions usually occur more readily in the opposite direction.

Coates, R. M.; Last, L. A. *J. Am. Chem. Soc.* **1983**, *105*, 7322–7326.

Answers to Problems

6-1.   Both of the orbitals have the same symmetry with respect to m_2 (S) and C_2 (A). Since orbitals alternate with respect to the symmetry of these elements, there must be another MO between them, which is A with respect to m_2, and S with respect to C_2. Also note that the lowest energy orbital

shown has no nodes, but the next higher one has two nodes.
There should be a MO, of intermediate energy, with one node.

6-2. Note: when no sign is given for the amplitude of the wave
function at an atom, there is a node at that atom. When
there is one node and an odd number of atoms, that node
must lie on the central atom, so that the MO can be either
totally symmetric or totally antisymmetric with respect to
symmetry elements m_2 and C_2 (see Helpful Hint 6-3).

	m_2	C_2	nodes

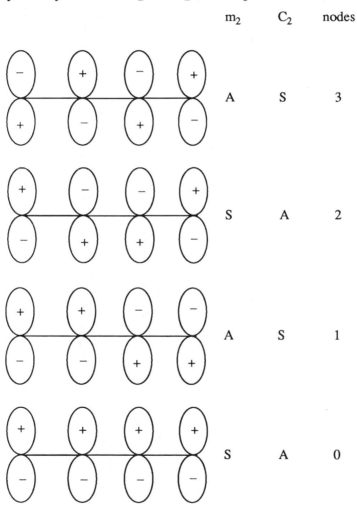

	m_2	C_2	nodes
	A	S	3
	S	A	2
	A	S	1
	S	A	0

m₂   C₂   nodes

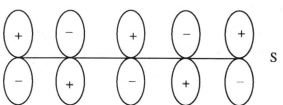

S   A   4

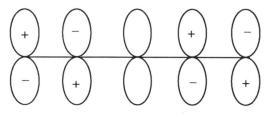

A   S   3

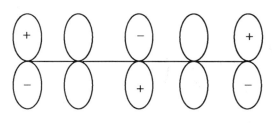

S   A   2

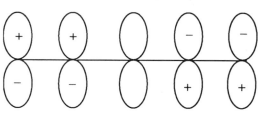

A   S   1

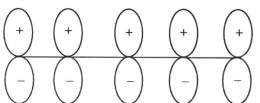

S   A   0

6-3. The pentadienyl cation, ⌒⌒CH₂⁺, has a total of 4 π electrons. Thus, starting at the MO of lowest energy we fill in the electrons (as up or down arrows). The lowest two MO's are filled with electrons. The MO with 1 node is the HOMO, and the MO with 2 nodes is the LUMO.

electrons

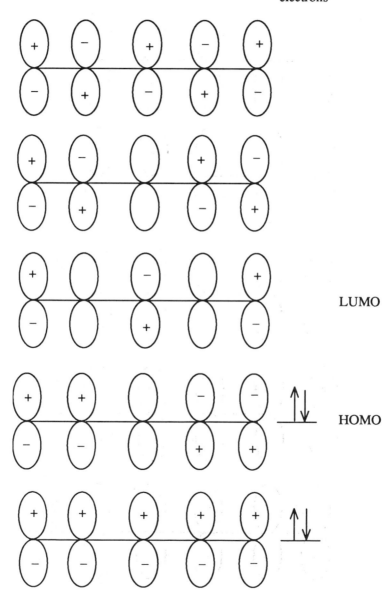

The pentadienyl radical, $\dot{C}H_2$, has a total of five electrons. Thus, there is one electron in the MO with 2 nodes which is the HOMO, and the LUMO is the next higher MO.

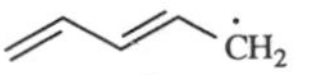

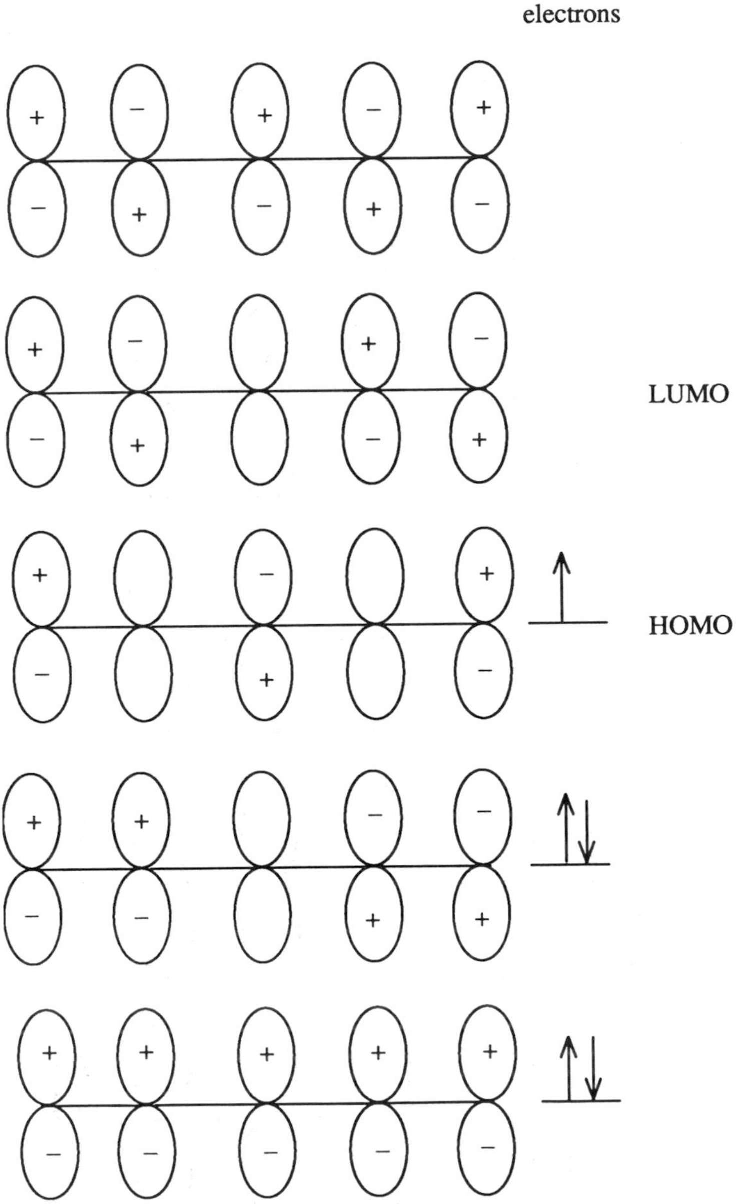

electrons

LUMO

HOMO

The pentadienyl anion, $\diagup\!\diagdown\!\diagup\!\diagdown\!_{\bar{C}H_2}$, has a total of 6 π electrons, two are counted for each π bond and two for the negative charge. Thus, the three MO's of lowest energy are filled with electrons. The HOMO is the MO with 2 nodes, and the LUMO is the MO with 3 nodes. The HOMO and the LUMO for the radical and the anion are the same.

electrons

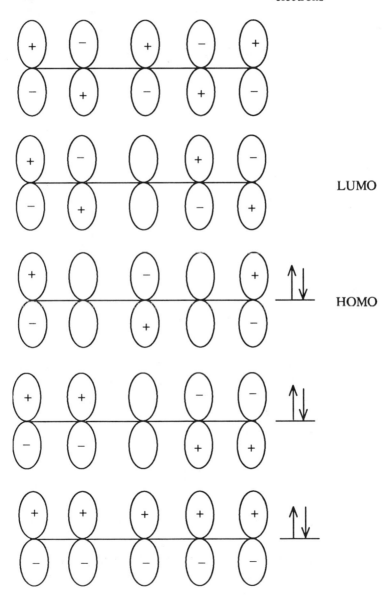

LUMO

HOMO

6-4.a. According to Helpful Hint 6-4, only those antarafacial components that contain 0, 4, 8, 12, etc. electrons are counted. Thus the $\pi 4_a$ process is counted, and the $\pi 2_a$ process is not counted. There is an odd number of counted processes; the reaction is thermally allowed.

6-4.b. According to Helpful Hint 6-4, only those suprafacial components that contain 2, 6, 10, etc. electrons are counted. Thus the $\pi 8_s$ process is not counted, but the $\pi 2_s$ is. There is an odd number; the reaction is thermally allowed.

6-4.c. Neither the $\pi 10_a$ nor the $\sigma 0_s$ processes are counted. There is an even number (0); the process is photochemically allowed.

6-5.a. For the stereochemistry shown, the process must be disrotatory. This is a $\pi 4_s$ process, and is thus "symmetry forbidden" thermally.

This is an example of a symmetry forbidden reaction which, nonetheless, actually takes place. The electronic effects of the substituents, R_1 and R_2, on the course of the reaction are complex.

6-5.b. In order to rotate the substituents into their positions in the product, a conrotatory mode is required. This is a $\pi 4_a$ process and is thermally allowed. Note that the diene, and not the cyclobutene, is considered in giving the designation and stereochemistry.

Note the unusual *trans* orientation of the ring carbons across the lower double bond in the product.

6-6.a.   According to Helpful Hint 6-4, a concerted, thermal reaction of a 6π electron system would be suprafacial. This would require a disrotatory mode of ring closing, which would lead to a *cis* orientation of the hydrogens.

6-6.b.   (1) This is a 6π electron process. (Only one of the π bonds in the furan ring is involved in the reaction). The photochemical reaction will be conrotatory (antarafacial), leading to *cis* methyl groups:

The thermal reaction will be disrotatory, leading to *trans* methyl groups:

(2) The photochemical ring opening of the thermal product will be conrotatory. (Remember, the total number of electrons

counted in a process is that of the open-chain compound.)
Thus, the product will be an isomer of the starting material
for the original process.

6-6.c.   First, rotate about single bonds, to get the ends of the
tetraene system into proximity, for reaction. Such rotations
change the conformation of the molecule, but not the stereochemistry. However, you may *not* change the stereochemistry, by rotating about any of the double bonds, to get
the ends into position. All of the double bonds have *cis*
substituents. Since this is an 8π electron system, the thermally allowed process will be conrotatory (antarafacial).
Such a rotation gives the product with the two methyl
groups *trans*.

6-6.d.   As in the previous example, the starting material must be
placed in a conformation in which the ends of the π system
are in proximity, for reaction. If it is not clear to you which
electrons are necessary for the transformation, drawing arrows for the redistribution of electrons can be helpful.

In order to get the product, all 12 π electrons must be involved. Thus, the thermal reaction should be conrotatory, and the hydrogens will be *trans*.

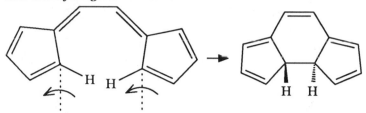

6-7. The reaction only involves 6 electrons: the conjugated diene and the σ bond between the carbons common to the 4- and 6-membered rings. Thus, this is a 6π electron electrocyclic transformation. The two electrons of the cyclobutene π bond are not involved in the process. The rotation necessary to form the product is shown below:

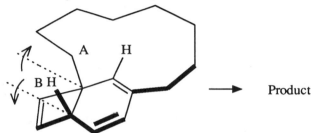

Product

Thus carbon A of the loop must move behind the plane of the page and hydrogen B must move in front of the plane of the page. This disrotatory, $\pi6_s$, reaction is thermally allowed.

6-8. Ionization occurs to a 2π electron allylic cation. To be thermally allowed, therefore, the motion must be disrotatory. Only the disrotatory motion, which moves the hydrogens outward, will occur. This places the developing p orbitals opposite the bromide leaving group for the second compound only. The bromide ion then reacts with the cation to produce the product.

6-9.a.   Both the cycloaddition, to give the intermediate, and the cycloreversion of the intermediate, to give the product, are [4 + 2] reactions. The electrons involved in each process are highlighted in the following structures:

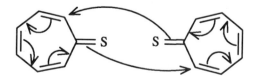

In the retrocycloaddition, only two of the π electrons of the N_2 product are used. The other two π electrons are perpendicular (orthogonal) to the first two and cannot take place in the reaction. Thus, by considering the products of the reaction, this is a $[\pi 4_s + \pi 2_s]$ process. On the other hand, the reaction of the intermediate would be designated as a $[\sigma 2_s + \sigma 2_s + \pi 2_s]$ process.

6-9.b.   [8 + 8] By using arrows, it can be seen that all 8 π electrons in each structure are utilized in the reaction which leads to the product indicated.

6-9.c.   [6 + 4] It is not necessary to use the π electrons of the C=O.

6-9.d.   [4 + 2]

6-10.a. [$_\pi 4_s + _\pi 2_s$] The phenyls of the diene and the carbonyls of the heterocycle are not involved in the ring-forming process, and thus the π electrons in these substituents are not counted. The two hydrogens, which are shown up (*cis*) in the product, are *cis* in the starting material. Also, the two phenyl groups of the diene end up on the same side of the 6-membered ring. Both of these stereochemical consequences are the result of suprafacial processes. This can be demonstrated readily by using models.

6-10.b. [$_\pi 2_s + _\pi 2_s$] The best way to understand the stereochemistry of the process is to look at models. Placing the appropriate termini together, in a model of the starting material, shows that the stereochemistry of the product is produced when both components react suprafacially. The π electrons of the C=O group and the dimethoxy-substituted π bond are not counted, because they are not involved in the reaction. This is a thermal reaction; thus, a nonconcerted pathway is expected.

6-11. [$_\pi 4_s + _\pi 4_s$] The nitrone component contains four electrons: the two π electrons of the double bond, and the two π electrons on the oxygen which overlap with them. The diene component also contains 4 electrons. The nitrone component contains three atoms, and the diene component contains four atoms; a [4 + 3] adduct is formed. The authors cited propose that this adduct is formed by a diradical mechanism. For more on this reaction, see Problem 7-2 b.

6-12.

6-13. The following shows the actual regiochemistry observed.

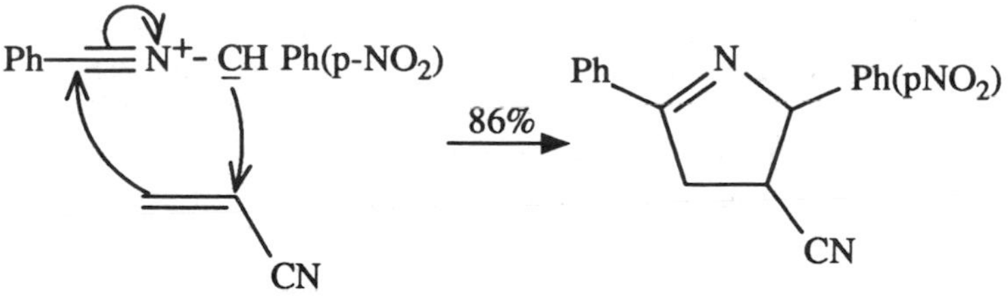

However, you should have also written the following reaction as a possibility:

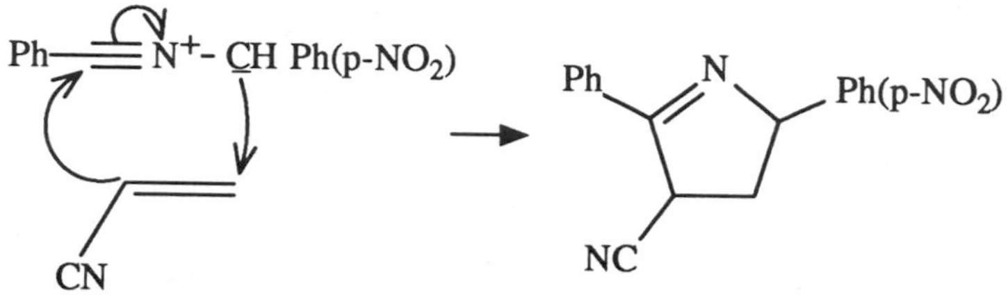

As indicated in the text, many of these reactions show high regioselectivity, which depends on electron distributions which can only be obtained through MO calculations.

6-14. A possible mechanism is intermolecular proton transfer from one molecule of starting ester-imine to another to give a 1,3-dipole, which then cycloadds with the acetylene.

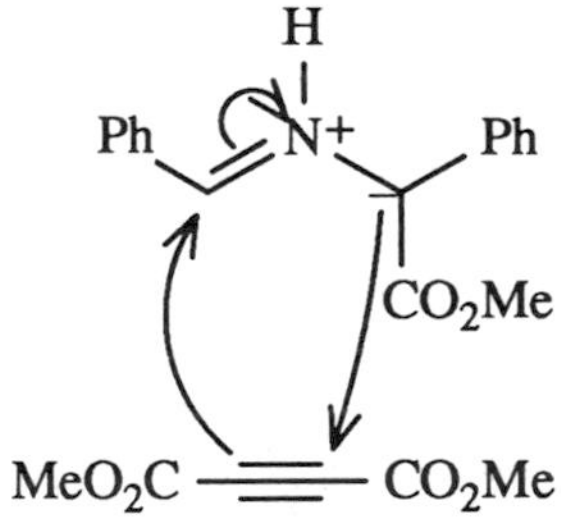

6-15.a. This is a [1,5] sigmatropic shift. The highlighted σ bond has been broken, the π system has shifted, and a new σ bond has been formed at the other end of the π system. Numbering atoms, from each end of the original σ bond, through the terminal atoms of the new σ bond gives:

Thus the hydrogen has moved from the 1-position to the 5-position on the chain. The "1" in the designation, [1,5] indicates that at one end of the new σ bond is an atom (the hydrogen) which was also at one end of the old σ bond. The "5" indicates that the other end of the new σ bond is formed at the 5 position along the carbon chain, atom number 5. To be thermally allowed, this would be a $[_\pi 4_s + _\sigma 2_s]$ reaction.

6-15.b. This is a [3,3] sigmatropic shift. It is an example of the Claisen rearrangement, a [3,3] sigmatropic rearrangement of a vinyl ether. It is also a $[_\pi 2_s + _\pi 2_s + _\sigma 2_s]$ reaction.

6-15.c. [2,3] or $[_\pi 2_s + _\pi 2_s + _\sigma 2_s]$

6-15.d. This is a [3,3] sigmatropic shift or $[_\pi 2_s + _\pi 2_s + _\sigma 2_s]$ reaction. It is an example of the Cope rearrangement, in which one of the carbons has been replaced by nitrogen. The σ bond broken and the new σ bond formed are highlighted in the structures.

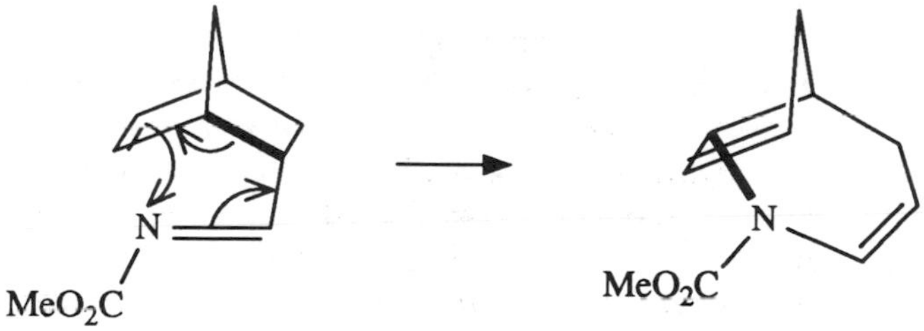

6-16.a. This rearrangement could take place by two different mechanisms. One is a [1,5] sigmatropic shift ($[_\pi 4_s + _\sigma 2_s]$) to give the product directly:

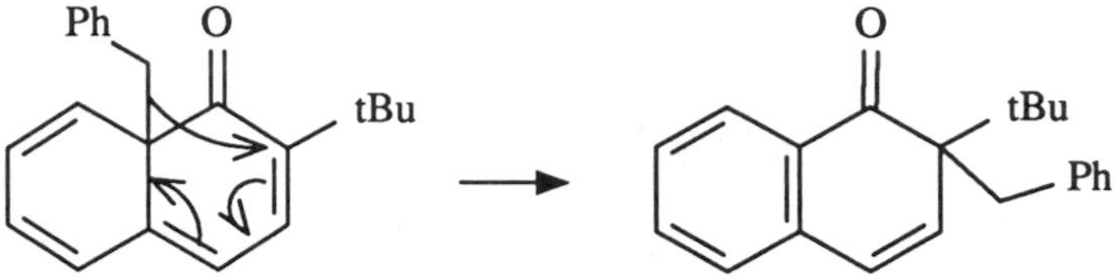

The other is two sequential [3,3] sigmatropic shifts:

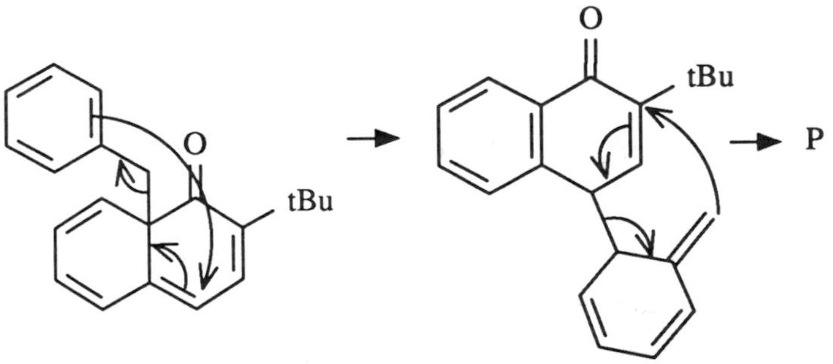

The authors of the cited paper favor the [1,5] rearrangement, because direct formation of the product is thermodynamically more favorable than proceeding through the intermediate from the first [3,3] sigmatropic shift.

6-16.b. This transformation was described by the cited authors as a [1,5] shift, followed by a "spontaneous" [1,7] shift.

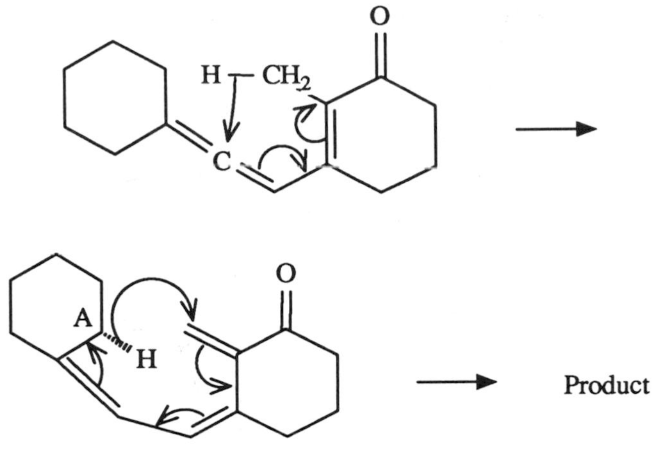

6-18

The [1,5] shift would be a $[_\pi 4_s + _\sigma 2_s]$ reaction. The [1,7] shift must be antarafacial in the π component to be thermally allowed, $[_\pi 6_a + _\sigma 2_s]$. The proton, located at carbon A in **6-18**, moves from the bottom side of the 6-membered ring to the top side of the π system at the other end. In general the antarafacial process, necessary for a thermal [1,7] sigmatropic shift, occurs with ease, when compared to the antarafacial process which would be required for a thermal [1,3] sigmatropic shift.

6-17.

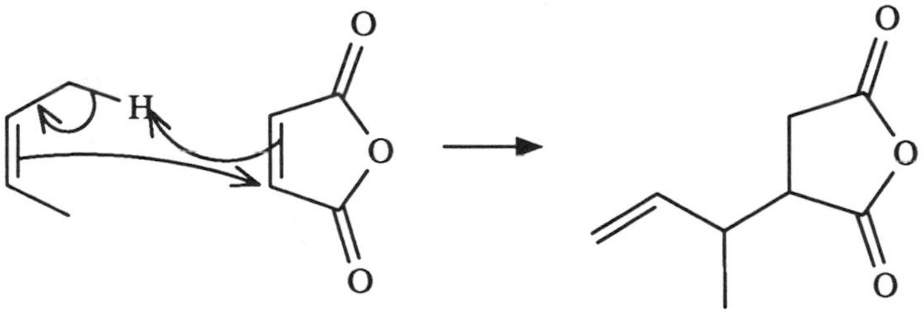

This is an example of a $[_\pi 2_s + _\pi 2_s + _\sigma 2_s]$ reaction.

6-18.a. The major pathway, a $[\pi 2_s + \pi 2_s + \sigma 2_s]$ reaction, is an ene reaction.

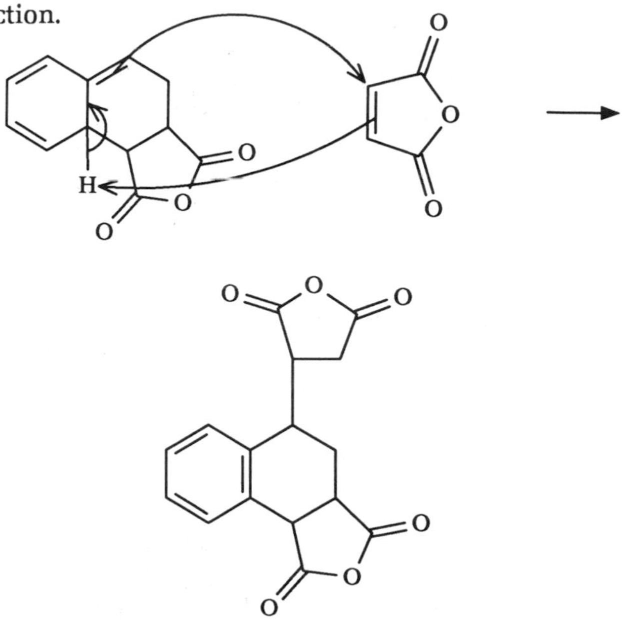

A stepwise mechanism could also be written for this reaction:

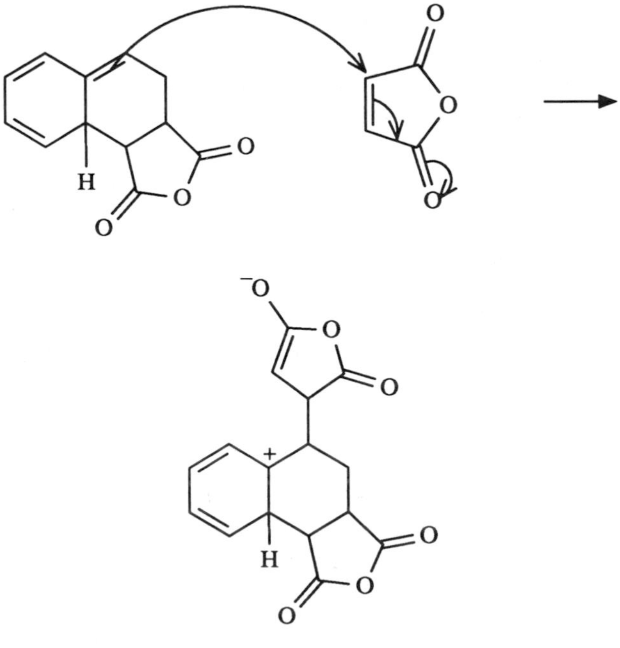

6-19

The initially-formed intermediate, **6-19**, is both an acid and a base. Appropriate intermolecular acid-base reactions, followed by tautomerization, can generate the product readily.

The product, formed in 11% yield, results from a Diels-Alder reaction, a $[_\pi 4_s + _\pi 2_s]$ cycloaddition.

6-18.b. The first step is a $[_\pi 4_s + _\pi 2_s]$ cycloaddition to give **6-20**.

6-20

The second step is a retrocycloaddition to give nitrogen and **6-21**, a $[_\pi 4_s + _\pi 2_s]$ process. Finally, **6-21** undergoes a [1,5] sigmatropic shift $[_\pi 4_s + _\sigma 2_s]$ to form the final product. In **6-21**, the bonds involved in the sigmatropic shift are highlighted.

6-20   **6-21**

Another possible decomposition of **6-20**, $[_\pi 4_s + _\pi 2_s]$, could give **6-22**, which then could open in a 6π-electron electrocyclic reaction $[_\pi 6_s]$ to give **6-21**.

6-20   **6-22**

6-21

The direct opening of **6-20** to **6-21** appears more favorable, on the basis of immediate relief of the strain of the 3-membered ring.

6-18.c. This reaction is a $[\pi 6_s + \pi 4_s]$ intramolecular cycloaddition. The paper cited noted that the reaction is "periselective"; that is, none of the [4 + 2] cycloadduct is produced.

The two alternative $[\pi 6_a + \pi 4_a]$ modes of reaction are extremely unlikely, because they give products with very high strain energy.

6-18.d. The product can be explained as the result of two consecutive [1,5] sigmatropic shifts, $[\pi 4_s + \sigma 2_s]$ reactions.

⟶   Product

6-18.e.   The product can be produced by two consecutive [3,3] sigmatropic shifts, $[_{\pi}2_s + _{\pi}2_s + _{\sigma}2_s]$ reactions.

The first reaction is a Cope rearrangement; the second reaction, rearrangement of a vinyl ether, is another example of a Claisen rearrangement.

6-18.f.   Step (a) looks like a [1,3] sigmatropic shift. However, because a concerted shift would have to be a $[_{\pi}2_a + _{\sigma}2_s]$ reaction, which is unfavorable sterically, the mechanism is probably not concerted. A radical chain reaction, initiated by a small amount of peroxide or oxygen, is possible.

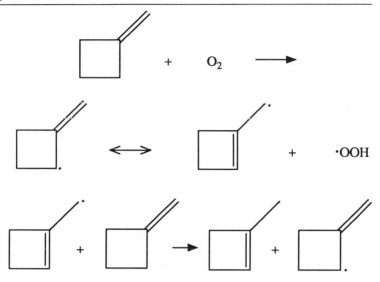

Step (b) is the ene reaction, $[_\pi 2_s + _\pi 2_s + _\sigma 2_s]$.

Steps (c) and (d) are electrocyclic ring openings, $_\pi 4_a$, processes. Because there are only hydrogens on the sp^3-hybridized carbons of the cyclobutene rings, there is no way to observe the stereochemical consequences of ring opening. Therefore, the arrows shown indicate only the rearrangement of electrons.

Step (c):

Step (d):

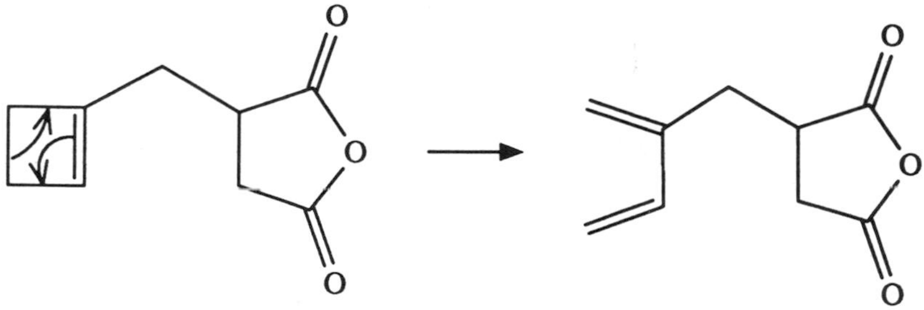

Steps (e) and (f) are $[_\pi 4_s + _\pi 2_s]$ cycloadditions.

6-19. In the equations below, R is the allyl group. The first step is very common for benzocyclobutenes: electrocyclic ring opening of the 4-membered ring. There are two possible products for this ring opening (each from a possible conrotatory mode), but only **6-23** has the correct orientation to continue the reaction. The last step is a [3,3] sigmatropic shift.

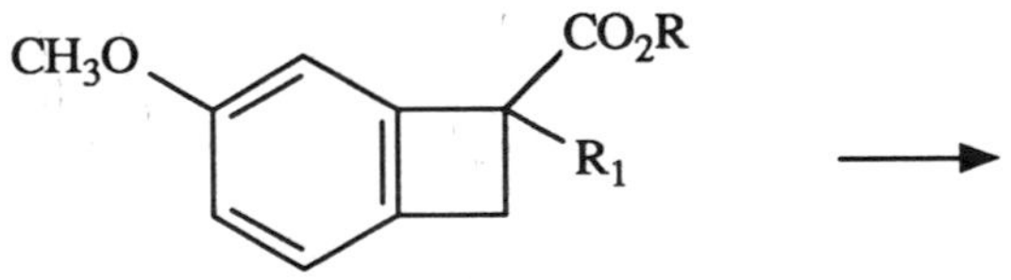

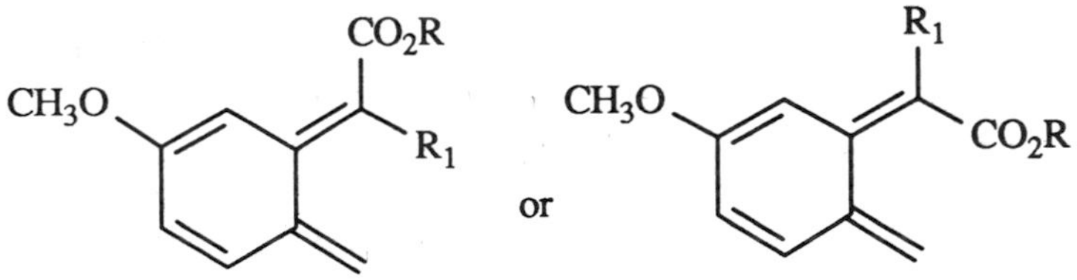

6-23

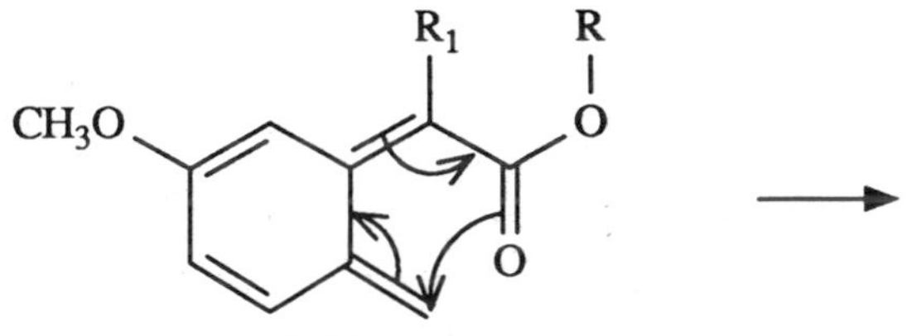

6-23

6-20. The fact that the reaction occurs readily suggests that it is not concerted, because a $[_\pi 4_s + _\pi 4_s]$ reaction is not a thermally allowed process. (See Helpful Hint 6-4.) While the stereochemistry of the product shows that anthracene has reacted suprafacially, the oxygen and nitrogen at the ends of the other 4π electron component preclude determining the stereochemistry of addition at this component. However, antarafacial reaction of this component is unlikely, because it would require a large amount of twist in the transition state. Thus, this transformation probably proceeds through an intermediate, which is likely to be charged. A positive charge could be stabilized by conjugation with two aromatic rings of the anthracene component. A negative charge could be stabilized by N and/or O, as well as the electron-withdrawing chlorines. Therefore, reaction of anthracene on either oxygen or nitrogen gives a stabilized intermediate:

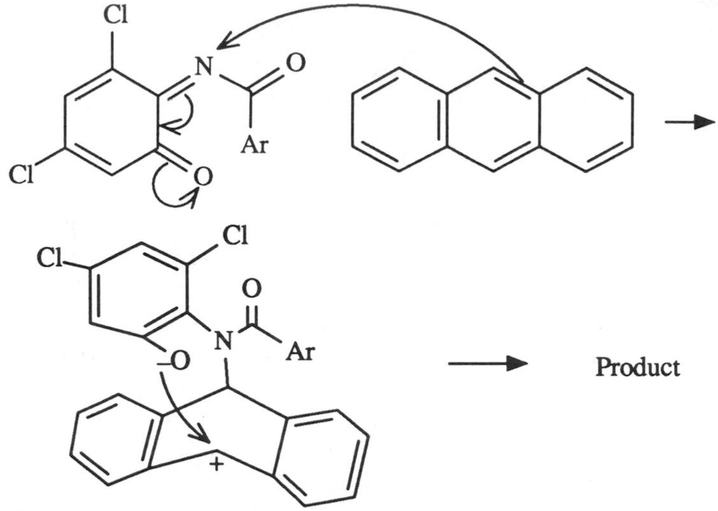

6-21. The first ring opening is a conrotatory process typical of a
4π electron thermal reaction. That is, in **6-24** there is an
anion and a π bond, a total of four electrons:

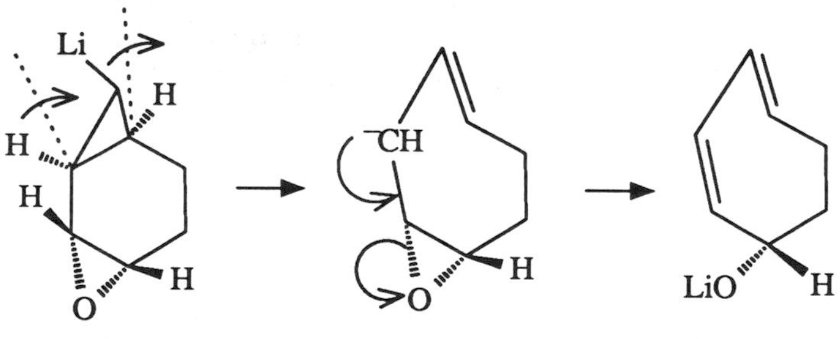

| | **6-24** | **6-15** |

Formation of **6-17** follows, from conrotatory ring closure of
the heptadiene:

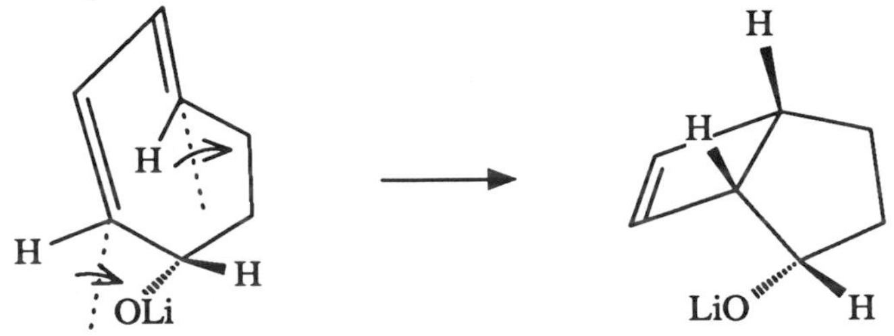

6-17

Notice that the conrotatory motion moves the hydrogens up and the carbons down. Ring closure is favored, because of the steric strain associated with a *trans* double bond in a 7-membered ring.

The isobenzofuran acts as a 4π electron component in a [4 + 2] cycloaddition to trap the intermediate. As usual, both components must react suprafacially. Thus, because the fusion between the 6- and 7-membered rings in **6-16** is *trans*, it must be the *trans* double bond in **6-15** which reacts in this trapping reaction.

6-16

Chapter 7

A Mixed Bag—Additional Problems

This chapter includes additional problems related to the material in Chapters 3 through 6. Some of the mechanisms will be mixed; for example, there might be a pericyclic reaction followed by a hydrolysis in either acid or base. If a reaction appears to be pericyclic, be sure to determine whether the reaction is symmetry allowed or symmetry forbidden under the reaction conditions. If it is symmetry forbidden, a nonconcerted reaction pathway through radical or charged intermediates will be the most likely mechanism.

Problem 7-1. Choose one principle or mechanism from each of Chapters 3 through 6, and find a reaction from the recent (past 5 years) literature which illustrates that principle or mechanism. Write a detailed step-by-step mechanism for each reaction. Journals to look at include: J. Org. Chem., Tetrahedron Lett., Synthesis, Synthetic Comm. Sometimes a synthetic sequence can be found which answers the entire question!

Problem 7-2. Show how the following transformations could occur.

441

a.

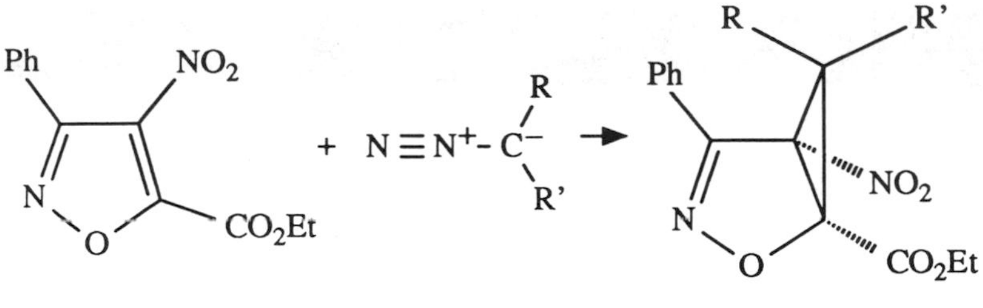

7-1

Nesi, R.; Giomi, D.; Papelao, S.; Bracci, S. *J. Org. Chem.* **1989**, *54*, 706–708.

b.

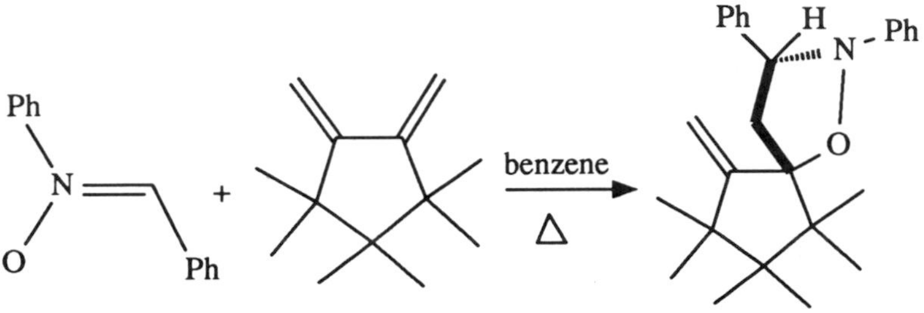

7-2

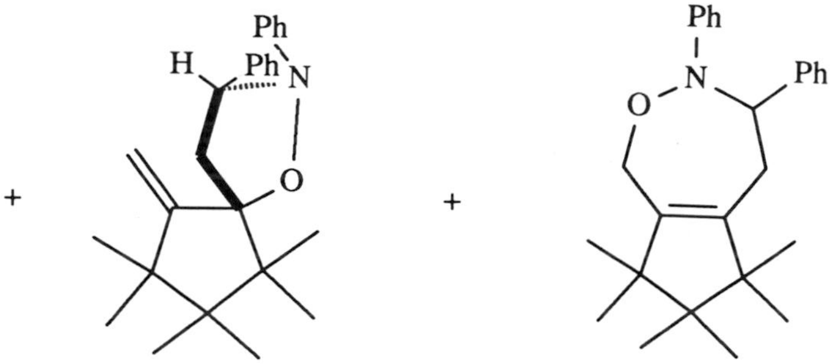

7-3   **7-4**

Baran, J.; Mayr, H. *J. Am. Chem. Soc.* **1987**, *109*, 6519–6521.

c.

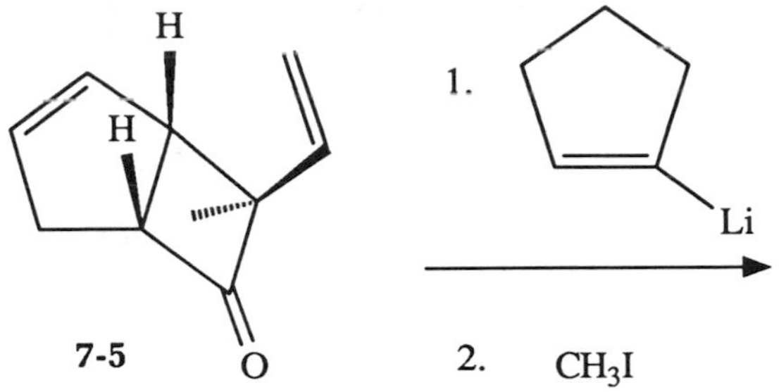

7-5

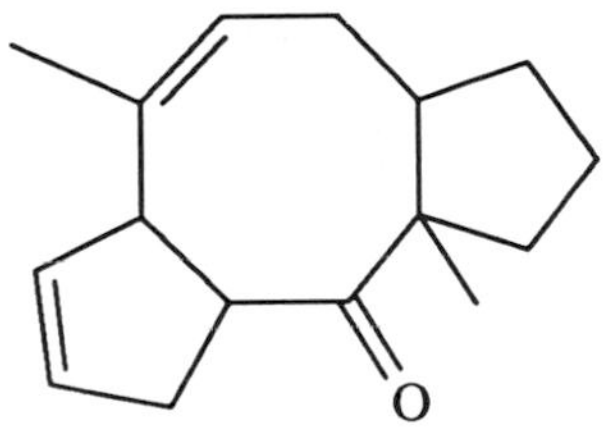

1.

2. CH₃I

7-6

Paquette, L. A.; Andrews, D. R.; Springer, J. P. *J. Org. Chem.* **1983**, *48*, 1147–1149.

Problem 7-3. What is the designation for the sigmatropic shift in the transformation of 7-7 to 7-8 in the following reaction sequence?

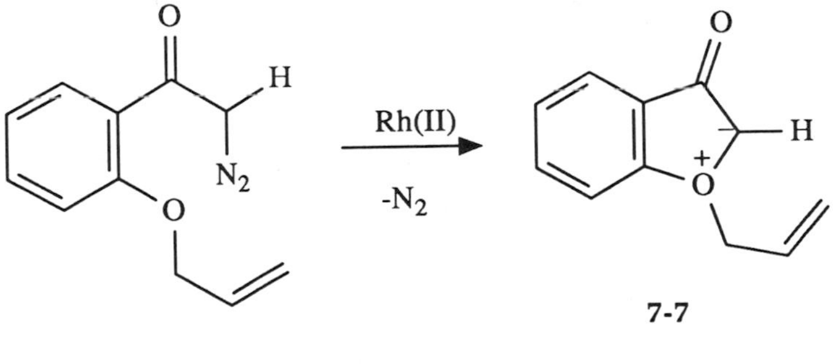

Rh(II)

-N₂

7-7

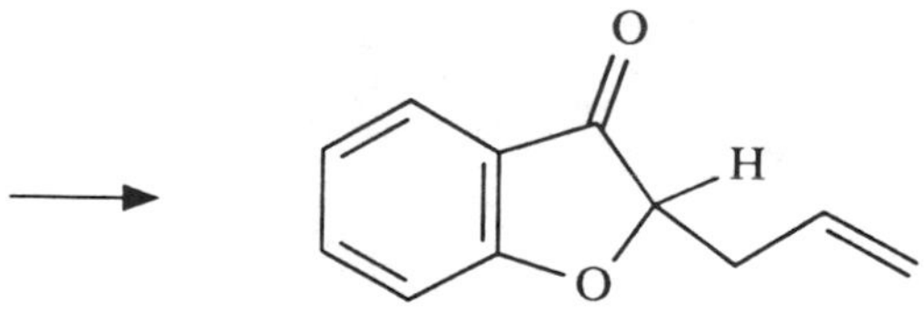

7-8

Pirrung, M. C.; Werner, J. A. *J. Am. Chem. Soc.* **1986,** *108,* 6060–6062.

Problem 7-4. Write a step-by-step mechanism for the following transformation. Other products for this reaction are covered by Problem 4-14 b.

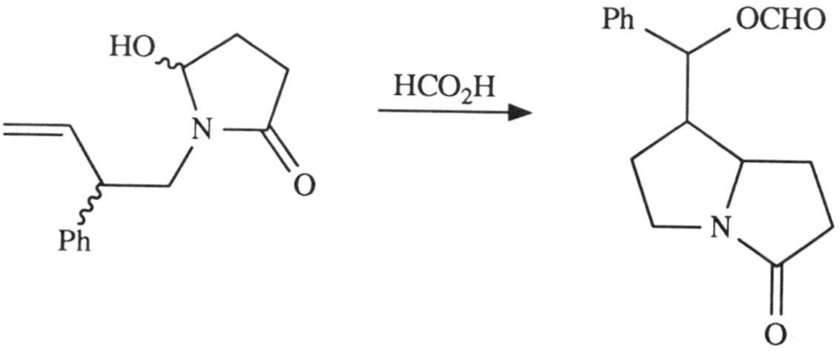

Ent, H.; de Koning, H.; Speckamp, W. N. *J. Org. Chem.* **1986,** *51,* 1687–1691.

Problem 7-5. Problem 5-11 showed the transformation of dibenzyl sulfide to a number of products, by reaction with n-butyllithium, followed by treatment with methyl iodide. Some tetramethylenediamine was included in the reaction mixture to coordinate the lithium. Another product, produced by this reaction, is shown below; write a reasonable mechanism for its formation.

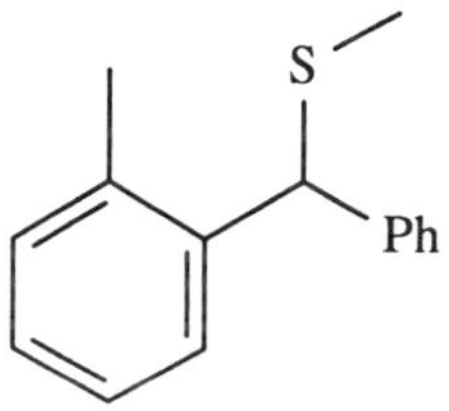

7-9

Problem 7-6. Using mechanistic principles, as well as the molecular formulas, determine structures for 7-10 and 7-11.

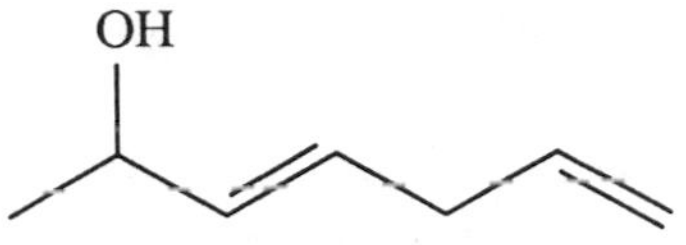

1. KH/THF

2. Cl$_3$CCN

3. trace MeOH

→ C$_9$H$_{12}$Cl$_3$NO

93%

7-10

xylene
Δ

→ C$_9$H$_{12}$Cl$_3$NO

72%

7-11

Roush, W. R.; Straub, J. A.; Brown, R. J. *J. Org. Chem.* **1987**, *52*, 5127–5136.

Problem 7-7. (a) Write a step-by-step mechanism for the formation of 7-14, and (b) propose a reasonable structure for the isomeric product, 7-15.

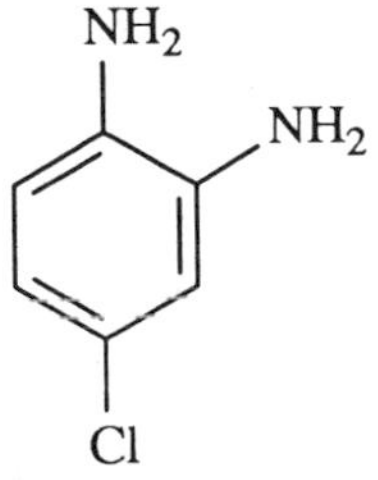

7-12 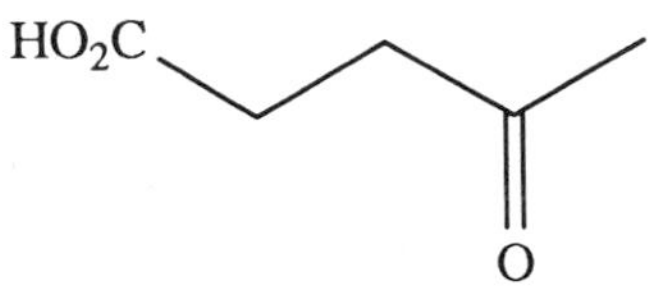   **7-13**

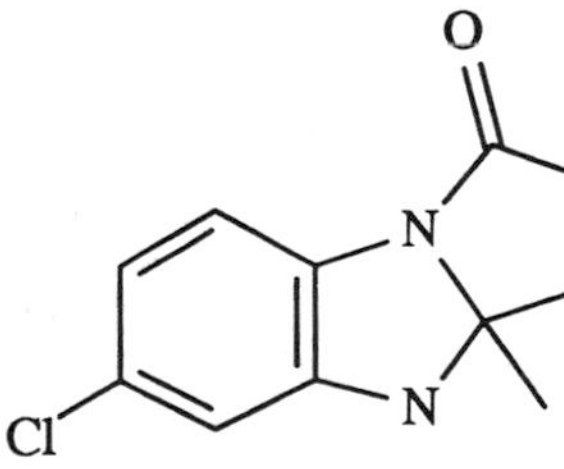

+   isomer

7-14   **7-15**

The reaction is run in refluxing toluene with azeotropic removal of water.

Chimirri, A.; Grasso, S.; Monforte, P.; Romeo, G.; Zappala, M. *Heterocycles* **1988**, *27*, 93–100.

Problem 7-8. Write a mechanism for the following reaction, a reduction-rearrangement in which zinc is a reducing agent. The first steps of a mechanism could involve protonation of the nitrogen, and reaction with zinc.

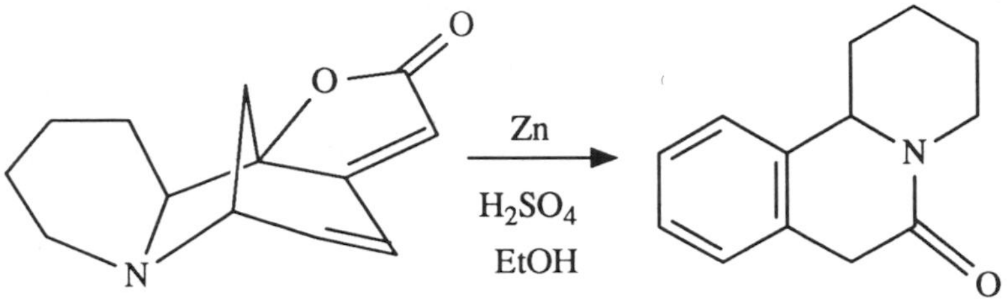

Problem 7-9. Give mechanisms for the formation of both products.

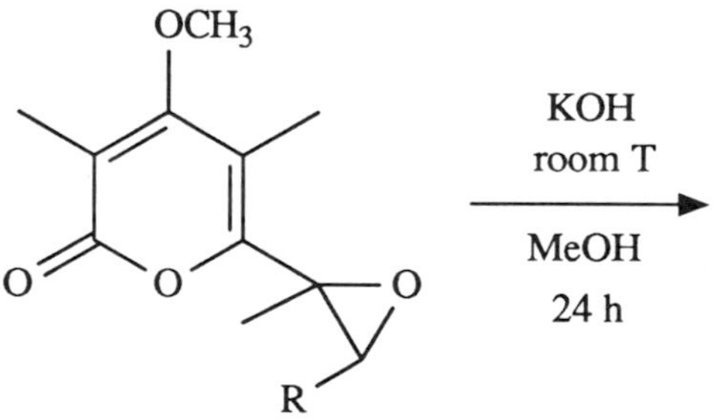

7-16

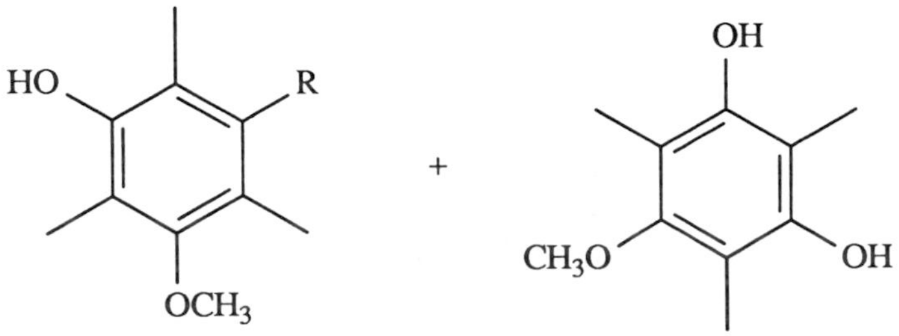

7-17   **7-18**

Ganguli, M.; Burka, L. T.; Harris, T. M. *J. Org. Chem.* **1984**, *49*, 3762–3766.

Answers to Problems

7-2.a.   There are several reasons why it is unlikely that loss of nitrogen from the diazo compound, followed by direct addition of the carbene to the double bond, is the mechanism for the reaction. (1) Even though the carbene has two alkyl substituents, it is still quite electrophilic, and would not be expected to react with one of the double bonds to the exclusion of the other. (2) There is no catalyst or uv light to aid in the generation of the carbene.

A more probable mechanism could be 1,3-dipolar addition to the electrophilic double bond to give 7-19, followed by loss of nitrogen from the intermediate to give the product.

7-19

The loss of nitrogen might occur with the intermediate formation of ions or radicals. In the heterolytic process a negative charge will be developed on carbon, because of resonance stabilization by the nitro group and the nitrogen in the oxazole ring.

7-2.b.   The formation of **7-2** and **7-3** can result from 1,3-dipolar cycloaddition of the nitrone to one double bond. This is a $[\pi 4_s + \pi 2_s]$, thermally-allowed reaction.

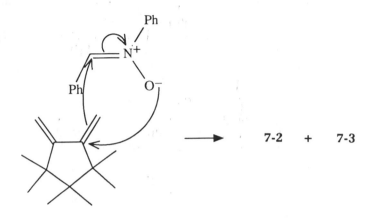

$$\longrightarrow \qquad \textbf{7-2} \quad + \quad \textbf{7-3}$$

Depicting the reaction as above, does not indicate clearly how the two isomers arise. This is best seen showing the MO's which are interacting. The HOMO of the 4 electron component, the nitrone, and the LUMO of the 2 electron component are used. The stereoisomeric compounds, **7-2** and **7-3**, arise from reaction at either the top face or the bottom face of the nitrone. Just that part of the diene which is reacting, one of the π bonds, is shown in the next two equations.

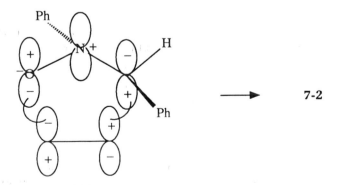

$$\longrightarrow \qquad \textbf{7-2}$$

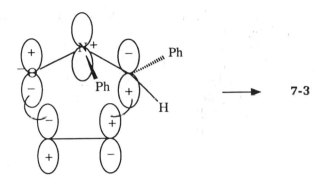

Concerted formation of **7-4** would be a $[\pi 4_s + \pi 4_s]$ cycloaddition which is not thermally allowed. Thus, a mechanism to this product probably involves a diradical intermediate:

Formation of an intermediate ion pair, **7-20**, is also possible. However, in a nonpolar solvent like benzene, radical intermediates, which require less solvation than ions, seem more likely.

7-20

Formation of **7-2** and **7-3** is also possible through these intermediates.

The following reaction mechanism was ruled out by the cited authors: a [4 + 2] cycloaddition of the diene with the nitrone to give 3° amine oxide, **7-21**, which then thermally rearranges to the product. (Thermal rearrangement of a 3° amine oxide to an alkylated hydroxylamine is called a Meisenheimer rearrangement.)

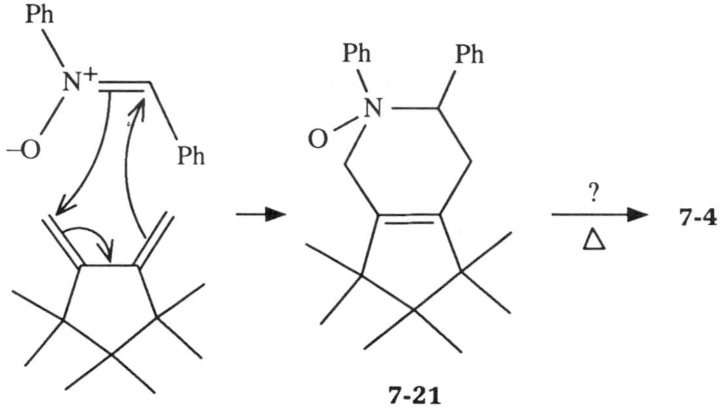

7-21

When **7-21**, synthesized by an alternate route, was subjected to the reaction conditions, it gave **7-22**, not **7-4**. Therefore, **7-21** cannot be an intermediate in the mechanism to **7-4**.

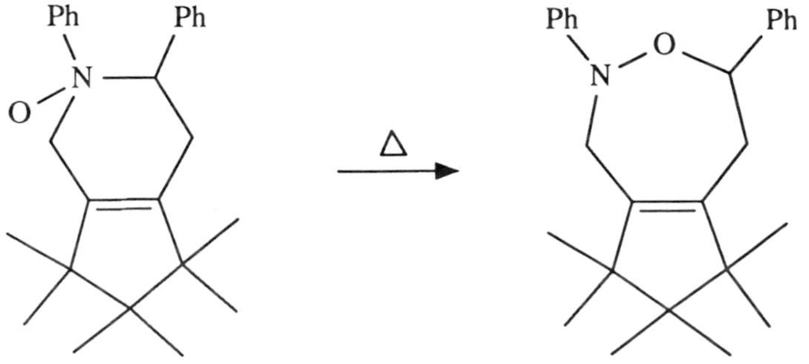

7-21   **7-22**

7-2.c.   The lithium reagent must react with the carbonyl carbon of
7-5 to give **7-23** as an intermediate. Since neither of the
double bonds in the starting material is conjugated with
the carbonyl, 1,4-addition of the lithium reagent is not possible. Comparison of the starting material, and **7-23**, with
the product simplifies the problem considerably.

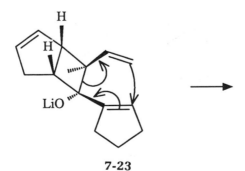

7-23

Note the following:

(1) The double bond in the 5-membered ring of the starting
ketone, **7-5**, is retained in the product.

(2) The other two double bonds in **7-23** have rearranged.
The allyl double bond and the double bond in the new
cyclopentenyl group are suitably situated for a [3,3]
sigmatropic shift.

(3) There is one methyl group in the starting material and
two methyl groups in the product. The second methyl must
be introduced by an S_N2 reaction of methyl iodide. One of
the methyls in the product is bonded to a carbon α to the
carbonyl group. Thus, the nucleophile in the S_N2 reaction
is most likely an enolate ion.

7-23

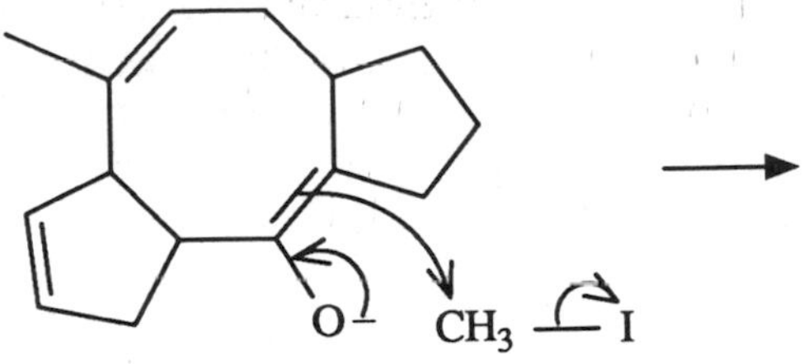

7-6

7-24

Ring opening to allyl anion **7-25** would be less likely, because **7-25** would be considerably less stable than enolate anion **7-24**.

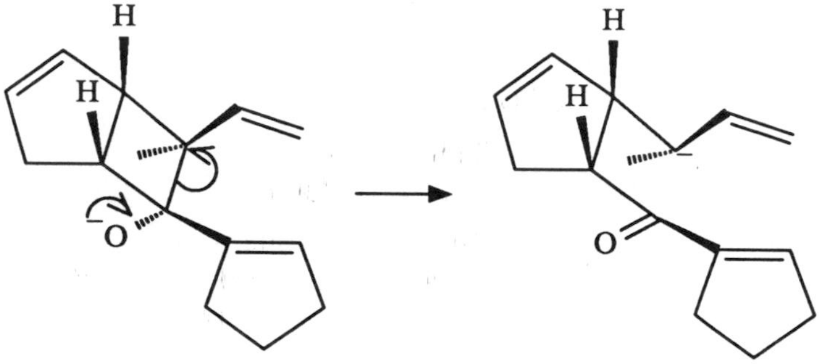

7-25

7-3. The usual numbering scheme shows that this is a [2,3] sigmatropic shift, involving a total of 6 electrons.

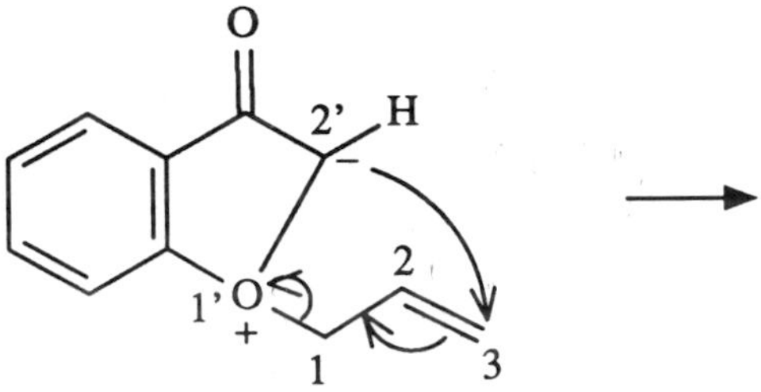

7-8

Another potential mechanism would be a [1,2] sigmatropic shift of the allyl group. However, if concerted, this would be a [$_\pi 2_s$ + $_\sigma 2_s$] reaction which is not thermally allowed. For purposes of illustration ONLY, suppose that a [1,2] sigmatropic shift takes place. There are two imaginable ways to write the flow of electrons for this reaction. The usual arrow pushing technique cannot be correct, because it leaves a pentavalent, 10 electron carbon, which is not possible.

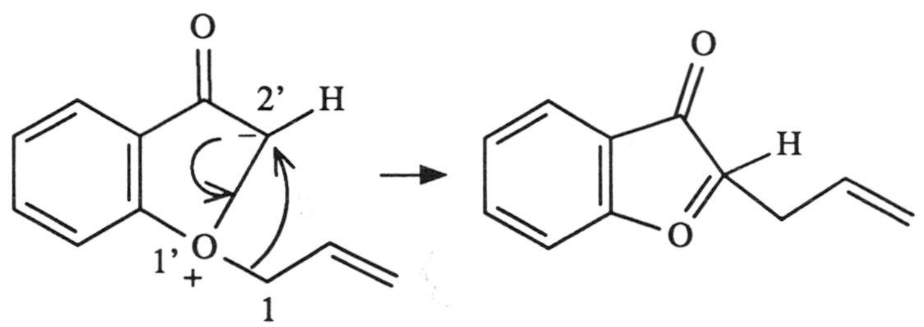

The method of writing electron flow, as we have been doing it in this book, ACTUALLY MEANS THE SHIFT OF A SINGLE ELECTRON FROM ONE ATOM TO ANOTHER. That is, in the above mechanism, the two electrons that were on C-2' are shared with oxygen in the product. C-2' is now perceived as having just one of those original two electrons; that is, a single electron has shifted to the oxygen.

The other way to write the flow of electrons for a hypothetical [1,2] shift would be transfer of two electrons from carbon 2' to the oxygen, as illustrated below. But, as just re-emphasized, reaction mechanisms are not written in this way.

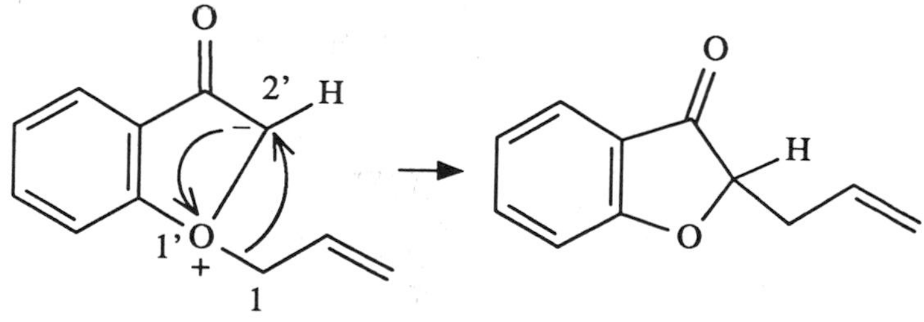

For further reading on this concept of the single electron shift, see "The Single Electron Shift as a Fundamental Process in Organic Chemistry: The Relationship between Polar and Electron-Transfer Pathways", Pross, A. *Acc. Chem. Res.* **1985**, *18*, 212–219.

7-4.   Protonation of the hydroxyl group followed by loss of water gives an intermediate, **7-26**, which can undergo a [3,3] sigmatropic shift to **7-27**.

As in the case of an analogous radical cyclization (see Section 5-6-2), the *exo* ionic cyclization of **7-27** is favored here. The reasons for this preference are subtle. With other substitution patterns, there might be steric and/or electronic factors which would make the cyclization to the 6-membered ring more favored.

7-27

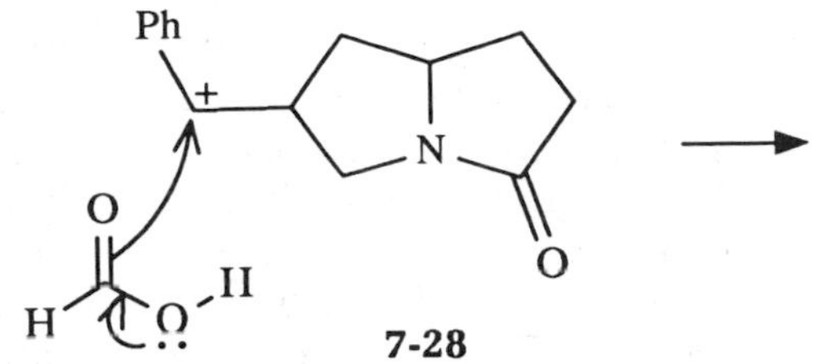

7-28

In the nucleophilic reaction of the carboxylic acid with the cation, **7-28**, the carbonyl oxygen, not the C-O oxygen, acts as the nucleophilic because only then is the resulting intermediate resonance stabilized.

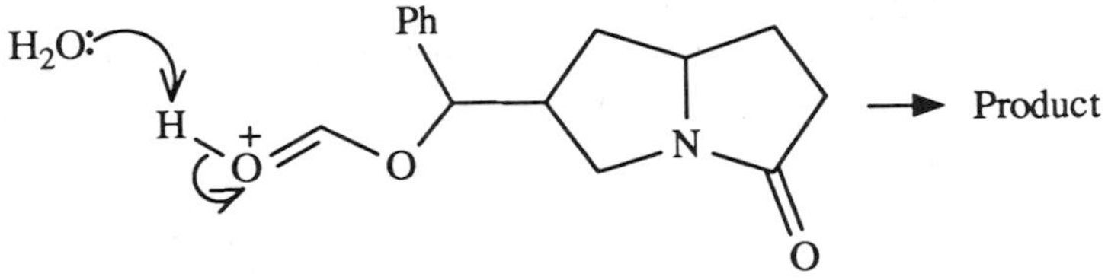

Product

7-5. The first step in the reaction would be removal of the most acidic proton in dibenzyl sulfide, **7-29**, by n-butyllithium to give anion **7-30**. This anion then undergoes a [2,3] sigmatropic rearrangement to **7-31**. This rearrangement utilizes a σ bond, the anionic electrons, and one π bond of one of the aromatic rings.

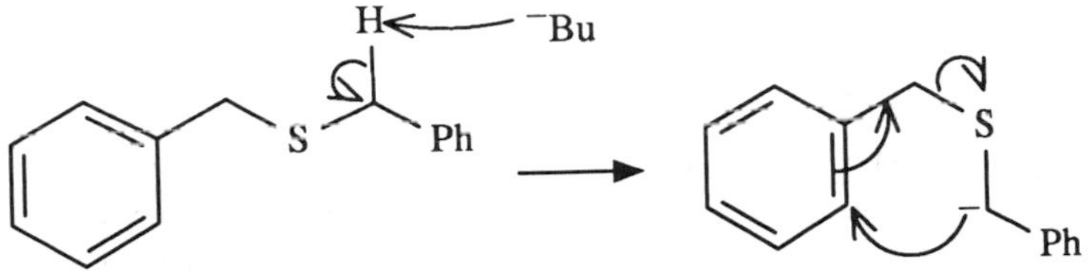

7-29   **7-30**

The intermediate, **7-31**, can be converted to the more stable benzylic anion, **7-32**, by a tautomerization.

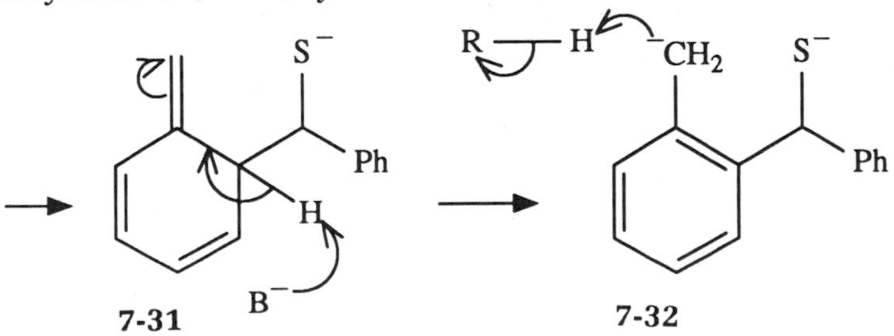

7-31   **7-32**

The base, which removes the proton from **7-31**, could be butyllithium, a molecule of anion **7-32**, or the anion from tetramethylenediamine (TMEDA). Anion **7-32** can then remove a proton from a molecule of **7-31**, or from a molecule of TMEDA, to produce **7-33**. It is unlikely that **7-32** picks up a proton from butane because of the high pK_a value of butane, relative to the amine or **7-31**. (Compare the following pK_a's from Table 1-2: diisopropylamine, 36 or 39; ethane, 50; toluene, 43; and ammonia, 41.)

The paper, from which this problem comes, claims that **7-31** is "rapidly converted by proton shift" to **7-32**, but as we saw in Chapter 6, [1,3] concerted thermal shifts of hydrogen are unlikely.

The last step is an S_N2 reaction of the nucleophilic sulfur anion with methyl iodide to give the methylated sulfur.

7-33

7-6. The molecular formula of **7-10** indicates that trichloro-acetonitrile has reacted with the starting material. Using Helpful Hint 1-2, 3 rings and/or π bonds are calculated. Since the starting alcohol has 2 π bonds and the nitrile has 2 π bonds, a nucleophilic reaction with the nitrile is likely. Potassium hydride should remove the most acidic proton from the starting alcohol, the proton on the oxygen. There are two possible condensation reactions for the nitrile. One is nucleophilic reaction of the oxyanion with the carbon of the nitrile functional group. This carbon is activated by nitrogen, and also by the strongly electron-withdrawing trichloromethyl group. The other possibility, S_N2 displacement of chloride, is ruled out by the fact that there are three chlorines and 4 π bonds in the product. After the nucleophilic reaction, a trace of methanol is needed to form a neutral product by protonation of the anion.

7-10

The second reaction is an isomerization. That is, the molecular formula of **7-11** is the same as that of **7-10**. Since xylene is an unreactive aromatic hydrocarbon, a reasonable assumption is that it functions only as the solvent. Therefore, the first thing to look for is a concerted reaction, and indeed a [3,3] sigmatropic reaction is possible.

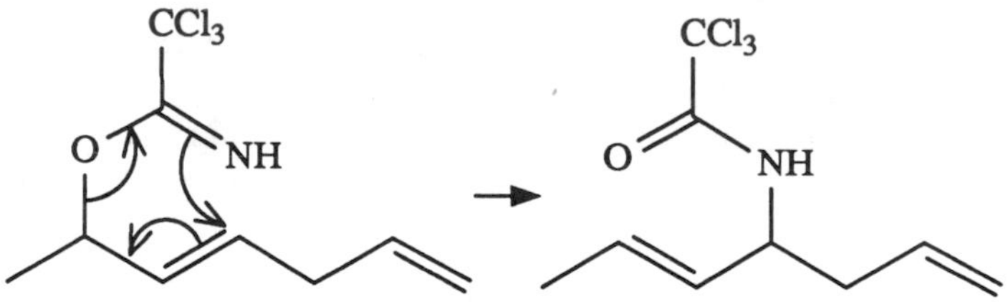

7-10   7-11

7-7.a.   Two types of condensations occur: an amine with a car-
         boxylic acid, and an amine with a ketone. The latter should
         occur more readily (and is written first in the mechanism
         below), because the hydroxyl of the carboxylic acid group
         reduces the electrophilicity of its carbonyl group to below
         that of a ketone, by a resonance effect. Because two rings
         are formed in this reaction, either the smaller 5-membered
         one or the larger 8-membered one could be formed first. Be-
         cause of entropy, the smaller ring should be favored.
             The first step is protonation of the ketone carbonyl
         group by the carboxyl group. This might be either an inter-
         molecular or intramolecular reaction.

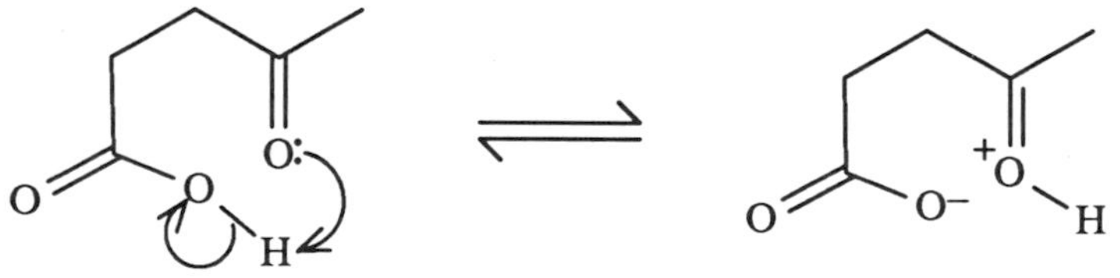

Then the nitrogen of the aniline can react, as a nucleophile, at
the electrophilic carbon of the protonated carbonyl group:

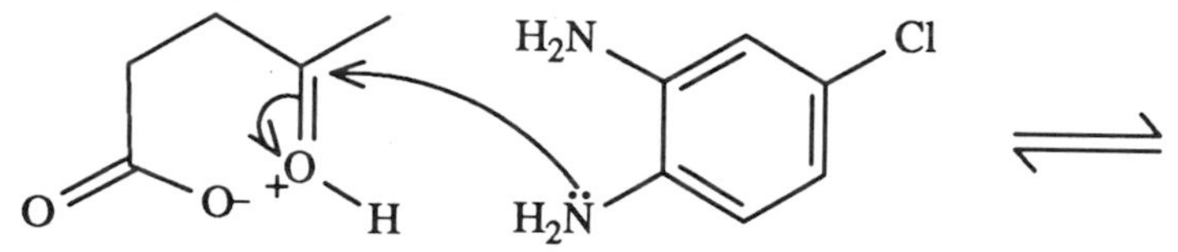

Subsequently, the positively charged nitrogen undergoes
deprotonation, and then the hydroxyl group undergoes
protonation.

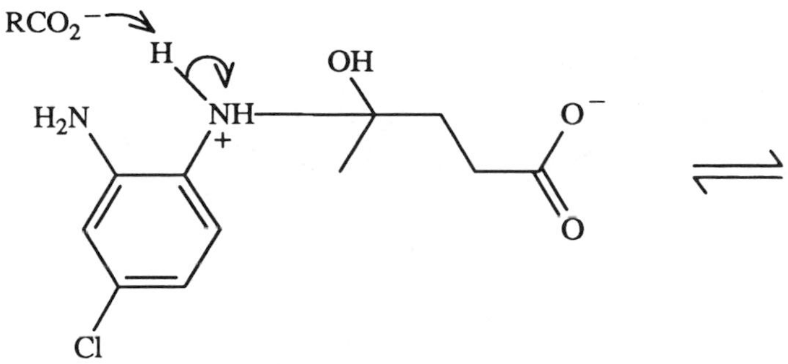

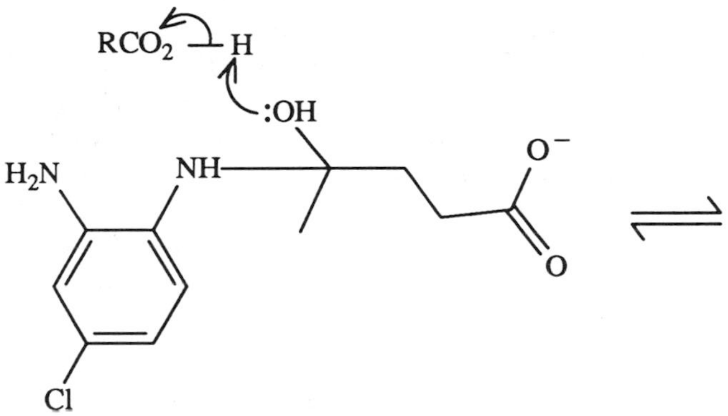

Loss of a molecule of water leads to an iminium ion, **7-35**, which is susceptible to nucleophilic reaction at carbon:

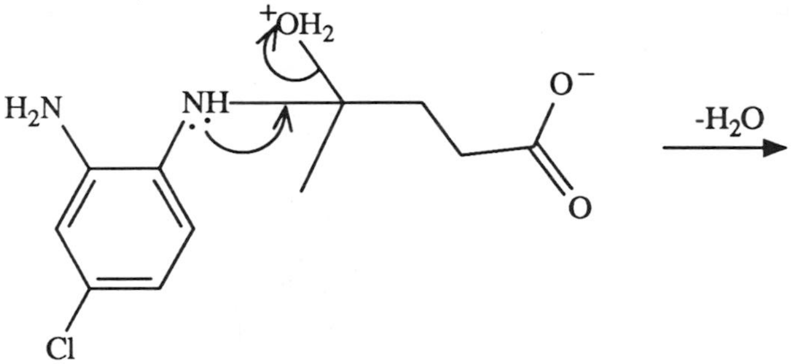

7-34

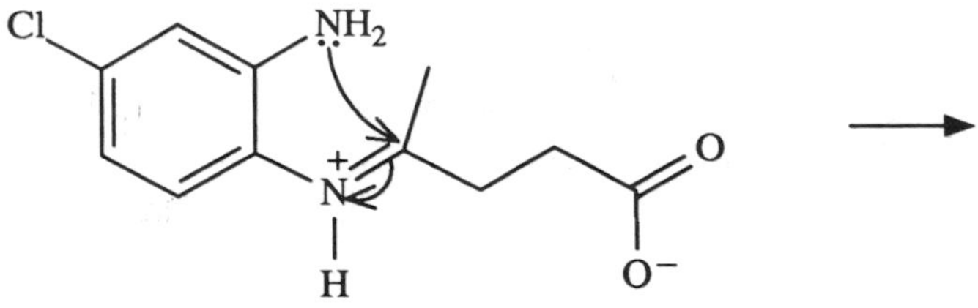

7-35

After deprotonation of the positively charged nitrogen, one of the nitrogens acts as a nucleophile with the electrophilic carbon of the neutral carboxy group. In the structures that have been written up to this point, the carboxylate anion has been indicated. However, this anion and the neutral carboxylic acid can be written interchangeably, because they would both be present under the reaction conditions. (As stated above, the condensation of the free NH$_2$ group with the carboxylic acid to give an 8-membered ring, followed by formation of the 5-membered ring, is less favored because of entropic considerations; that is, because of the low probability for productive collision of the ends of an 8-atom chain.)

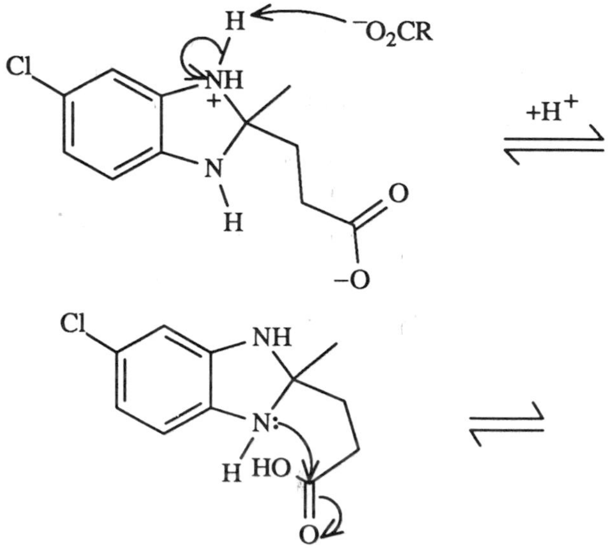

7-36

Protonation of oxygen and deprotonation of nitrogen occurs.

Then protonation of one of the hydroxyl groups, elimination of water, and deprotonation gives the product.

Because this water, as well as the water formed earlier in the mechanism, is removed as it is produced in the course of the reaction, the reaction is driven to completion.

Additional Comments

(1) The following equilibrium, which undoubtedly occurs under the reaction conditions, is unproductive; that is, it is not a reaction of the amine that is on a pathway to the product.

(2) There are several possible mechanisms for the loss of water from the protonated aminol, **7-34**. A partial structure representing the possibility depicted earlier is:

But another possibility is the simultaneous loss of a proton from the nitrogen:

Finally, the proton might be removed from the nitrogen before elimination of water takes place:

This last mechanism is an example of an E_{1CB} elimination, which is not common unless the intermediate anion is especially stabilized by strongly electron withdrawing groups. Although the anion, produced in this reaction, is resonance stabilized by the ring, it is probably not stable enough to be produced without the driving force of simultaneous formation of the double bond. Also, bases used in E_{1CB} reactions are usually much stronger than the carboxylate anion.

(3) Estimations of the basicity of carboxylate ion **7-37**, formed from **7-13**, and of the starting aniline, **7-12**, suggest that either one may be used as the base in the mechanisms written.

7-37

From Table 1-2, the pK_a of p-chloroanilinium ion is 4.0, and that of acetic acid is 4.76. This means that the acetate ion is more basic than p-chloroaniline, but by less than a factor of 10. However, the basicity of **7-12** would be enhanced by the second amino group, and the basicity of **7-37** would be decreased by the keto group. Thus, the basicities of these two bases may be approximately the same.

(4) The base in the mechanism would not be **7-38**, formed by removal of a proton from one of the amino groups. The equilibrium to form **7-38** is completely on the side of starting material. This can be calculated from values given in Table 1-2. The hydronium ion, H_3O^+, ($pK_a = -1.7$) is at least 10^{20} times more acidic than the substituted aniline (pK_a of m-chloroaniline is 26.7).

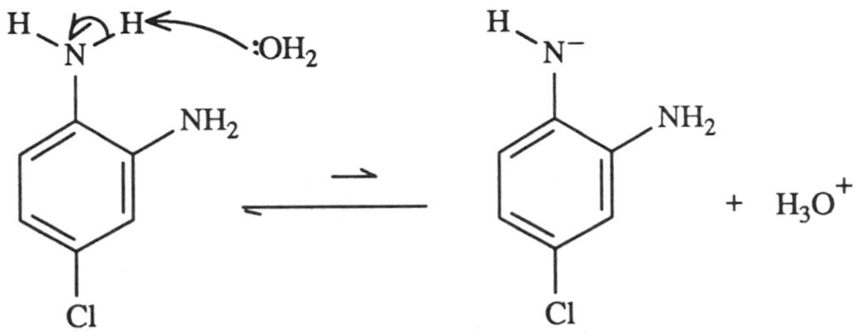

7-38

7-7.b.   Formation of the isomeric product, **7-15**.

If the other nitrogen in **7-36** attacks the carboxyl carbon, the following isomeric product would be formed:

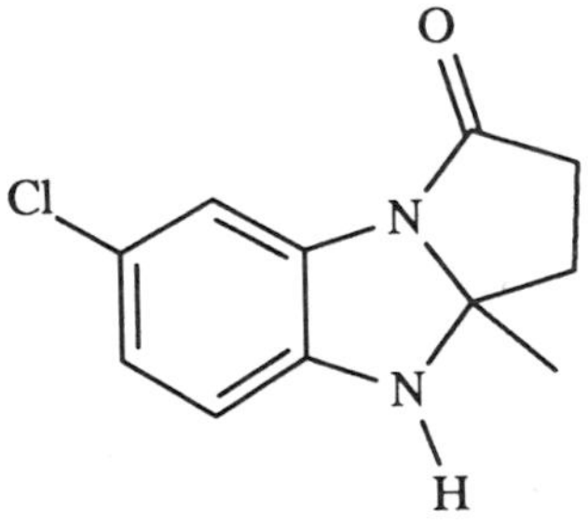

7-15

The nitrogens in the starting amine are not identical, so this isomer also could be produced by the nitrogen, *meta* to chlorine, acting as the original nucleophile. The mechanism then would be written just as it was in (a). See the cited paper for experimental evidence which supports just one of these two possibilities.

7-8. Protonation of the nitrogen is followed by reduction by zinc. This is a two-electron reduction, in which zinc is oxidized to Zn^{2+}. The remainder of the mechanism involves acid-catalyzed reactions.

7-39

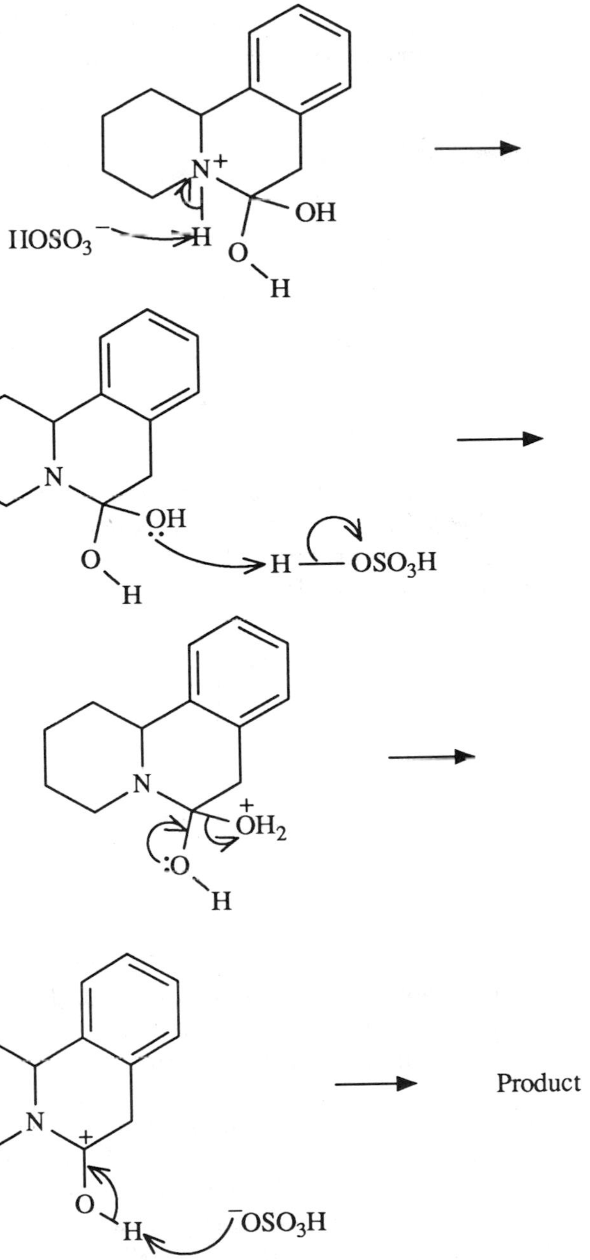

HOSO$_3^-$ ⋯→ H O OH

H—OSO$_3$H

$\overset{+}{O}H_2$:O H

$^-$OSO$_3$H

Product

An alternative route from **7-39** involves forming the remaining six-membered ring before the five-membered ring opens:

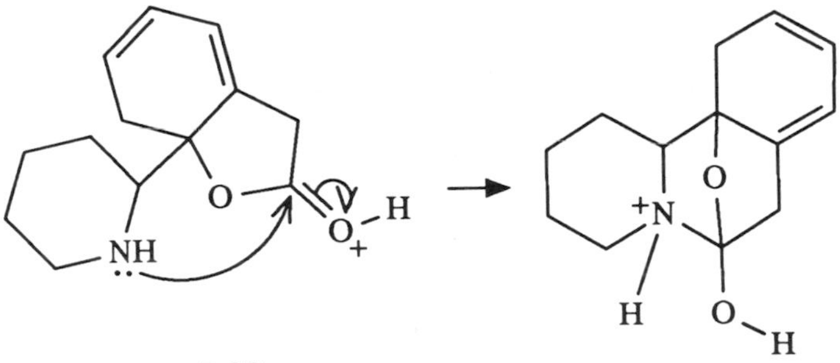

7-39

Subsequent acid-catalyzed steps will also give the product.

If you failed to use zinc, a reduction would not be effected, and the product of acid-catalyzed reactions of the starting material would contain two less hydrogens, as in **7-40**.

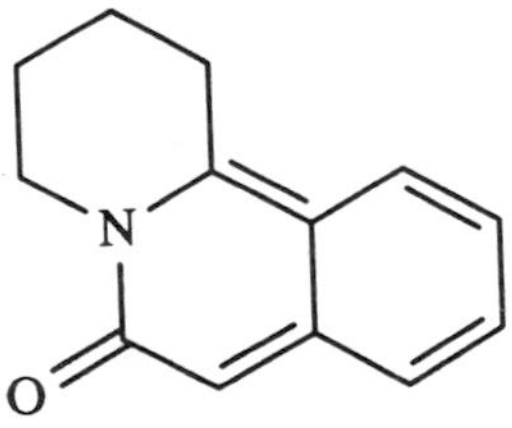

7-40

7-9.   Application of Helpful Hint 2-13 can aid in organizing thoughts about how formation of the products might occur. Formation of the first product involves a loss of one carbon. It appears most likely that this would be CO_2 or CO. It is probably CO_2, because CO is usually lost only in strong acids or, under thermal conditions, in reverse Diels-Alder reactions or free radical decarbonylation of aldehydes.

Once the atoms in the starting material, **7-16**, have been numbered, the carbons attached to the R group and the methoxy group in product, **7-17**, can be numbered.

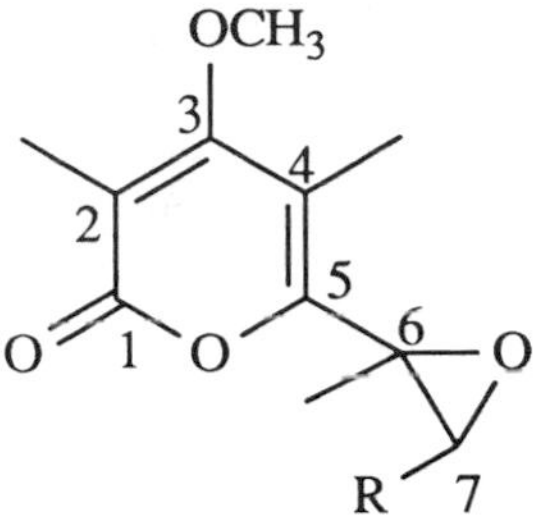

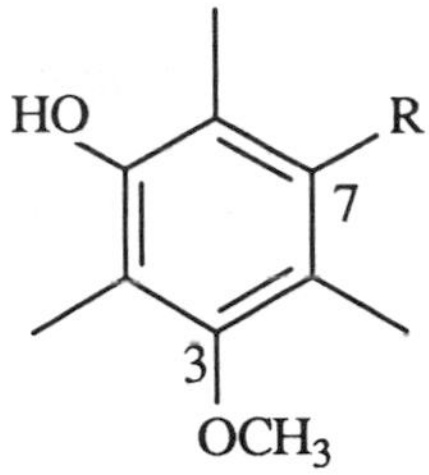

7-17

There are two possible numbering schemes for the remaining atoms in **7-17**:

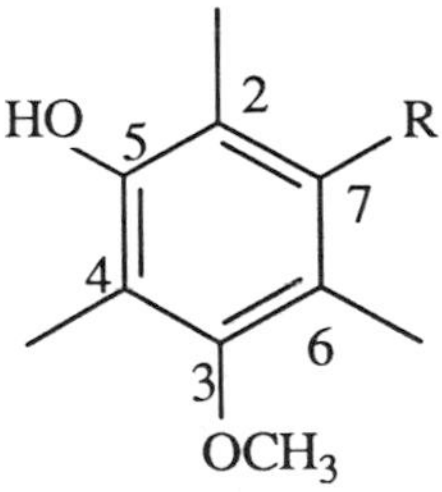

7-41 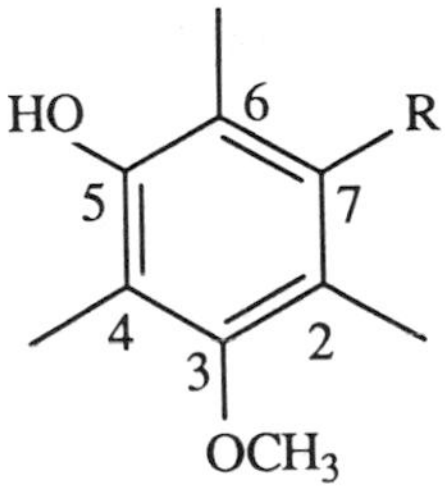   **7-42**

Getting to **7-42** would involve much less bond making and breaking than getting to **7-41** and thus, applying Helpful Hint 2-14, can serve as a roadmap for what is occurring: C_2 becomes bonded to C_7, and C_1 is lost. All other carbons remain attached to the same atoms.

Because the reaction conditions are basic, the next thing to do is search for protons in the molecule which are acidic and carbons susceptible to reaction with a nucleophile. At this point we will look at possible nucleophilic reactions and will return to acidic protons later in the discussion. Two obvious electrophilic carbons are those of the epoxide. Moreover, if the C_2-C_7 bond is formed by nucleophilic reaction of an anion, at C_2, on carbon C_7, ring opening of the epoxide would also occur in this step. This ring opening would be a driving force for this reaction. Thus, we should search for steps which would lead to an anion at C_2. This is also probably not obvious. Electrophilic positions in **7-16** other than the epoxide carbons, include the carbonyl carbon, C_1, and the positions β and δ to it, C_3 and C_5.

One other comment should be made before we start writing a mechanism. The acidities of methanol (pK_a = 15.5) and water (pK_a = 15.7) are practically identical; so methoxide or hydroxide ion as base, and methanol or water as acid, can be used interchangeably in the mechanism.

$$^-OH \quad + \quad HOCH_3 \quad \rightleftharpoons \quad HOH \quad + \quad ^-OCH_3$$

We need to start playing around with the nucleophilic reactions suggested above. Let's start by adding hydroxide ion to the carbonyl group. Then a ring-opening reaction can occur.

This ring opening reaction leads to **7-43**, a resonance-stabilized anion:

7-43-1   **7-43-2**

7-43-3   **7-44**

In the basic medium the carboxylic acid will be rapidly converted to the carboxylate anion, **7-44**. Rotation in this intermediate brings carbons 2 and 7 close enough to react with each other. After ring-opening of the epoxide, the resulting oxyanion can remove a proton from solvent to give the corresponding alcohol.

7-44

The carboxylate ion is vinylogous to the ketone carbonyl and thus, by analogy to the decarboxylation of the carboxyanion of acetoacetic acid, should decarboxylate readily.

The resulting anion then undergoes base-promoted elimination of water to give the anion of a phenol, **7-45**. This
adds a proton, on workup, to give the neutral product, **7-
17**. The driving force for this elimination is the formation
of an aromatic system. Hence, this reaction occurs with facility, even though the leaving group is a poor one.

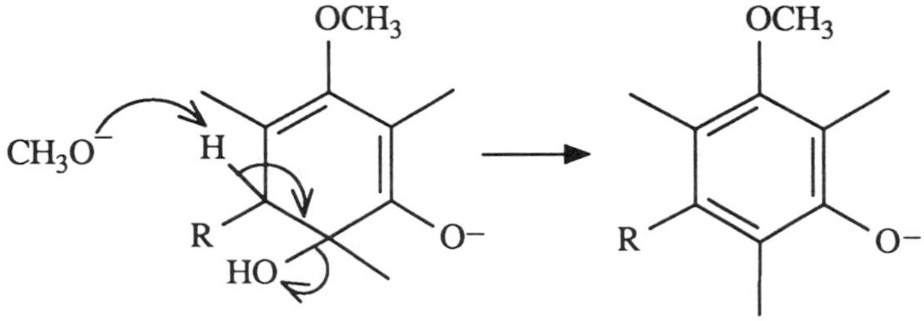

7-45

Another possible mechanism, from **7-44** to product, involves a different sequence of reactions. Thus, **7-44** might
pick up a proton to give **7-46**.

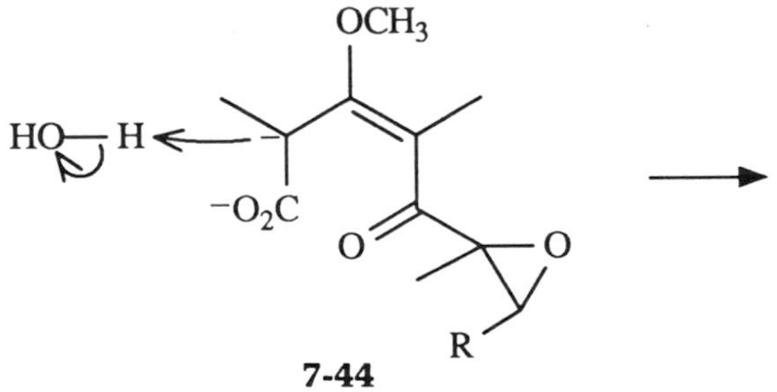

7-44

This intermediate can then decarboxylate to give an anion,
which can then ring close with the epoxide.

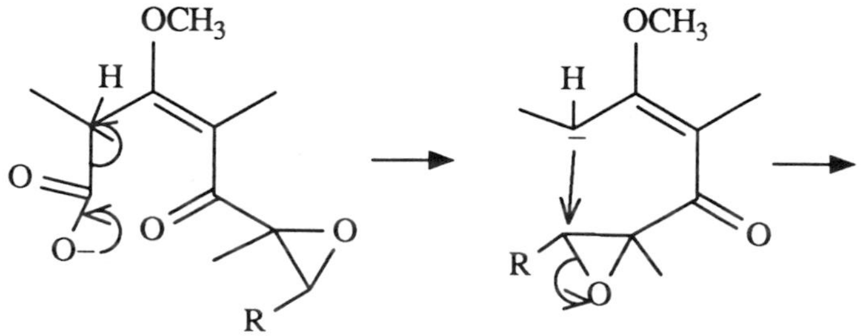

7-46

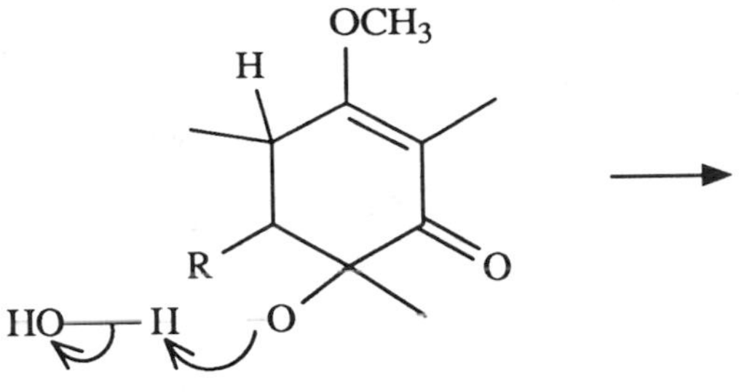

7-47

Protonation of **7-47** by water or methanol gives an intermediate, which can undergo base promoted elimination of
water and tautomerization to give product **7-17**.

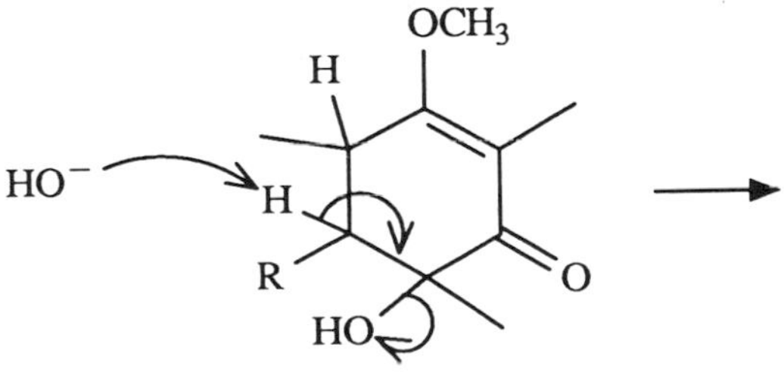

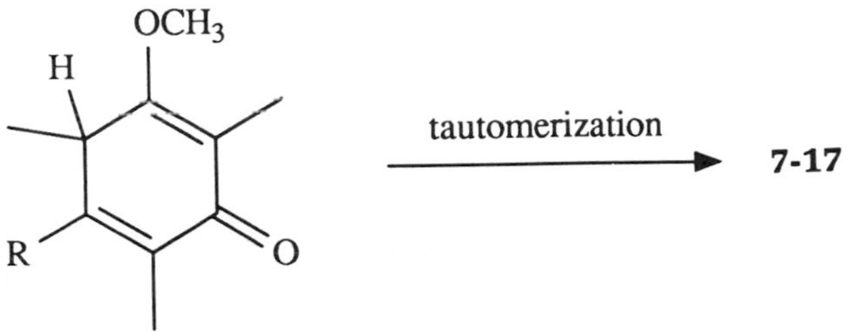

tautomerization
7-17

Another possible mechanism for the formation of **7-17** begins
with nucleophilic addition of hydroxide to C$_5$ of **7-16**. The
resulting anion reacts with the epoxide ring, which opens
giving **7-48**.

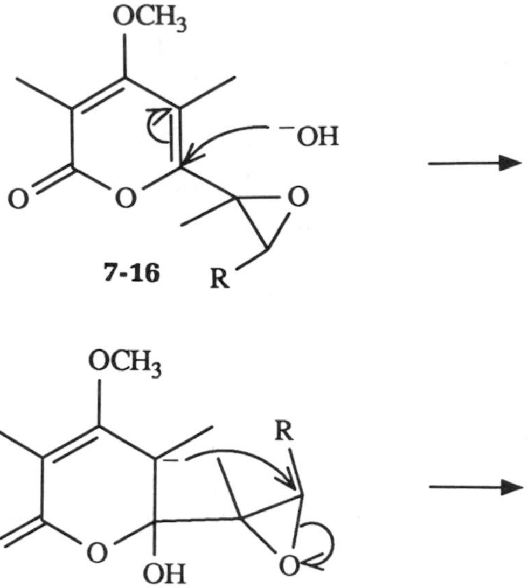

7-16

Intermediate **7-48** picks up a proton on one oxygen and loses a proton from another oxygen.

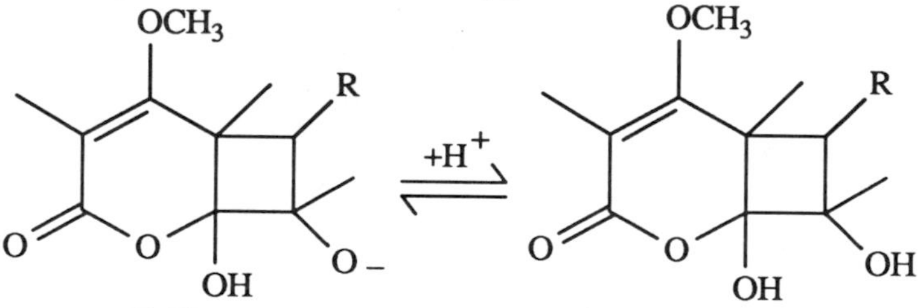

7-48

The 6-membered ring of the new alkoxide ion, **7-49**, then opens, with decarboxylation, to give an anion, **7-50**, which recyclizes on the carbonyl group, giving **7-51**.

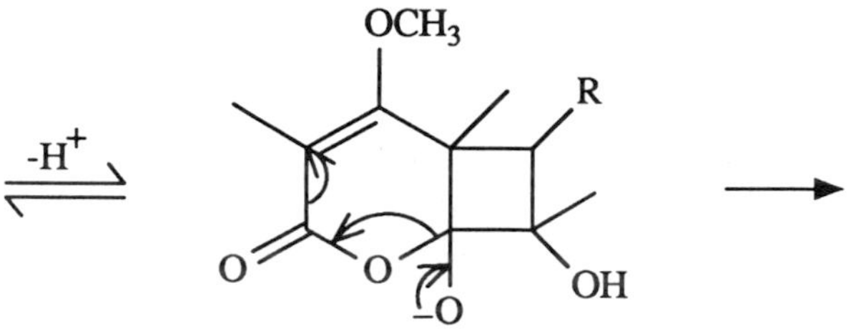

7-49

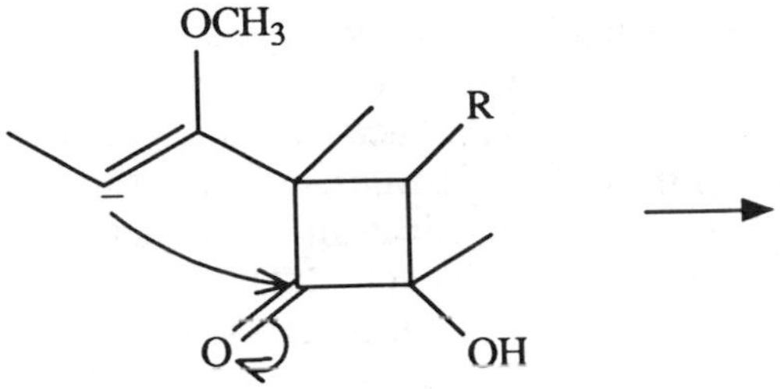

7-50

The remaining steps from **7-51** to product are: protonation, electrocyclic ring opening, and elimination of water.

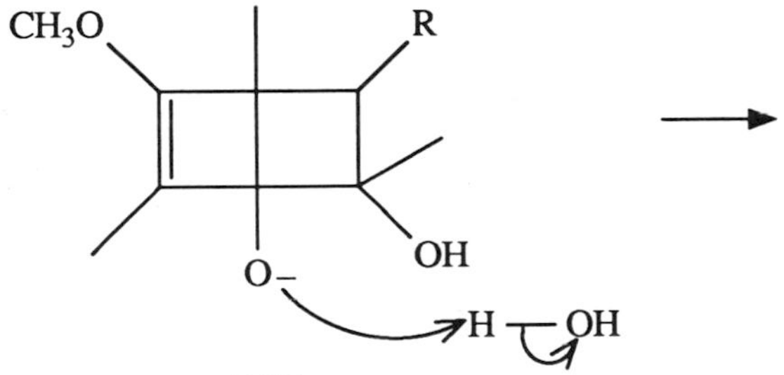

7-51

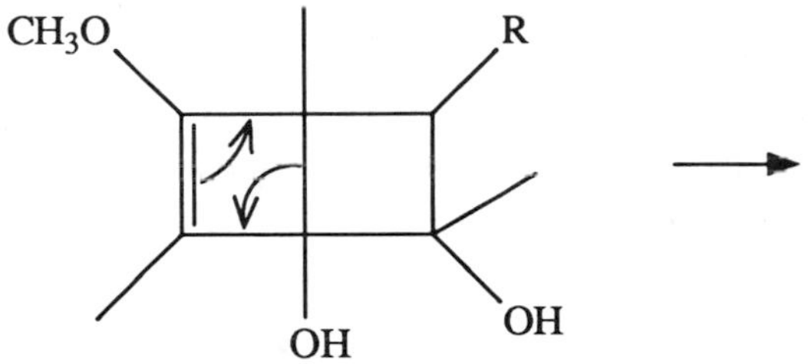

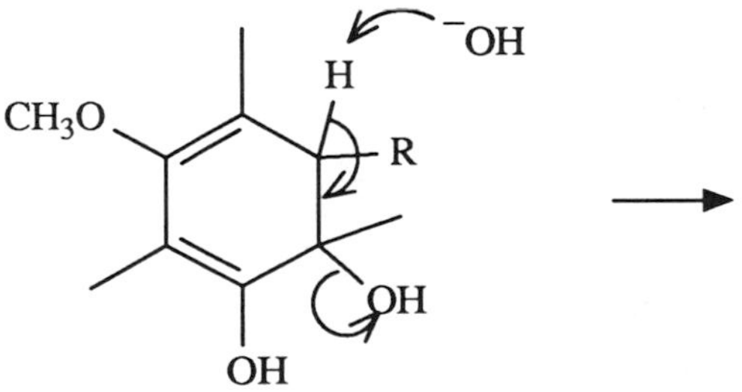

7-17

This mechanism has several disadvantages, when compared to those previously given. Intermediate carbanion **7-50** is relatively unstable, because its negative charge is localized. In contrast, the carbanions of the other mechanisms are stabilized by delocalization. Another intermediate of this mechanism, the bicyclo[2.2.0]hexenyl system, is not very stable, because of high strain energy.

Mechanism for formation of **7-18**.

Analysis by Helpful Hint 2-13 of the relationship between **7-16** and **7-18** gives the following:

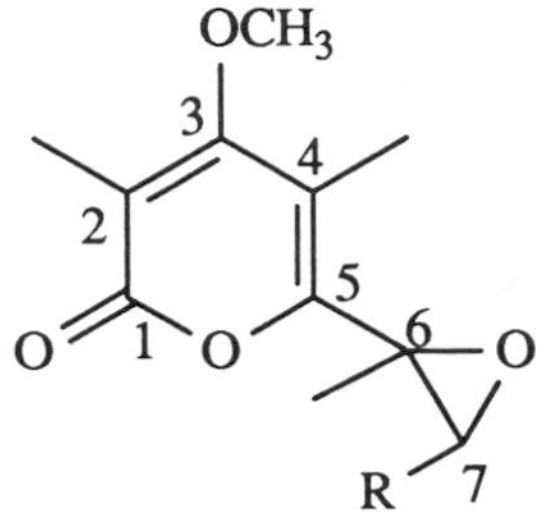

7-16 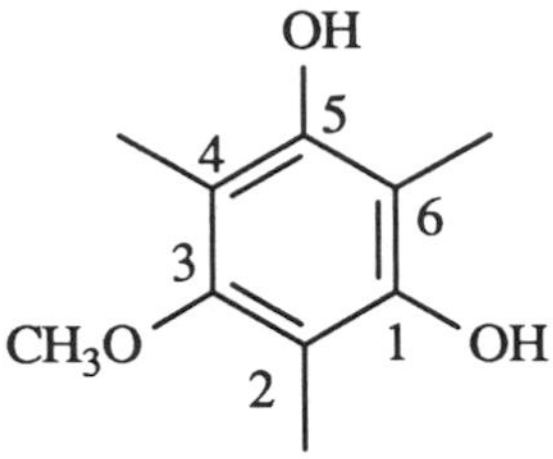   **7-18**

This numbering scheme in the product gives the fewest changes in bond making and bond breaking, and shows that the only new connection is between C_6 and C_1. C_7 and the R group attached to it are not present in the product. The cited paper draws the following mechanism for this transformation:

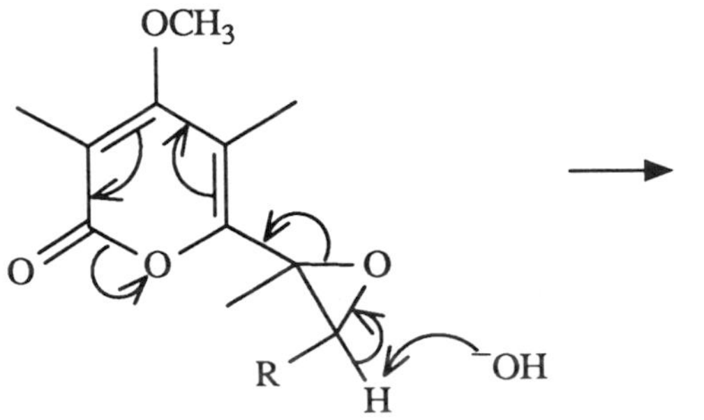

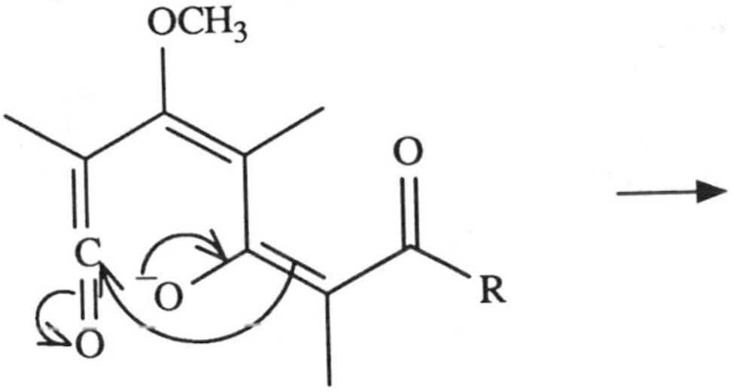

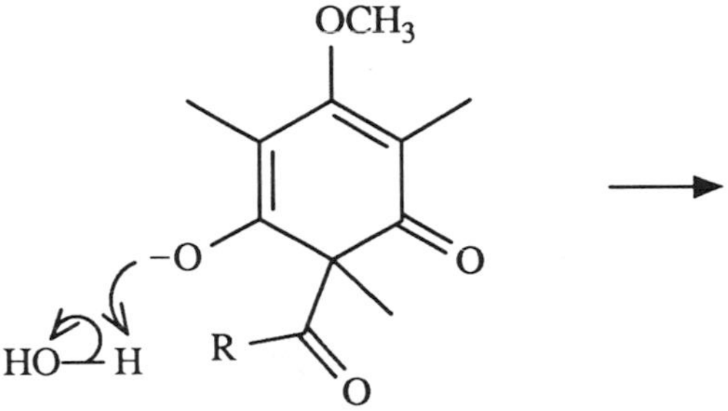

7-52

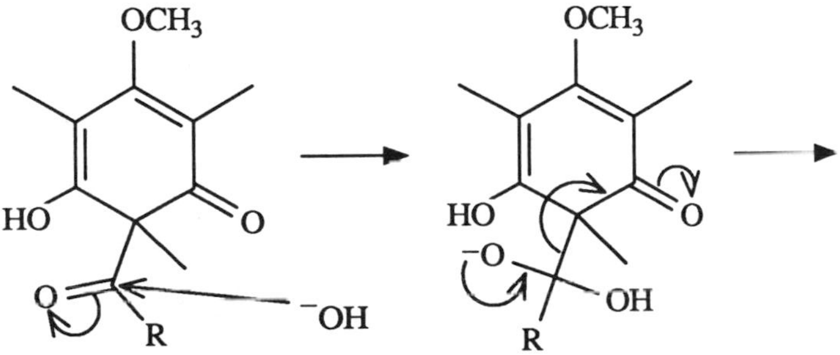

The phenolate would then pick up a proton during acidic workup to give the product phenol.

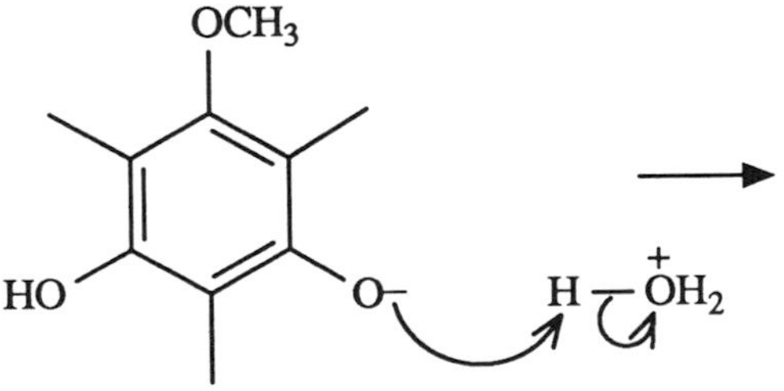

7-18

An alternative mechanism for the formation of intermediate **7-52** might be a nucleophilic addition of hydroxide to C_5 of **7-16**, followed by ring opening and subsequent modification of the epoxide:

The following reaction was written by a student.

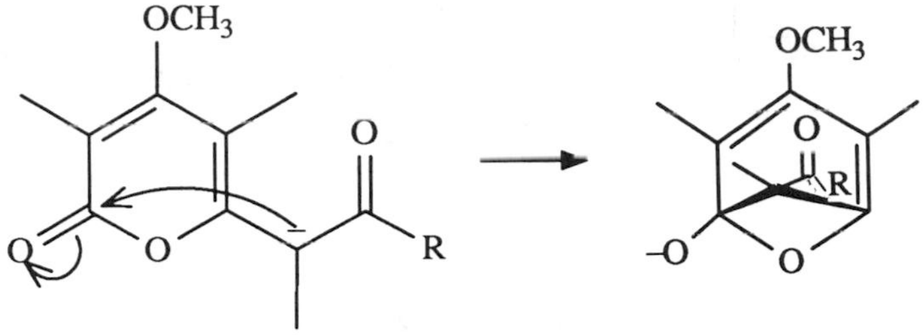

7-53   **7-54**

This mechanistic step suffers from three problems. First, anion **7-53** is attacking itself! This anion is a resonance hybrid; one of the resonance forms, **7-53-2**, shows that there is already considerable electron density at the carbonyl group:

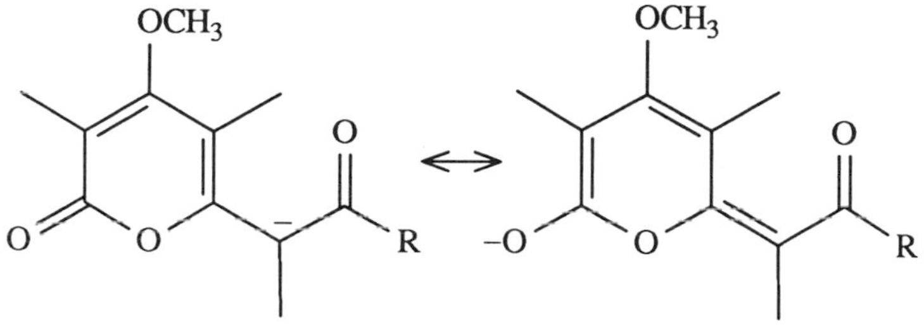

7-53-1   **7-53-2**

Second, all of the atoms involved in resonance stabilization must be coplanar. Thus, the exocyclic carbanion is not in a position geometrically favorable for attack on the carbonyl group. Third, in **7-54** there is a double bond at a bridgehead which involves considerable strain energy.

Another student wrote the following, as a first step in the reaction:

7-55

Anion **7-55** is not stabilized by resonance in any way. The
acidity of the proton on the epoxide ring, relative to that of
normal hydrocarbons, is enhanced by the higher s character in the C-H bond and by the adjacent oxygen. Nonetheless, neither of these factors enhance its acidity enough to
bring it close to the pK_a's of methanol or water. The
mechanism given in the cited paper starts with removal of
this proton, (see page 476) but makes it part of a concerted
reaction in which the epoxide ring is opened. In this case,
the release of strain energy and the stability of the resulting
resonance-stabilized anion provide considerable driving
force.

Index